Engineering Materials

This series provides topical information on innovative, structural and functional materials and composites with applications in optical, electrical, mechanical, civil, aeronautical, medical, bio- and nano-engineering. The individual volumes are complete, comprehensive monographs covering the structure, properties, manufacturing process and applications of these materials. This multidisciplinary series is devoted to professionals, students and all those interested in the latest developments in the Materials Science field, that look for a carefully selected collection of high quality review articles on their respective field of expertise.

Indexed at Compendex (2021)

More information about this series at http://www.springer.com/series/4288

Meiqi Hu • Bohong Gu

Impact Damages of Braided Composites

Springer

Meiqi Hu
School of Textiles
Donghua University
Shanghai, China

Bohong Gu
School of Textiles
Donghua University
Shanghai, China

ISSN 1612-1317 ISSN 1868-1212 (electronic)
Engineering Materials
ISBN 978-981-16-5702-3 ISBN 978-981-16-5703-0 (eBook)
https://doi.org/10.1007/978-981-16-5703-0

National Natural Science Foundation of China

This Springer imprint is published by the registered company Springer Nature Singapore Pte Ltd.
The registered company address is: 152 Beach Road, #21-01/04 Gateway East, Singapore 189721, Singapore

For
our constant loves
and
our family members and colleagues who
always give us supports

Foreword

Braiding, an ancient textile method, has now become an advanced technology for manufacturing complex-shaped high-performance composite materials frequently used in aircraft, high-speed vehicles, sports, and civil facilities. According to the Merriam-Webster dictionary, braiding is defined as three or more fiber strands interweaved to form a regular diagonal pattern along its length, and it is usually a narrow fabric of intertwined threads used especially for trimming. With the invention of various man-made fibers, especially carbon fibers, braiding has been extended to prepare braided preform for composite material manufacturing. The braided preform is impregnated into matrix to be a stiff material after matrix consolidation. This stiff material is called braided composite. Based on the various braided preform shapes, the braided composite could be an integrated complex-shaped mechanical part without any inner connection. This seamless mechanical part will have higher damage resistance compared with other connected mechanical parts. Therefore, the braided composites have been extensively used in aerospace, mechanical, and civil engineering.

Impact is a very common phenomenon during the service life of braided composites. The braided composites will have global deformation and local damages accompanied with thermal generation and local temperature rise. The temperature rise will alter mechanical behaviors of matrix and furthermore the entire braided composites. The mutual influence between temperature and mechanical behaviors is thermo-mechanical coupling. The thermo-mechanical coupling is a key to characterize impact damage of the braided composite and to optimize the impact resistance.

In order to reveal impact damage mechanisms, the relationship between structure and impact behaviors should be known. There are thousands of million fibers in the braided composite. There are also several structure levels from fiber to braided composite, for example, fibers, fiber tows, matrix-impregnated fiber tows, unit cells, and structure parts. Micro-level is for 10^{-6} m level in the braided composite, such as diameter in a single fiber. Meso-level is for 10^{-3} m level, such as diameter of a fiber tow or size of a unit cell. Macro-level is 10^{0} m level, such as width or length of the braided composite. The structure level in the braided composite is called multiscale structure. The multiscale structure concept will be employed in analyzing impact damages of the braided composite.

One of the authors of this monograph, Prof. Bohong Gu, who was my master's degree student 30 years ago, has focused on the impact damages of the braided composite for more than 10 years. He has supervised more than ten PhD students recently for knowing how the thermo-mechanical coupling and multiscale structure will influence impact damages for carbon fiber/epoxy 3-D braided composite material. In this monograph, the methodology of combining thermo-mechanical coupling and multiscale structure for analyzing impact damages will be presented for 3-D braided composite beams and tubes. The theoretical framework, test results, and finite element analyses will be given. It is expected that the book could be a guideline or reference for high-impact damage tolerance braided composite design.

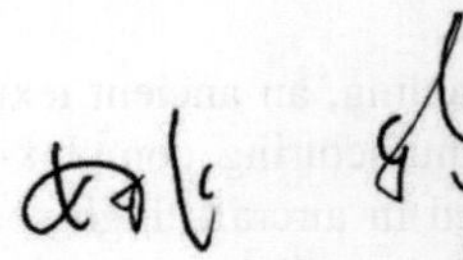

Xi'an Polytechnic University, Senior Member of Chinese Academy of Engineering, Xi'an, China

Yao Mu

Preface

Three-dimensional (3-D) braided composite materials are composite materials prepared with fiber tow braiding and composite material forming, which can realize the integration of material structure to manufacture complex structural parts and reduce the number of assembly connections. The 3-D braided composite material is one of the unique textile composites with integrated braided preform structure. The material has been widely used in aerospace, high-speed vehicles, and important civil facilities. Impact loading and high/low temperatures are very common in such kinds of applications. To investigate impact damage behaviors under different temperature environments is an important topic and also a tough challenge for braided composites.

The objectives of this book are to bridge the gap among materials science, impact dynamics, and textile engineering and to provide an insight on impact damages of the braided composites under room- and high-temperature environments. With the understanding of damage mechanisms, we hope that the braided composites could be optimized for structural designs in aerospace, high-speed vehicles, and important civil facilities in a more efficient way.

It is only during the last 35 years that the subject of braided composites has become to occupy a significant part of the textile and mechanics literature. This is not to say that the braided materials have not been studied in one form or another by engineers and scientists one or two centuries ago. A quick review of the indices of the "*Strength of Materials from Galilei to Lord Kelvin*" will show many references under textile-like categories, such as wires, threads, and helical springs. All the subjects described therein provide the basis of the methodologies in today's applications of braided composites, which include but not limited to braiding techniques, composite curing, and multiscale structure characterizations.

This monograph is in a multidisciplinary field among mechanical engineering, textile engineering, and impact dynamics. In this book, we will report thermo-mechanical coupling constitutive equations and impact damage distributions of 3-D braided composite materials under impulsive loadings. The model was verified with transverse impact tests under different impact velocities and different temperatures on split-Hopkinson pressure bar apparatus. We present a thermo-mechanical coupled multiscale geometrical model of the 3-D braiding composite beams and tubes for analyzing damage mechanisms under various impact velocities. Impact deformation and damage morphologies have been described in test observations with high-speed cameras, micro-CT (computer tomography), and finite element analyses (FEA). The FEA model could largely reduce time, personnel, and resource cost, providing strength prediction and feasibility study for thermo-mechanical coupling analyses and applications of various composite structures in extreme environments. All the impact damages are shown in figures for unveiling the relationships between microstructure and failure modes. This provides an intuitive and visible way for designing braided structures with high-impact damage tolerance.

This monograph is intended for investigators, engineers, and graduate students who are interested in textile composite materials; investigators working on impact dynamics of composite structure design; and engineers working on impact-proof structure design.

Shanghai, China
July, 2021

Meiqi Hu
Bohong Gu

Acknowledgments

In the Laboratory of Dynamic Strength of Textile Composite, which is located in the College of Textiles, Donghua University, Shanghai, China, more than 30 graduate students work with us together year by year. They are diligent, smart, and enthusiastic young people with strong interest in designing, manufacturing, testing, and modeling of various textile composite materials. We have worked together for more than 10 years like a big family. We should acknowledge all the lab-mates and students for their hard work.

The College of Textiles (CoTs) was founded in 1951 and has the longest history and the strongest power in Donghua University, Shanghai, China. The CoTs obtained the first qualification for conferring three-level academic degrees in China. The CoTs is the main body of national key discipline of Textile Science and Engineering (TSE). Now the TSE holds the top and A+ position in China. We acknowledge the friendly and competitive academic environment in the CoTs. We love the CoTs environment for encouraging us to find more interesting topics in impact damages of textile composites, and always.

We have been granted financial supports from the Chinese Government for many years, especially from the National Natural Science Foundation of China (NSFC). We have had opportunities to work in this area based on the constant financial supports. We acknowledge the NSFC for providing a tolerant environment to us in the recent 10 years. Our progress depends on NSFC supports.

Contents

Abbreviations

CAE	Computer-aided engineering
CTE	Coefficients of thermal expansion
CZM	Cohesive zone model
DMA	Dynamic mechanical analysis
DSC	Differential scanning calorimetry
FEA	Finite element analysis
FEM	Finite element method
PID	Proportional integral derivative
PTFE	Polytetrafluoroethylene
RTD	Resistance temperature detector
RUC	Representative volume cell
SEM	Scanning electron microscope
SHPB	Split-Hopkinson pressure bar
TG	Thermogravimetric
UC	Unit cell
VARTM	Vacuum-assisted resin transfer molding
VUMAT	Vectorized user-defined material
μ-CT	Micro-computer tomography

Chapter 1
Literature Review

Textile composites are made from textile preforms and matrix, in which the textile preforms are impregnated in matrix. The textile composite materials have been widely used in manufacturing high-speed vehicles, aircrafts, etc., where the impact resistance is a critical issue. Understanding impact damage mechanisms is fundamental for designing composites with high impact resistance. The impact damage mechanisms and design methods based on isotropic or homogeneous materials are not applicable to textile composites, while the heterogeneity of textile composite structures and the anisotropy of mechanical properties challenge the current methods for analyzing impact damage mechanisms and designing impact-resistant structures.

The local materials around impact area are under adiabatic state. Temperatures of the local materials rise during the transient loading. Thermal sensitivity of polymer matrix will induce the change of mechanical behaviors of the composite materials. The change of the mechanical behaviors will also alter the temperature rise. This is thermo-mechanical coupling effect. The thermo-mechanical coupling effect is an important factor for designing high impact resistant composites.

Fabrics consist of structure levels from fibers, yarns, and fabrics. Textile composites also consist of several structure levels from fiber, yarn, resin-impregnated yarn, unit cell, and macroscopic material. This is called multiscale structure levels. The multiscale structure level is a basic feature for textile composite design, including structure, stiffness, and strength designs. This book will focused on impact damage mechanisms of a unique textile composite, 3-D braided composite. The braided composite is a composite material in which braided preform is impregnated in matrix. The composite can realize the integration of material structure to manufacture complex structural parts and reduce the number of assembly connections. The material has been widely used in aerospace, high-speed vehicles, and important civil facilities.

M. Hu, B. Gu, *Impact Damages of Braided Composites*, Engineering Materials,
https://doi.org/10.1007/978-981-16-5703-0_1

1.1 Textile Composites

Textile composite is one kind of advanced fiber-reinforced composite materials. Using the traditional textile processing technology, the fiber tows are interlaced into a certain structure or shape to form a preform, and then the preform is impregnated in matrix for curing. The textile composite materials are made from high-performance fibers (such as carbon fiber, glass fiber, ceramic fiber, basalt fiber, aramid fiber, etc.) and functional resins (such as epoxy resin, phenolic resin, polyester resin). Compared with metals, the composites have superior behavior of light weight, high specific strength, high specific stiffness, corrosion resistance, etc. and have been widely used in aerospace, automotive, marine, construction, and other types of engineering and protection field.

2-D laminates have low interlaminar shear strength and toughness due to the lack of fibers along the thickness direction (or called Z-direction). Since the 1960s, relying on the needs of the aerospace industry, 3-D textile composites have been developed. There are fibers or yarns along Z-direction in the 3-D textile composites. The 3-D textile composites have superior integrity and formability to overcome the disadvantages of 2-D laminates such as easy delamination [1]. 3-D textile composites are mainly classified into 3-D woven (orthogonal woven, angle-interlock woven) composites, 3-D knitted composites, 3-D stitched composites, and 3-D braided composites [2].

1.2 3-D Braided Composites

Documented braiding dates back as far as 4000 years to ancient Chinese women's hair braids [3], and early traditional braiding methods were mainly used to make shoelaces, ropes, and other household items. The industrialization of braiding technology began in the eighteenth century with Thomas Walford, who described in his 1748 patent "a machine for the distribution of different ropes in mixed arrangements" and is regarded as the originator of the braiding machine. Unlike weaving machines and knitting machines, which could produce wide widths with relative ease, braiding machines generally produced very narrow widths that could only be used for ropes, decorative tapes, etc. This has to some extent influenced the development of braiding technology.

The first systematic description of the structure and formation of braided fabrics began in 1953 [4, 5]. The rapid development of 3-D braiding technology started in 1982 with the introduction of the "Omniweave" concept by Florentine, who depicted in his patent a braiding machine with rectangular or circular arrangement of the spindles, in which the spindles are moved in longitudinal and transverse tracks to form a 3-D braided fabric [6]. Phoenix [7] published the first article on the structural mechanism of tubular braids. Starting in the 1980s, a large body of literature on 3-D braids began to deal with braided composites [8].

Fig. 1.1 Application of fiber-reinforced composites. (**a**) Rocket motors, (**b**) wind turbine blades, (**c**) automobiles, (**d**) aircraft, (**e**) satellites

Braiding technology [8] makes the yarn continuously interwoven to form an integrated spatial structure and offers a near net-shape manufacturing advantage. According to the process, technology can be divided into 2-D braiding technology and 3-D braiding technology. The 2-D braided fabric thickness is difficult to meet part of the preform thickness requirements. 3-D braiding technology has been gradually developed since the late 1960s and mainly includes two-step, four-step, and solid braiding techniques [9, 10], as well as six-step [11] and multistep [12] braiding techniques derived from the four-step method. The 3-D braiding process allows the manufacture of complex shaped parts [13, 14], such as tapered parts, openings, and bifurcated parts, with typical cross-sectional shapes including round, I-shaped, and T-shaped.

The fiber tows in 3-D braided preforms are interwoven in an integrated structure. 3-D braided composites are near to a monolithic structure and overcome the shortcomings of traditional laminated composites which are easy to delaminate under impact. There is not delamination in 3-D braided composite under transverse impact. And the impact damage mechanism and damage modes are obviously different from laminates [2]. 3-D four-step braiding technology also has excellent self-adaptability; it not only can weave a rectangular cross-section fabric but also can weave a circular tube, L-beam, T-beam, I-beam, and other shaped cross-section fabrics or even variable cross-section fabrics. The 3-D braided composites can be divided into two types: 3-D four-directional braided composites and 3-D five-directional braided composites. The 3-D five-directional braided composites have improved the fiber volume fraction and axial mechanical properties due to the addition of axial yarns (fifth-directional yarns) along the braided gap in the preform braiding process [15]. It has become an important structural material in traditional and emerging engineering fields such as aerospace, biomedical, transportation, and ballistic protection [16–23], as shown in Fig. 1.1. In the abovementioned fields, the composites are inevitably subjected to high- and low-temperature environments and multiple impact loads, with the risk of structural damage and failure. In order to ensure the safety and reliability of the braided composites in service, it is important to quantify the internal damage distribution and accumulated plastic deformation under multiple impact damage, as well as the thermo-mechanical coupling effects, to optimize the material design.

1.2.1 3-D Braided Composite Beams

The 3-D braided geometry is an important factor affecting the overall mechanical properties of this type of composite. Researchers have used various methods to obtain the spatial structure of 3-D rectangular braids and to establish mathematical relationships between the braiding parameters and prediction of the mechanical properties.

Ko [24] presented the first 3-D braided preform unit cell model in the Fourth International Conference on Composite Materials (ICCM-4) held in Tokyo, Japan, in 1982. This unit cell model consists of four straight line segments diagonal to the unit cell cube, and the four straight line segments intersect with the center point. In 1993, Du and Ko [25] made further development based on the geometry of 3-D braided composites such as fiber tow orientation and fiber volume fraction after the four yarns in this unit cell are truncated by six planes, so there is no longer a complete yarn in any unit cell.

Yang et al. [26] developed the fiber inclination model to treat 3-D braided composite unit cell as an assembly of tilted unidirectional plates and predict the in-plane elastic properties of 3-D braided composites based on classical laminate theory. The model can be applied not only to 3-D braided composites but also to other 3-D textile structure composite mechanical property calculations such as 3-D angle-interlock woven composites [27].

Li et al. [28] regarded 3-D braided yarns as cylinders without radial deformation and obtained the interior unit cell model of the 3-D braided preform based on the calculation of structural parameters such as fiber tow orientation angle, fiber volume fraction, and preform profile size. Also, they found that the 3-D braided surface structure differs from the internal structure.

Wang [29, 30] had reported structural features of 3-D braided preforms and yarns in composites. Wang [29, 30] started from a four-step 1×1 braiding method and used a geometric topology method to construct a structural model of 3-D braided preforms. This is the first time that the 3-D braided preforms were divided into three different unit cells: internal unit cell, surface unit cell, and corner unit cell. In 2001, Wang et al. [31] used a digital-element simulation method to study the 3-D braided preform structure. She equated the yarn as a chain of digital rod units similar to a peg connection, and when the distance between digital units is zero, the chain of units becomes a fully flexible body. It is also defined that the two yarns start to contact when the distance between two nodes is the yarn diameter. In this method, the simulated yarns have mechanical properties such as tension, compression, and friction between each other, so that the 3-D braided preform simulated best reflects the real configuration. The advantage of this method is that it is possible to observe the path of any yarn in the 3-D braided preform. And this is very difficult to obtain in both test and existing theoretical analysis. Later, Zhou et al. [32] and Miao et al. [33] developed the multichain digital-element simulation method to study the 3-D braided preform structure at the mesoscale by using fibers instead of yarns as the object of study. In addition to the yarn structure of 3-D braided composites, Sun et al. [34] used this method to analyze the effect of fiber volume fraction and braiding angle on the mechanical properties of 3-D braided composites.

Kalidindi et al. [35] used a new isostrain and isostress weighted average method to calculate the mechanical property response of 3-D braided composites based on 3-D braided composite representative volume element (RVE) and compared the effect of different braiding parameters on the mechanical properties of 3-D braided composites.

Chen et al. [36] used test observation and analytical model to divide the 3-D braided composites into three unit cells: internal unit cell, surface unit cell, and corner unit cell. He went further than Wang et al.'s [30] in terms of surface and corner unit cells. The yarns can be connected into a whole after the combination of unit cells, and also the yarn morphology is closer to the actual results.

Zeng et al. [37, 38] presented simple models for 3-D braided composites with transverse and longitudinal cracks, respectively. The stress distribution was derived by repeating the regression equation defined by the quadratic ordinary differential equation in a unit cell. Numerical calculations showed that Young's modulus and Poisson's ratio of 3-D braided composites decreased with increasing braiding angle.

Tang et al. [39, 40] established the fiber volume fraction model based on three unit cell models of 3-D braided composites by normalizing the braided pitch length as an important parameter for 3-D braided composites. The braiding angle and fiber volume fraction were also expressed as a function of the braided pitch length.

Zeng et al. [41] discretized the 3-D braided composites into cubic elements, and the discretized elements were further divided into three types: matrix elements, yarn elements, and hybrid elements. This new finite element method can be used to predict the mechanical properties and local stresses of 3-D braided composites. Also, this method can be applied to calculate the mechanical properties of 3-D braided composites with different structural parameters and geometrical sizes.

Gu [42] developed the four-step 3-D braided composite realistic meso-structure model and used it in ballistic impact finite element calculations. The meso-structure model is based on the assumption that the same set of yarns in the 3-D braided preform moves with the same trajectory, differing only in the starting point. It is also assumed that the yarns are straight inside the 3-D braided preform and only bend and change direction at the surface. In the finite element calculation of the meso-structure model, the energy absorbed by the preform and the resin matrix during the ballistic impact can be obtained separately, so that the difference in mechanical properties between the preform and the matrix can be distinguished.

Kang et al. [43, 44] used a computer aided design (CAD) method to establish the unit cell model of 3-D braided composite. The path of the yarn in the 3-D braided preform was described by a cubic spline curve. They also predicted the axial and transverse tensile mechanical properties of 3-D braided composites.

Zhang et al. [45] studied the meso-structure of 3-D braided preforms with complex rectangular cross-sections based on a four-step braiding method and developed a unit cell model. The number and volume of each cell and the fiber volume fraction were derived from the unit cell model. And a design system was established to select the number and arrangement of spindles according to the size of the 3-D braided preforms, yarn linear density, fiber density, and fiber volume fraction.

Li et al. [46–48] established three meso-structure unit cell models, internal unit cell, surface unit cell, and corner unit cell, based on the braiding process of 3-D four-directional and 3-D five-directional rectangular braided preforms, respectively, and derived the relationships between structural parameters such as yarn orientation angle, fiber volume fraction, and stacking coefficient. The stiffness and strength properties of the 3-D five-directional braided composites were also derived based on the three unit cell theories. The finite element method was also used to develop the unit cell model and predict the stiffness and strength properties of the 3-D braided composites. Comparison of test results, theoretical derivations, and finite element results showed that the latter two accurately predicted the mechanical properties of the 3-D braided composites.

1.2.2 3-D Braided Composite Tubes

Researchers have used various methods to obtain the spatial structure of 3-D circular braided tubes. The mathematical relationships between the braiding parameters and mechanical properties have been established for impact strength design.

Since the 1950s, scholars [5, 28, 49–54] have studied the relationship between the braiding process and the meso-structure geometry of fabrics based on the tubular fabric braiding process. Mathematical relationships between yarn orientation, fiber volume fraction, preform profile size, and mechanical operating conditions (including braiding rate, friction between yarns, yarn sliding on the mandrel, etc.) were derived.

Wang et al. [55] performed a topological analysis of 3-D circular braided preforms and used the control volume and control surface methods to classify the structures into interior unit cell and inner and outer surface unit cells. Wang [56] further described the geometric relationships of tubular braided structures based on this method and treated the yarn assemblies in tubular braided composites as curved cross laminates based on laminate theory.

Chen et al. [57] studied the spatial yarn interweaving structure by defining a representative control domain method based on the motion law of 3-D circular braiding carriers. The unit cell model of the preform structure was developed, which contains six types of unit cells, four internal and two surface unit cells. It was found that the geometric properties of the unit cells were related to their positions in the cross-section of the preforms. Zhao [58] used the stiffness averaging method on this basis to calculate the elastic property parameters of 3-D circular braided composites. The results showed that both the braiding angle and the fiber volume fraction had significant effects on the elastic modulus.

Sun [59] used a digital-element method to dynamically simulate a 3-D circular braiding process to obtain a meso-structure of the braided fabric. The yarn in this method resembles a long chain formed by rod elements connected by pins. Compared with the topological method, this method is more accurate in describing the meso-structure geometry of the 3-D braided circular preforms.

Li et al. [60] studied the meso-structure of 3-D five-directional circular braided composites, established the internal unit cell model, and derived the mathematical relationship between the unit cell geometric properties and the braiding process. It was shown that not only the 3-D five-directional braided structure has good mechanical properties of 3-D four-directional braided structure, but also the axial mechanical properties of the circular tube were improved by the addition of axial yarn.

Kang et al. [44] established 3-D circular braided preform geometry by CAD system and analyzed the geometry parameters and interrelationships between parameters. A third-order spline curve was used to describe the spatial path of yarn to obtain the complex internal braided structure of the braided preform and to determine the representative volume unit. Lee [61] and Song [62] also used the spline method to describe the yarn path of the 3-D circular braided structure.

Hwan et al. [63] developed a space spring model, where the fiber tow in a unit cell is composed of spring elements with only axial stiffness and the matrix consists of solid elements. The model was experimentally verified to be effective in predicting the elastic modulus of 3-D braided composite tubes.

Wang et al. [64] established the geometry model of a 3-D circular braided composite tube based on free deformation theory. The three geometric mapping equations of the circular braided structure subunit cells were derived from the rectangular braided subunit cells. The validity of the geometric model was verified with computer tomography observations.

Ma et al. [65] explored the variable cross-section 3-D five-directional circular braiding technique to model the fabric unit cell in different regions and predict the variable cross-section fabric geometry and fiber volume fraction distribution.

1.3 Impact Damages on Textile Composites

1.3.1 3-D Braided Composite Beams

The mechanical properties and impact damage mechanisms of 3-D braided composites have been a hot topic since 2000. Figure 1.2 shows the number of papers published in the field of mechanical properties of braided composites in the Science Citation Index Expanded (SCI-Expanded) database from 1980 to 2020. With the expansion of advanced composites into various fields of society, the mechanical research on braided composites has been increasing year by year.

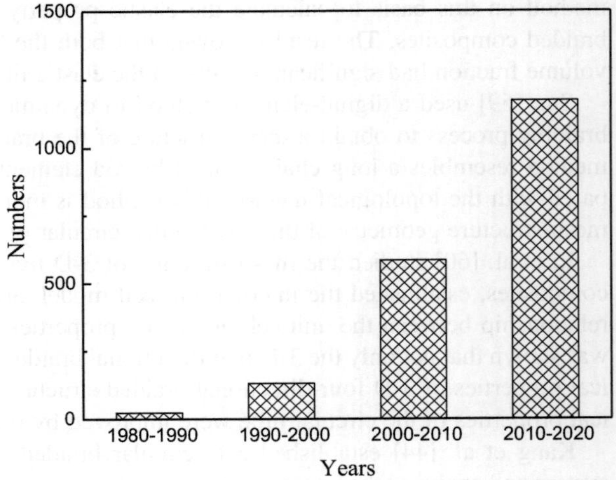

Fig. 1.2 Number of papers published in the field of mechanical properties of braided composites in the Science Citation Index (SCI) database from 1980 to 2020

By experimentally investigating the ballistic penetration behavior of 3-D braided composites, Flanagan et al. [66] found that the damage morphology of 3-D braided ultrahigh-molecular-weight polyethylene (UHMWPE) composites after ballistic shot holes transitioned from a tensile damage mode to a shear damage mode at medium and high impact velocities. Shear blockage occurred at velocities corresponding to the reduction in wreckage mass when the kinetic energy of the bullet exceeded the inherent strain energy of the material at the critical velocity. Jenq et al. [67, 68] applied quasi-static behavior to ballistic penetration analysis of braided composites and found that the test ballistic limit of the four-step braided specimens was approximately 7% lower than that of the two-step braided specimens due to the absence of axial yarns in the four-step braided specimens. In addition, the high strain rate effect of GFRP composites with off-axis fiber-reinforced structures is not significant, and thus the toughening effect is reduced under the action of strain rates. Yang et al. [69] experimentally investigated the deformation and damage mechanism of 3-D braided carbon fiber-reinforced plastics (CFRPs) under ballistic impact loading by combining C-scan and DIC techniques to obtain the entire transient displacement field and strain field during impact, locating damage initiation and extension. It was found that the energy absorption was approximately linearly related to the incident velocity.

The effect of impact velocity on the damage behavior of 3-D braided composites has also gradually attracted the attention of scholars. Shaker et al. [70] studied the damage mechanism of 3-D braided aramid fabric-reinforced epoxy resin composites under low- and high-velocity impacts and found that under low-velocity impact loading, the energy absorption tolerance of the structure is most important and the geometry determines the impact response of the target. And conversely under high-velocity impact loading conditions, the bullet on the target produces a local form response, and geometric parameters such as width and length have little effect on the impact response and produce a greater degree of damage to the target. At the point of impact, the damage to the structural integrity of the composite by a small bullet loaded by high-velocity impact is usually greater than that by low-velocity drop hammer impact loading. Zhang et al. [71] described the impact damage morphology of 3-D braided composites at different impact energies phenomenologically and found that the impact damage was jagged, which is manifested by fiber fracture and resin cracking. The damage area and energy absorption increased with increasing impact velocity. Zhao et al. [72] tested the dynamic compression behavior of 3-D braided composites with different braiding angles in longitudinal, transverse, and thickness directions using a split Hopkinson pressure bar and showed that the compression properties exhibited strain rate strengthening effects in all directions and were characterized by brittle damage at high strain rates. The composites with braiding angles of 20° and 45° are the most sensitive to longitudinal strain rate, followed by the thickness direction, and the least sensitive to transverse strain rate. The composites with braiding angle of 30° were the most sensitive to strain rate in longitudinal compression, while the strain rate sensitivity was lower in the transverse and thickness directions. Zhang et al. [73] used frequency domain analysis to study the energy absorption characteristics and damage mechanism of 3-D braided composites under transverse and low-velocity impacts. The results showed that with the increase of impact velocity, the peak value of impact load, displacement, and total energy absorption increased and the failure mode changed from resin cracking to fiber fracture. The results of frequency domain analysis show that the stress wave amplitude increases with the increase of impact velocity, but the corresponding frequency is independent of the impact velocity.

The effect of braiding structure on the mechanical properties of composites should not be ignored. Liu et al. [74] compared the mechanical properties of 3-D four-directional/five-directional braided quartz-reinforced silica composites. The results showed that the tensile strength, flexural strength, and shear strength of five-directional composites were higher than those of four-directional composites, the strength and toughness were significantly enhanced, and the distribution of yarns in the preform affected the failure mode of the composites. Cui et al. [75] studied the mechanical properties, real-time progressive damage law, and fracture mechanism of 3-D five-directional braided composites with different braiding angles. It was found that the smaller the braiding angle, the more sensitive the longitudinal mechanical properties of the composites, especially the composites still have a certain post-peak load capacity after longitudinal impact, showing a nonlinear trend of increasing and then decreasing.

Most of the numerical simulations are based on the unique unit cell model of 3-D braided composites, i.e., corner unit cell, surface unit cell, and interior unit cell [76]. Zhang et al. [77] defined the unit cell model of 3-D braided composites, derived the finite element constitutive equations and failure criterion, wrote the vectorized user-defined material (VUMAT) subroutine, and incorporated it into the commercial finite element software ABAQUS/Explicit to calculate the transverse impact properties. Wu et al. [78] numerically simulated the three-point bending fatigue behavior of 3-D braided composite using the three unit cell model and developed a user-defined material subroutine to describe the stiffness matrix and fatigue damage evolution of the unit cell, and the results showed that stress concentrations and large deflections occurred at the middle and the bottom of both ends of the composite plate. As the stress increased, the composite deformation increased and accumulated more damage, which leads to accelerated fatigue failure. Zhang et al. [79] developed solid structural models of 3-D braided composites with interior, surface, and corner unit cells and simulated the mechanical response of 3-D braided composites with finite element method under periodic boundary conditions to give the deformation and stress distribution of the three unit cell models. Then they studied in detail the effects of braiding angle and fiber volume fraction on the elastic constants of the composites. Zhang et al. [80] developed a three unit cell-based finite element model to evaluate the penetration process of 3-D braided composites under high-speed impact loading. A 3-D rate-dependent constitutive model was used to determine the behavior of the components in the three unit cell. A transient degradation scheme was proposed based on the failure criterion of the yarn and matrix, which was encoded with VUMAT. It is shown that the penetration process is summarized in three stages: hole punching, fiber breakage, and bullet penetration (damage extension) stages. The residual velocity is linearly related to the bullet impact velocity, and the energy absorption increases with the impact velocity, but not linearly.

With the gradual development of the finite element method (FEM), full-scale meso-structure models are increasingly used to simulate the impact process, which can more realistically reflect the mesoscale damage of materials. Zhou et al. [81–85] used test and FEA to study the transverse impact behavior, deformation, and damage distribution of 3-D braided composite tubes. The effects of impact velocity, number of braided layers, and braiding angle on the impact resistance of the composite tube were analyzed to reveal the transverse impact damage mechanism. At lower impact velocities, cracks nucleate in the weak region and propagate along the path of least resistance, leading to matrix cracking and interfacial debonding. For high impact velocities, cracks propagate along a relatively straight path, and the fibers fracture under high shear and tensile stresses. Stresses propagate faster at small braiding angles than at large braiding angles, and the orientation of the braided yarn has a significant effect on crack extension. Zhao et al. [86] studied the complex failure behavior of triaxial braided composites under in-plane transverse loading by quasi-static tests and finite element method with a 3-D progressive damage model combined with an interfacial cohesive model to simulate the initiation, accumulation, and extension of material damage. It was found that the transverse tensile strength and transverse compressive strength increased with the increase of interfacial strength. The specimen thickness had a significant effect on the overall response of the braided composites, mainly due to warpage and interfacial delamination caused by the free edge effect, and increasing the thickness could suppress the out-of-plane warpage but could not completely eliminate it. Lyu et al. [87] established a full-scale meso-structure model of 3-D braided composite based on real structural features. To investigate the transverse impact response characteristics and damage mechanism of the composites at different impact velocities, it was found that after four impact cycles, the peak loads at three impact velocities (13.6 m/s, 17.8 m/s, and 22.8 m/s) decreased by 29.1%, 21.5%, and 32.3%, respectively. The stress amplitudes decreased and then increased along the length and thickness direction and decreased continuously along the width direction. With the increase of impact velocity, the failure mode changed from fiber tow debonding with resin matrix to fiber tow fracture and resin cracking. Li et al. [88, 89] established a parametric finite element model of 3-D five-directional braided composites, which considered the cross-sectional deformation and surface contact of yarns during the braiding process, and studied the deformation damage mechanism of the composites by transverse quasi-static and high strain rate compression tests. It was found that the composites show obvious strain rate strengthening effect and dynamic toughness phenomenon.

However, the above studies mainly focus on the mechanical analysis and structural effects during the damage of composites at room temperature, including ballistic impact, low-velocity impact, quasi-static compression, high-velocity impact compression, tension, shear, and bending tests. The unit cell method and FEM of meso-structure model were used to analyze the mechanical response. It is found that the thermo-mechanical coupling effects has not been considered in the impact damage model. The above studies have limited the accurate assessment of the service reliability of 3-D braided composite components under multiple impacts and high-/low-temperature environments.

1.3.2 3-D Braided Composite Tubes

The mechanical properties of braided composite tubes mainly include axial static and dynamic crush properties [90–100], tensile properties [101], fatigue properties [102–104], transverse static compression and impact properties [105–107], and bending and torsion properties [108, 109]. The above studies mainly focus on the axial properties of 2-D biaxial and triaxial braided composite tubes, while the mechanical properties of 3-D braided composite tubes are not sufficiently studied.

Chiu et al. [94] and Wu et al. [98] investigated the axial static crush behavior and high strain rate dynamic impact compression behavior of 3-D braided composite tubes by test methods, respectively. The results showed that both the reinforcing fiber type and the braiding parameters had a significant effect on the axial compression failure mode of the 3-D braided composite tubes. In the static crush experiments [94], the damage of carbon fiber/epoxy resin composite tube started with the compressed surface, followed by bending until the final failure, while the Kevlar/epoxy resin composite tube exhibited a stable progressive folding damage process. High strain rate dynamic compression behavior [98] showed that the peak of axial compressive stress and compressive stiffness of 3-D braided composite tubes decreased as the braiding angle increased. The failure modes of the specimen with 25° braiding angle included shear damage, progressive damage, and eventual fragmentation. The specimens with 35° braiding angle are dominated by fiber breakage and flexural damage. The specimens with 45° braiding angle are dominated by matrix cracking.

Zeng et al. [117] and Wu et al. [98, 99] also used the unit cell model and meso-structure finite element model to simulate the axial dynamic compression damage process of 3-D braided composite tubes, respectively. Compared with the unit cell model, the meso-structure model can obtain a more intuitive view of the stress distribution and the respective damage of the preform and matrix.

Wu et al. [101] tested the tensile properties of 3-D braided composite tubes. The specimens were subjected to tensile stresses and compressive stresses of the clamp at both ends of the tube, and the fracture morphologies of the specimens were all local fractures.

Calme et al. [105] studied the behavior of 3-D braided carbon fiber/epoxy resin composite tubes subjected to transverse compression by flat plates. Transverse compression damage occurred initially on the outer surfaces of both sides of the circular tube, with cracks extending from the interfiber defects along the thickness direction and the outer surface yarns being pulled off. The whole damage behavior is progressive, and the load-displacement curve exhibits nonlinear characteristics.

Potluri et al. [108] studied the bending and torsional properties of biaxial and triaxial braided composite tubes and investigated the effect of braiding angle and number of braided layers on the mechanical properties. It was shown that the bending stiffness and torsional stiffness of triaxial braided composite tubes in the presence of axial yarns were greater than those of the biaxial specimens. The bending and torsional stiffness of the composite with two layers was greater than that of the composite with single layer. The braiding angle increased from 31° to 65°, the bending stiffness tended to decrease and then increase, and the torsional stiffness increased.

Gideon et al. [103] investigated the damage properties of 3-D braided composite tubes under quasi-static compression and compressive fatigue loading and explored the effect of different braiding angles on the compressive mechanical response. The results showed that specimen with the braiding angle of 45° had longer fatigue life. The braiding angle is an important parameter affecting the fatigue damage mode of tubes, and the specimens with small braiding angle are mainly broken, while the specimens with large braiding angle show a progressive damage process. The damage mechanisms mainly include resin cracking, fiber breakage, interface debonding, and buckling, and the damage extension direction is consistent with the braiding angle direction. This author [104] also developed the three unit cell model for finite element analysis of the quasi-static compression behavior of the specimens.

3-D braided composites are inevitably subjected to impact in aerospace and defense applications. The instantaneous impact process often brings catastrophic consequences to human beings. The impact responses are also popular topic as the applications extended. Main researches include comparing the impact resistance of 3-D braided composites with other textile composites [110–112], comparing the mechanical response under quasi-static and dynamic loading conditions [67], studying the dynamic mechanical properties of materials under different loading modes (e.g., impact tension [113, 114], compression [98, 115, 116], shear [117, 118], transverse impact [71, 77, 119]), and studying the factors affecting the dynamic properties of 3-D braided composites (e.g., strain rate effect [115], temperature effect [120], structural effect [121]). The dynamic response of 3-D braided composites has also been simulated with various finite element models (e.g., unit cell model [77], meso-structure model [42], multiscale model [122]). The studies have mainly focused on rectangular braided composites, and there are not enough studies on the dynamic mechanical properties of tubular braided composites [17, 94, 98, 99].

Ko et al. [110–112] have been studying the impact properties of 3-D braided composites since 1986 and comparing them with conventional 2-D laminates. The results showed that the 3-D braided composites have higher delamination resistance and impact damage tolerance and lower notch sensitivity compared to conventional 2-D laminates due to the enhancement in the thickness direction.

Jenq et al. [68] studied the ballistic impact behavior of 3-D four-step braided glass fiber/epoxy resin composites in 1998. The quasi-static and ballistic impact properties of the target plate were tested with the same warhead. Quasi-static experiments were used to describe the progressive damage mode of the material, while dynamic tests investigated the target plate penetration process and obtained a ballistic limiting velocity of 69 m/s. The finite element analysis software named MARC was used to simulate the dynamic impact process, and the modeling was based on the 3-D braided composite constitutive relationship and quasi-static penetration model. The ballistic limit of the four-step braided composite target plate was reduced by about 7% compared to the two-step braided composite specimen previously studied by Jenq et al. [67], due to the fact that no axial yarn was incorporated in this four-step braided composite.

Flanagan et al. [66] experimentally studied the penetration failure mode of textile composites under high-velocity impact loading. Different types of fibers (Spectra®, Kevlar®, and Twaron®) and composites with different fabric structures (woven, braided, and nonwoven) were compared and analyzed for failure modes in the impact velocity range of 200–1100 m/s. The results show that the 3-D Spectra® braided composites have the best penetration resistance and structural integrity after impact relative to other types of composites. The delamination resistance of 3-D braided composite is better than that of woven composite.

Feng et al. [123] studied the impact properties of 3-D braided carbon fiber/epoxy resin composites by explosive impact experiments in 2000. The damage to the specimens under impact was caused by the high temperature, high velocity, and high pressure generated during the explosive blast, and the specimens were blown off along the direction of the charge strip, suffered ablation and grooving on the front side, and had significant resin chipping and cracking along the braiding angle direction on the back side.

Shen [124] conducted high-speed symmetric collision experiments and projectile penetration experiments on 3-D braided composites using a hydrogen cannon in 2002. The shock wave velocity of 12,000 km/h and shock wave pressure of more than 1200 MPa were measured in the symmetric impact experiments, which far exceeded the static bending strength. The impact damage mechanism is based on matrix cracking and fiber breakage, and there are obvious damage areas around the perforations. This 3-D braided structure improves the impact resistance of the material at the cost of certain area damage.

Gu et al. [42, 125, 126] have developed various models (quasi meso-structure finite element model, refined quasi meso-structure model, and meso-structure model) since 2004 to predict the ballistic impact behavior of 3-D braided composites. The quasi meso-structure finite element model [125] combines the fabric structure and a modified classical laminate theory, which consists of four inclined plates. The ballistic penetration behavior of individual plates is first calculated by finite element analysis, and then the ballistic impact behavior of the composite is obtained based on the principle of energy conservation. The refined quasi meso-structure model [126] is composed of four thin plates and fiber tows arranged alternately tilted and parallel between the thin plates, and the simulation results can observe matrix cracking and fiber breakage. This model can roughly predict the ballistic impact damage of the real case of the 3-D braided composite. The meso-structure model [42] is based on the real meso-structural geometry, including the matrix and the braided preform, which more accurately predicts the failure modes of the material and more intuitively reflects the effect of the braided structure on the dynamic properties of the composite.

Sun et al. [67, 115] used the split Hopkinson pressure bar (SHPB) apparatus to test uniaxial impact tensile and impact compressive behaviors of 3-D braided composites and compared with quasi-static test results. The results showed that the 3-D braided composites exhibited strain-rate sensitivity in both the tensile and compressive directions. In the impact tensile tests, with the increase of strain rate, the tensile modulus and failure stress increase, while the failure strain decreases. The composites showed brittle fracture under high strain rate loading conditions. The results of compression experiments showed that the out-of-plane compressive stiffness of the composites was higher than in-plane under quasi-static loading conditions; the in-plane compressive stiffness was greater than out-of-plane compressive stiffness under high strain rate compression, and the in-plane compressive stiffness was more sensitive to strain rate.

Li et al. [89, 121, 127] studied the strain rate effect of longitudinal and transverse impact compression of 3-D five-directional braided carbon fiber-reinforced phenolic composites since 2009 and analyzed the effect of braiding angle and fiber volume fraction on impact compression behavior. For longitudinal dynamic compression, the composites with smaller braiding angle or larger fiber volume fraction had better dynamic compression behavior. For transverse dynamic compression, the 3-D five-directional braided composites were more sensitive to strain rate as the braiding angle and fiber volume fraction increased.

Zhang et al. [71, 77] tested the transverse impact properties of 3-D four-step braided composite rectangular plates using the SHPB test setup and developed the unit cell model of the composite and derived the elastic-plasticity constitutive equations, combined with VUMAT, to calculate the transverse impact mechanical response of 3-D braided composite.

Yan [128] investigated the low-velocity impact behavior of 3-D five-directional braided composites. The effects of impact velocity and braiding angle on the impact response of 3-D braided composites were investigated. The results showed that as the impact velocity increased, the damage degree of material, peak load, and displacement increased. As the braiding angle increased, the damage degree of material decreased, the peak load increased, and the displacement decreased, indicating that the specimens with large braiding angle had better impact damage resistance.

In recent years, mesoscale models [114, 116, 129] and multiscale models have also been developed [122] to analyze the dynamic damage mechanisms of 3-D braided composites, as well as the mechanical response of materials to impact compression [120, 130, 131] and transverse impact properties [119] under temperature fields.

1.4 Multiscale Structures

3-D textile composites are structural materials, and different weaving structures correspond to different types of textile composites. In turn, the stress-strain relationship depends on the morphology and spatial orientation of fiber tows. The accurate prediction of the mechanical properties of 3-D textile composites is still a challenge. The structure of 3-D textile composites can be well described with the meso-structure. The meso-structure of textile composites such as woven, knitted, and braided composites have periodicity, so there is a general way to analyze effective properties of different types of textile composites.

In order to better develop and utilize 3-D textile composites, it is necessary to grasp the influence of the meso-structure on the macroscopic properties of composites, i.e., the composite properties should be studied at the multiscale level. How to bridge the structure, constituents' properties, effective stiffness, and strength is one of the core objectives of the composite design. In recent years, with the gradual development of meso-mechanics and the deepening understanding of asymptotic homogenization theory, the connection between the multiscale structures and mechanical behaviors has been established [132].

Multiscale approach [132–135] is to study the coupling of physical problems on different spatial scales or timescales. It is one of the important branches of the study of complex systems, which can study different mechanisms of the same material at different scales. The approach has profound scientific connotation and research value and is one of the important tools for solving various complex material and engineering problems. For solving various scale-related discontinuity problems, such as the properties of composite and heterogeneous materials, and for considering the macro-/micro-physical properties of materials, multiscale methods are quite effective [132]. The multiscale mechanic problem studied in this book is in the range of material multiscale mechanics, more precisely in the length range of 10–10^0 m and in the time range of 10^{-9}–10^6 s. That is, the material is divided into multiple scales, i.e., microscale, mesoscale, and macroscale. The microscale is used to predict the mesoscale material properties, and then the mesoscale predicts the macroscale material properties. The method specifically embodied in textile composites is to generalize from fiber and matrix properties to yarn, then calculate the unit cell properties from yarn and matrix properties, and finally predict the overall properties of textile composites from the unit cell properties. The basic idea of the multiscale analysis method is to equate the original nonhomogeneous material with the global homogeneous material and can satisfy that the constitutive relationship between them is close or even identical. The method focuses on establishing a homogeneous model. The multiscale method uses multistep homogenization and combines with FEA to carry out calculations, taking into account the multiscale characteristics of textile composites, which can analyze the equivalent modulus and deformation of the material from the microscopic scale as well as the structural response from the macroscopic scale. This method has a high calculation accuracy, reduces the structural complexity of textile composites, and can greatly shorten the modeling process and reduce the amount of calculation with high calculation efficiency. The multiscale homogenization method is currently a more popular method for analyzing the macroscopic mechanical properties of composites internationally. However, the method suffers from the problems of obtaining microscopic stresses from macroscopic stresses and reasonable transition between different scales.

Finite element methods with meso-structure nonhomogenization [136–138] can be used to predict the effective properties of textile composites by modeling them in detail. Due to the differences between fiber and matrix, composites exhibit nonhomogeneous properties, and finite element modeling on the basis of meso-structure to study the mechanical properties of composites is also known as nonhomogenization method. The meso-structure nonhomogenization FEM is capable of predicting the stress and strain fields inside the textile composite in detail, which leads to failure mechanism analysis. However, the meso-structure nonhomogenization FEM requires the mesoscale modeling of textile composites. If the textile structure is complex, establishing a finite element model that can fully show the specific structural characteristics is a tedious process, especially for braided composites with high fiber volume fraction. It is difficult to establish a model that truly reflects the cross-sectional shape of fiber tows, and only some approximations can be made to the geometric cross-sectional shape of the fiber tows of the textile composites [31, 33, 116, 139].

The mechanical properties of 3-D braided composites are complex and influenced by many factors, such as braiding angle, fiber volume fraction, porosity, etc., which affect the mechanical properties of the braided structure. The research on the influence of braided structure on the constitutive relationship and strength of composites is still in the developing stage. Therefore, in order to apply 3-D braided composites more safely, effectively, and rationally as well as to better serve the integration of material structure design function, a large number of test and theoretical studies are still needed to reveal the deformation and damage laws. The mechanical properties of 3-D braided composites have been richly investigated with test methods [71, 140, 141], analytical methods [142], systematic methods

[143], multiscale methods [133], micro-/meso-structural finite element methods [116, 126], and unit cell methods [78, 79, 144].

Earlier mechanical models for predicting complex engineering structures were based on homogenization methods (multiscale methods, unit cell methods, etc.). In fact, the multiscale approach allows the mechanical properties calculated by the microscopic model to be used as mesoscopic material properties, which in turn can be transferred to macroscopic material properties, and eventually to calculate the macroscopic mechanical response of the whole material. This is exactly the mechanical prediction method currently needed for composites with complex structures. However, due to the complex structure of 3-D braided composites, the multiscale mechanical prediction model is still in the exploration stage so far. For example:

Gu [145] used fiber-inclined unit cell model to simulate the ballistic impact behavior of 3-D braided composites. According to the acceleration-time history of bullets and the failure modes of composites predicted by finite element method, it can be seen that this model can roughly predict the impact damage and energy absorption trend of 3-D braided composites in real situation.

Zhang et al. [77] used a homogenized unit cell model of 3-D braided composite combined with a user-defined subroutine to simulate the transverse impact properties of 3-D braided composites and finally compared with test results found that the method was effective in predicting the mechanical properties of the material.

Sun et al. [146, 147] similarly used a homogenized unit cell model of 3-D braided composite combined with a user-defined subroutine to simulate the three-point bending fatigue behavior and low-velocity impact behavior of 3-D braided carbon fiber/epoxy resin composites in 2012 and 2013, respectively.

Tang et al. [148] proposed an extended damage model for 3-D braided composite stress distribution, failure analysis, and multiscale analysis of property degradation, where the entire analysis was performed at three scales, fiber/resin scale, fiber tow scale, and entire composite macroscale, while the effect of braiding parameters on the extended failure behavior was explored.

Yu [149] developed finite element models at two scales to predict stiffness and strength properties of four-step 3-D braided composites, discussing in detail the effect of braiding angle and fiber volume fraction on strength properties.

Flesher [150] developed a dynamic crush model to characterize the failure properties and energy absorption of braided composites, and the analysis was performed at three scales, including the fiber tow scale, the braided composite unit cell scale, and the entire composite scale.

Deng [151] proposed a multiscale correlation model to obtain not only the stress/strain fields at each scale of the 3-D braided composites simultaneously but also the local failure modes, the damage sequences, and the subsequent failure extensions. Recently, multiscale analysis methods have been widely used to predict the visco-elastic properties [152] and axial and non-axial tensile properties [153] of braided composites.

Ji et al. [133] used multiscale finite element approach to study the variation of elastic properties of braided composites at different scales and investigated the effect of fiber volume fraction on the elastic properties of biaxial and triaxial composites.

However, most current multiscale analysis methods are based on averaging unit cell finite element methods to predict the quasi-static mechanical properties of braided composites, and few multiscale studies have targeted the dynamic mechanical properties of materials at high strain rates. Literature [77, 147] used unit cell methods to study the dynamic mechanical properties of 3-D braided composites under high strain rate loading in 2012 and 2013, respectively. Due to the requirement of computational efficiency, the homogenization model did not consider the yarn geometry of 3-D braided composites. In fact, due to the interaction between yarns, the cross-section of yarn in 3-D braided composites is quite complex, and the stresses in the yarns are not uniformly distributed when subjected to applied loads. The damage and failure of the composite may occur in a specific region. However, these problems can be solved by a nonhomogenization meso-structure model.

Gu [42] developed a meso-structure model of 3-D braided composite to study the ballistic impact properties. This model assumes that the yarns inside the preform are straight and bending occurs only at the surface of the preform. Compared with other models established on continuous medium, the meso-structure finite element model is more accurate in predicting the failure modes and energy absorption mechanism of 3-D braided composite.

Fang et al. [154] established a representative volume element to test the tensile damage extension behavior of 3-D four-directional braided composite using an anisotropic damage model based on Murakami-Ohno damage theory. This damage extension model is controlled by the energy required for material fracture and the equivalent displacement. According to the different damage modes of yarn transverse tension, compression and shear, axial tension and compression, matrix cracking, and so on, the different failure modes and damage extension of braided composites are obtained. The numerical results indicate that the main failure modes of braided composites are yarn transverse tension, shear failure, and matrix cracking. The authors of this book [141] selected representative volume elements of 3-D braided composites to study the effect of interfacial properties on the uniaxial tensile mechanical properties of 3-D four-directional braided composites. The model considers not only the damage model of the yarn and matrix but also the interface damage model. The results show that the interfacial damage is an important factor causing the nonlinearity of the stress-strain curve and that the interfacial damage can reduce the elastic modulus of the composite but has little effect on the strength of the braided composite.

In literature [88, 155], finite element models of 3-D five-directional braided composites predict the effective elastic properties of braided composites in high precision for considering the spatial conformation, the cross-sectional shape, and the surface contact during the braiding process of the braided yarns and axial yarns.

Zhang et al. [116] used meso-structure model to simulate the quasi-static and dynamic compression properties and failure modes of 3-D braided composites. The local stress distribution, stress state, and damage extension process can be obtained by the finite element model. It was found that the surface yarns and edge yarns of 3-D braided composites play an important role under quasi-static and high strain rate compressive loading.

1.5 Thermo-mechanical Coupling

Thermo-mechanical coupling properties and damage mechanisms of textile composites have been investigated to find temperature effects, thermal-oxidative aging properties, and thermo-mechanical coupling mechanisms from both test tests and numerical simulations. Figure 1.3 shows the number of papers published in the field of thermo-mechanical properties of textile composites in the SCI database from 1980 to 2020. The development of new generation of national defense weapons and high-end industrial equipment introduces more stringent requirements for service environment adaptability, which puts forward higher requirements for thermo-mechanical behaviors and reliability evaluation of lightweight composites. Therefore, how to characterize mechanical properties of composites under multi-field coupled environment and how to establish accurate mechanical models of composites under real service conditions are the new challenges nowadays.

The mechanical properties of the composites were studied at different ambient temperatures. Halvorsen et al. [156] tested the low-velocity energy impact behavior of glass fiber and glass fiber/aramid sandwich structure composites and found that different sandwich structures had a significant effect on the mechanical properties of the composites at different temperatures. It is concluded that the dominant failure modes changed from brittle to ductile in different temperature ranges. Zhang et al. [120] studied the compressive properties of 3-D carbon fiber/epoxy resin braided composites at the temperature of 23–210 °C and the strain rates of 1200–2400 s^{-1}. It was found that the compressive modulus degraded rapidly at 120 °C, which was related to the glass transition temperature of epoxy resin. The coupling effect of strain rate and temperature reduced the effect of strain rate on the compressive modulus. Failure strains are very sensitive to strain rate, especially at lower temperatures. Shear deformation is the main failure mode of materials under high strain

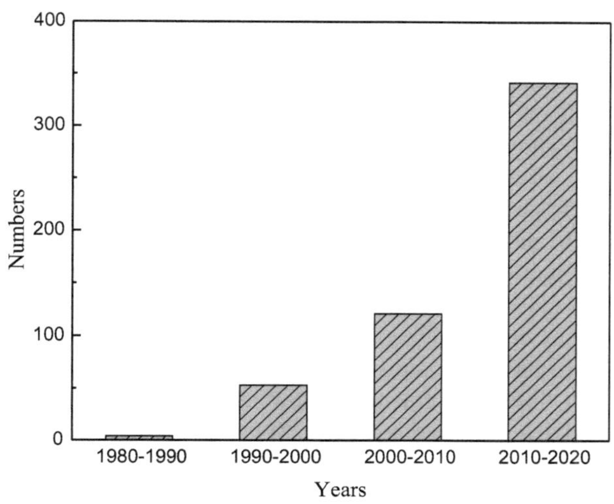

Fig. 1.3 Number of papers published in the field of thermodynamic coupling properties of textile composites in Science Citation Index (SCI) database from 1980 to 2020

rate, high compression load, and high temperature. Li et al. [89, 157–160] reported the effect of temperature on the bending properties of 3-D four-directional glass fiber/epoxy resin braided composites and the damage mechanism under high strain rate compressive impact loading. It was found that the compressive property at low temperature is significantly higher than that at room temperature. The degradation of the matrix properties at high temperature leads to a decrease in the dynamic mechanical properties of the composites with obvious plasticity characteristics. The material has the highest strength and modulus at room temperature, and the increase of strain rate improves the dynamic mechanical properties of the material. Li et al. [161] tested the bending properties of 3-D five-directional braided composites at room temperature, 90 °C, 110 °C, and 150 °C, and heated for 0.25 h, 10 h, and 30 h, respectively. It was found that the bending strength decreases with the increase of temperature and the effect of ambient temperature on the bending properties of the material is greater than the heating time. Zhang et al. [119] tested the transverse impact behavior of 3-D braided composite T-beams at different temperatures and impact velocities. It was found that the transverse impact properties were more sensitive to impact velocity than to temperature. As the temperature increased, the failure mechanism changed from brittle damage to ductile damage, with matrix cracking at the impact surface and fiber breakage mainly occurring at the rib and flange locations. Song et al. [162] tested the tensile strength of 3-D four-directional carbon fiber/epoxy resin braided composites after aging at 150 and 180 °C for 60, 120, and 180 h. It was found that with the increase of high-temperature accelerated aging time, the tensile strength decreased, but the decrease was less than that of the laminated composites, which was related to the damage degree of the matrix at high temperatures. Singh et al. [163] tested the three-point bending behavior of uncoated and sealed boron nitride-coated carbon fiber-reinforced silicon carbide composites at different temperatures. It was found that the bending strength of the sealed coated specimens was higher than that of the uncoated specimens at high temperatures due to the oxidation protection of the carbon fiber before loading and the different glass phase formations for sealing of microcracks, resulting in nearly 70% higher bending strength of the sealed coated specimens. At low-temperature field, Li et al. [158] conducted axial and transverse compression tests on 3-D braided composites with different braiding parameters at −196 °C. It was shown that the low temperature led to a significant increase in the compressive properties of the material, the brittle failure characteristics of the material became more pronounced, and the interfacial bonding ability was significantly enhanced.

Combining numerical simulations to investigate the effects of temperature and thermal-oxidative aging on the thermo-mechanical properties of the composites, Pan et al. [130, 131, 164–167] used test and numerical methods to study the compression behavior of 3-D braided basalt fiber tow/epoxy resin composites and biaxially warp-knitted composites at different temperatures and strain rates. It was found that fiber tows fracture at low temperatures neatly, with low fiber extraction rate and fine resin fragmentation. Microstructural models were used to reveal the thermo-mechanical behavior of the composites, and it was found that the inelastic heat generated by the matrix was greater than that of the preform and the distribution and accumulation of inelastic heat led to the formation of shear bands. The effect of fiber tow orientation and preform structure on the thermal stress distribution was revealed. Zhang et al. [167] studied the impact response behavior of carbon nanotube (CNT)-reinforced functional gradient composites with cylindrical shells. The effective properties of the material were modeled by the extended mixing law. The model adopts Reddy's high-order shear deformation theory and takes into account the thermal effect. Fourier series expansion and Laplace transform were used in the solution process, and the analytical expression of the transverse displacement was given. The results of the numerical analysis show that the impact response amplitude of CNT-rich composites on the top and bottom surfaces of the shell is lower than that of CNT-rich composites on the middle surface and unidirectional composites. The increase in the volume fraction of carbon nanotubes decreases the impact response amplitude, and the increase in temperature increases the impact amplitude of cylindrical shells. Liu et al. [168] studied the dynamic damage and failure mechanism of thermal-oxidative aging 3-D braided composites under compression conditions and found that as the aging temperature increases, the interfacial damage is more severe, and the epoxy tends to rubberize. For the low-temperature aged composites, the impact resistance degradation is mainly due to the weakened epoxy behavior, while the interfacial debonding is the main degradation mechanism for the high-temperature aged composites. The FEA results showed that the plastic deformation-induced temperature and thermal stress of aged composite were lower than those of unaged composite during impact. Zhang et al. [169–172] investigated the effect of thermal-oxidative aging on the compression properties of pure epoxy resin and 3-D carbon fiber/epoxy resin braided composites. The effect of thermal-oxidative aging on local properties of aged resin was studied by nano-indentation test. For the pure epoxy resin, degradation occurs only in the surface region, and the interior is protected from oxidation by an external oxide layer. For the composite, the matrix shrinkage induced during thermal-oxidative aging leads to interfacial damage, and interfacial cracks provide a path for oxidation expansion, leading to degradation. The compression behavior of low-temperature thermal-oxidative aged composites is similar to that of pure resin, so resin aging is the dominant aging mode. High-temperature thermal-oxidative aging leads to resin degradation and interfacial cracking of the composites, resulting in significant degradation of the compressive properties.

The research on the thermo-mechanical properties of textile composites mainly focuses on the mechanical tests at different temperatures and the mechanical properties of composites after thermal-oxidative aging, including low-velocity impact, high-velocity impact compression, bending, and tensile tests. Combined with finite element methods, the influence of temperature on the failure mode and thermal stress distribution of the composite is analyzed; however, the adiabatic effect has not fully considered.

The thermo-mechanical coupling property of composites has also gradually attracted attention. Pan et al. [173–175] used a fully coupled thermo-mechanical response model to study the compressive properties of 3-D braided composites at different temperatures. Under the influence of the preform structure, the adiabatic temperature rise formed sawtooth damage along the braiding angle in the composite. The preform was found to have a higher stress state and a lower heat generation, revealing the dynamic shear mechanism of the braided composites containing the asymptotic thermo-mechanical coupling response of the adiabatic temperature rise. Zhang et al. [176] established a thermo-mechanical coupling constitutive model which involves temperature effects and strain rate effects to reveal the coupled stress and adiabatic temperature rise distributions of 3-D braided composites under different temperatures and impact compression conditions. Shariat et al. [177] used combined Galerkin energy and finite element method to determine the thermoelastic coupling and impact control equations, which were solved by an updated Newton-Raphson-based procedure, and further investigated the impact behavior of rectangular composite plates. The results show that the temperature rise caused by impact and the temperature dependence of material properties increase the initial stress and final stress of the direct and inverse phase transition mechanisms, leading to larger hysteresis loops, strains, and stresses in the shape-memory alloy (SMA) wires. Liu et al. [178] developed a macroscale thermo-mechanical model to study the coupling effects of thermal-oxidative aging and braided structure on the high-velocity impact response of 3-D braided composites. The results showed that the adiabatic temperature rise, thermal strain, and thermal stress of the composites during impact increased with increasing braiding angle. The effect of thermal-oxidative aging time on the mechanical properties of the composites could be optimized by changing the braided structure to optimize the properties. The thermal-mechanical coupling properties of 3-D braided composites under impact compression were investigated [179], and "with thermal parameter model" and "without thermal parameter model" were established to calculate the thermal stresses induced by impact compression. The stress levels of resin and fiber tow in the model with thermal parameter were higher than those of the model without thermal parameter. Su et al. [180] proposed an effective general method for determining the optimized virtual field and extracting the elastic constants and coefficient of thermal expansion (CTE) of orthotropic anisotropic composites from the thermo-mechanical coupling deformation field and verified the effectiveness of the method by three-point bending tests. The results show that the optimized virtual field can effectively reduce the influence of noise error on the material parameters. Ye et al. [181] tested the quasi-static behavior of fiber-reinforced composite laminates in a high-temperature environment, using a representative volume element model at the microscopic scale and classical laminate theory at the macroscopic scale to describe the thermo-mechanical failure of the composites. It was found that higher ambient temperatures lead to stiffness degradation and failure strength reduction of the laminates, whether uniaxially loaded or biaxially loaded.

The current research on the analysis of thermo-mechanical coupling response of textile composites is still mainly at the macroscopic level, which can only reflect the macroscopic properties without considering the thermo-mechanical coupling mechanism at the microscopic level. In order to reveal the thermo-mechanical impact damage mechanisms, it is necessary to introduce the thermo-mechanical coupling effect under high-velocity impact.

References

1. Alyhassan, M.S., Hatta, H., Wakayama, S., et al.: Comparison of 2D and 3D carbon/carbon composites with respect to damage and fracture resistance. Carbon. **41**(5), 1069–1078 (2003)
2. Mouritz, A.P., Bannister, M.K., Falzon, P.J., et al.: Review of applications for advanced 3-D fibre textile composites. Compos. Part A. **30**(12), 1445–1461 (1999)
3. Chen, W.J.: The History of Textile Science and Technology in China (Ancient Part), pp. 21–24. Science Press, New York (1984) [In Chinese]
4. Brunnschweiler, D.: Braids and braiding. J. Text. Inst. Proc. **44**(9), 666–686 (1953)
5. Brunnschweiler, D.: The structure and tensile properties of braids. J. Text. Inst. Trans. **45**(1), 55–77 (1954)
6. Florentine, R.A.: US Patent 4,312,261, 26 Jan 1982
7. Phoenix, S.L.: Mechanical response of a tubular braided cable with an elastic core. Text. Res. J. **48**(2), 81–91 (1978)
8. Ko, F.: Braiding. In: Engineering Materials Handbook Volume 1: Composites, pp. 519–528. ASM International, Metals Park (1987)
9. Byun, J.H., Chou, T.W.: Process-microstructure relationships of 2-step and 4-step braided composites. Compos. Sci. Technol. **56**(3), 235–251 (1996)
10. Tsuzuki, M., Kimbara, M., Fukuta, K., et al.: 3-D fabric woven by interlacing threads with rotor driven carriers. Google Patents, 1991
11. Li, J., Chen, P., Elshiekh, A.: Construction and geometry of 6-step braided preforms for composites. In: International SAMPE Symposium and Exhibition: SAMPE Society for the Advancement of Material, pp. 826–833. SAMPE, Anaheim (1994)
12. Kostar, T.D., Chou, T.W.: Microstructural design of advanced multi-step 3-D braided preforms. J. Compos. Mater. **28**(13), 1180–1201 (1994)
13. Mungalov, D., Bogdanovich, A.: Complex shape 3-D braided composite preforms: structural shapes for marine and aerospace. SAMPE J. **40**(3), 7–21 (2004)
14. Mcconnell, R.F., Popper, P.: Complex shaped braided structures. U.S. Patent 4,719,837, 1988;1–19
15. Wu, D.L.: Three-cell model and 5D braided structural composites. Compos. Sci. Technol. **56** (3), 225–233 (1996)
16. Yang, Q.D., Rugg, K.L., Cox, B.N., Shaw, M.C.: Failure in the junction region of T-stiffeners: 3D-braided vs. 2D tape laminate stiffeners. Int. J. Solids Struct. **40**, 1653–1668 (2003)
17. Zeng, T., Fang, D., Lu, T.: Dynamic crashing and impact energy absorption of 3D braided composite tubes. Mater. Lett. **59**, 1491–1496 (2005)
18. Bilisik, K.: 3-D braiding for composites: a review. Text. Res. J. **83**, 1414–1436 (2013)
19. Soutis, C.: Carbon fiber reinforced plastics in aircraft construction. Mater. Sci. Eng. A. **412**, 171–176 (2005)
20. Friedrich, K., Almajid, A.A.: Manufacturing aspects of advanced polymer composites for automotive applications. Appl. Compos. Mater. **20**, 107–128 (2013)
21. Jia, Z., Li, T., Chiang, F., Wang, L.: An experimental investigation of the temperature effect on the mechanics of carbon fiber reinforced polymer composites. Compos. Sci. Technol. **154**, 53–63 (2018)
22. Gu, Q., Quan, Z., Yu, J., Yan, J., Sun, B., Xu, G.: Structural modeling and mechanical characterizing of 3-D four-step braided composites: a review. Compos. Struct. **207**, 119–128 (2019)
23. Ahmadian, H., Yang, M., Soghrati, S.: Effect of resin-rich zones on the failure response of carbon fiber reinforced polymers. Int. J. Solids Struct. **188–189**, 74–87 (2020)
24. Ko, F.: Three dimensional fabrics for composites-an introduction to the magnaweave structure. In: Proceedings of the ICCM-4 Japan Society Composite Material, pp. 1608–1609. Japan Society for Composite Materials, Tokyo (1982)
25. Du, G.W., Ko, F.K.: Unit cell geometry of 3-D braided structures. J. Reinf. Plast. Compos. **12** (7), 752–768 (1993)

26. Yang, J.M., Ma, C.L., Chou, T.W.: Fiber inclination model of 3-D textile structural composites. J. Compos. Mater. **20**(5), 472–484 (1986)
27. Cui, F., Sun, B.Z., Gu, B.H.: Fiber inclination model for finite element analysis of 3-D angle interlock woven composite under ballistic penetration. J. Compos. Mater. **45**(14), 1499–1509 (2011)
28. Li, W., Hammad, M., El-Shiekh, A.: Structural analysis of 3-D braided preforms for composites part I: the four-step preforms. J. Text. Inst. **81**(4), 491–514 (1990)
29. Wang, Y.Q., Wang, A.S.D.: Microstructure property relationship in 3-dimensionally braided fiber composites. Compos. Sci. Technol. **53**(2), 213–222 (1995)
30. Wang, Y.Q., Wang, A.S.D.: Geometric mapping of yarn structures due to shape change in 3-D braided composites. Compos. Sci. Technol. **54**(4), 359–370 (1995)
31. Wang, Y.Q., Sun, X.K.: Digital-element simulation of textile processes. Compos. Sci. Technol. **61**(2), 311–319 (2001)
32. Zhou, G., Sun, X., Wang, Y.: Multi-chain digital element analysis in textile mechanics. Compos. Sci. Technol. **64**(2), 239–244 (2004)
33. Miao, Y., Zhou, E., Wang, Y., Cheeseman, B.A.: Mechanics of textile composites: micro-geometry. Compos. Sci. Technol. **68**(7–8), 1671–1678 (2008)
34. Sun, X., Sun, C.J.: Mechanical properties of 3-D braided composites. Compos. Struct. **65** (3–4), 485–492 (2004)
35. Kalidindi, S.R., Franco, E.: Numerical evaluation of isostrain and weighted-average models for elastic moduli of 3-D composites. Compos. Sci. Technol. **57**(3), 293–305 (1997)
36. Chen, L., Tao, X.M., Choy, C.L.: On the microstructure of 3-D braided preforms. Compos. Sci. Technol. **59**(3), 391–404 (1999)
37. Zeng, T., Wu, L., Guo, L.: A damage model for 3D braided composites with transverse cracking. Compos. Struct. **62**(2), 163–170 (2003)
38. Zeng, T., Fang, D., Guo, L., Ma, L.: A mechanical model of 3D braided composites with transverse and longitudinal cracks. Compos. Struct. **69**(1), 117–125 (2005)
39. Tang, Z.X., Postle, R.: Mechanics of 3-D braided structures for composite materials – part I: fabric structure and fibre volume fraction. Compos. Struct. **49**(4), 451–459 (2000)
40. Tang, Z.X., Postle, R.: Mechanics of 3-D braided structures for composite materials – part II: prediction of the elastic moduli. Compos. Struct. **51**(4), 451–457 (2001)
41. Zeng, T., Wu, L.Z., Guo, L.C.: Mechanical analysis of 3D braided composites: a finite element model. Compos. Struct. **64**(3–4), 399–404 (2004)
42. Gu, B.H.: A microstructure model for finite-element simulation of 3D rectangular braided composite under ballistic penetration. Philos. Mag. **87**(30), 4643–4669 (2007)
43. Kang, T.J., Kim, S., Jung, K.: Prediction of mechanical properties of 3-D braided composite using CAD system. Text. Res. J. **78**(2), 174–182 (2008)
44. Kang, T.J., Kim, S., Jung, K.: Analysis of geometrical parameters using a CAD system for a 3-D braided preform. Text. Res. J. **78**(10), 922–935 (2008)
45. Zhang, W., Ding, X., Li, Y.L.: Calculation and design of parameters for four-step 3D braided preform with complex rectangular cross sections. J. Ind. Text. **38**(2), 139–150 (2008)
46. Li, D.S., Lu, Z.X., Chen, L., Li, J.L.: Microstructure and mechanical properties of 3-D five-directional braided composites. Int. J. Solids Struct. **46**(18–19), 3422–3432 (2009)
47. Li, D.S., Li, J.L., Chen, L., Lu, Z.X., Fang, D.N.: Finite element analysis of mechanical properties of 3D four-directional rectangular braided composites part 1: microgeometry and 3D finite element model. Appl. Compos. Mater. **17**(4), 373–387 (2010)
48. Li, D.S., Chen, L., Li, J.L.: Microstructure and unit-cell geometry of four-step 3-D rectangular braided composites. J. Reinf. Plast. Compos. **29**(22), 3353–3363 (2010)
49. Du, G., Popper, P., Chou, T.: Process model of circular braiding. Am. Soc. Mech. Eng. Mater. Div. (Publ.) MD. **19**, 119–133 (1990)
50. Du, G., Popper, P.: Analysis of a circular braiding process for complex shapes. J. Text. Inst. **85** (3), 316–337 (1994)

51. Zhang, Q., Beale, D., Broughton, R.: Analysis of circular braiding process, part 1: theoretical investigation of kinematics of the circular braiding process. J. Manuf. Sci. Eng. **121**(3), 345–350 (1999)
52. Zhang, Q., Beale, D., Broughton, R., et al.: Analysis of circular braiding process, part 2: mechanics analysis of the circular braiding process and experiment. J. Manuf. Sci. Eng. **121** (3), 351–359 (1999)
53. Li, W., Hammad, M., El-Shiekh, A.: Structural analysis of 3-D braided preforms for composites part II: the two-step preforms. J. Text. Inst. **81**(4), 515–537 (1990)
54. Potluri, P., Rawal, A., Rivaldi, M., et al.: Geometrical modelling and control of a triaxial braiding machine for producing 3D preforms. Compos. Part A. **34**(6), 481–492 (2003)
55. Wang, Y., Wang, A.: On the topological yarn structure of 3D rectangular and tubular braided preforms. J. Compos. Sci. Technol. **51**(4), 575–586 (1993)
56. Wang, Y.Q., Wang, A.S.D.: Spatial distribution of yarns and mechanical properties in 3D braided tubular composites. Appl. Compos. Mater. **4**(2), 121–132 (1997)
57. Li, C., Jialu, L., Xueming, L.: Analysis of 4-step 3-D tubular braiding structures. Acta Materiae Compositae Sinica. **20**(2), 76–80 (2003) [in Chinese]
58. Zhao, Y.L.: Analysis of structures and elastic properties of 3D tubular braided composites. J. Tianjin Polytech. Univ. **29**(3), 15–18 (2010) [In Chinese]
59. Sun, X.: Micro-geometry of 3-D braided tubular preform. J. Compos. Mater. **38**(9), 791–798 (2004)
60. Li, D.S., Lu, Z.X., Chen, L., et al.: Microstructure analysis and prediction of the elastic properties of 3d and 5-d tubular braided composites. Acta Aeronautica et Astronautica Sinica. **28**(1), 123–129 (2007) [In Chinese]
61. Lee, W., Kim, J.H., Shin, H.J., et al.: Constitutive equations based on cell modeling method for 3D circular braided glass fiber reinforced composites. Fibers Polym. **4**(2), 77–83 (2003)
62. Song, Y.S., Chung, K., Kang, T.J., et al.: Prediction of permeability tensor for three dimensional circular braided preform by applying a finite volume method to a unit cell. Compos. Sci. Technol. **64**(10–11), 1629–1636 (2004)
63. Hwan, C.L., Tsai, K.H., Chen, W.L., et al.: Predicting the elastic moduli of 3-D (four-step) braided tubes using a spatial spring model. J. Compos. Mater. **47**(8), 991–1000 (2012)
64. Wang, Y., Liu, Z., Liu, N., et al.: A new geometric modelling approach for 3D braided tubular composites base on Free Form Deformation. Compos. Struct. **136**, 75–85 (2015)
65. Ma, W., Zhu, J., Jiang, Y.: Studies of fiber volume fraction and geometry of variable cross-section tubular 3D five-direction braided fabric. J. Compos. Mater. **46**(14), 1697–1704 (2011)
66. Flanagan, M.P., Zikry, M.A., Wall, J.W., El-Shiekh, A.: An experimental investigation of high velocity impact and penetration failure modes in textile composites. J. Compos. Mater. **33**, 1080–1103 (1999)
67. Jenq, S.T., Mo, J.J.: Ballistic impact response for two-step braided 3-D textile composites. AIAA J. **34**(2), 375–384 (1996)
68. Jenq, S.T., Kuo, J.T., Sheu, L.T.: Ballistic impact response of 3-D four-step braided glass/ epoxy composites. Key Eng. Mater. **141**, 349–366 (1998)
69. Yang, Y., Zhang, L., Guo, L., Zhang, W., Zhao, J., Xie, W.: Dynamic response and research of 3D braided Carbon Fiber Reinforced Plastics subjected to ballistic impact loading. Compos. Struct. **206**, 578–587 (2018)
70. Shaker, M., Ko, F., Song, J.: Comparison of the low and high velocity impact response of Kevlar fiber-reinforced epoxy composites. J. Compos. Technol. Res. **21**(4), 224–229 (1999)
71. Zhang, Y., Sun, B., Gu, B.: Experimental characterization of transverse impact behaviors of four-step 3-D rectangular braided composites. J. Compos. Mater. **46**, 3017–3029 (2012)
72. Zhao, Z., Liu, L., Chen, W., Huang, X., Zhao, Z., Liu, L., et al.: Dynamic compressive behavior in different loading directions of 3D braided composites with different braiding angle. Lat. Am. J. Solids Struct. **15**(9), e109 (2018)
73. Zhang, Y., Wang, P., Guo, C.: Energy absorption behaviors of 3D braided composites under impact loadings with frequency domain analysis. Polym. Compos. **37**, 1620–1627 (2016)

74. Liu, Y., Chen, Z., Zhu, J., Jiang, Y., Li, B., Edmond, B.F.: Comparison of 3D four-directional and five-directional braided SiO2f/SiO2 composites with respect to mechanical properties and fracture behavior. Mater. Sci. Eng. A. **558**, 170–174 (2012)
75. Cui, C., Dong, J., Mao, X.: Effect of braiding angle on progressive failure and fracture mechanism of 3-D five-directional carbon/epoxy braided composites under impact compression. Compos. Struct. **229**, 111412 (2019)
76. Wang, Y.-Q., Wang, A.S.D.: Microstructure/property relationships in 3-Dly braided fiber composites. Compos. Sci. Technol. **53**, 213–222 (1995)
77. Zhang, Y., Jiang, L., Sun, B., Gu, B.: Transverse impact behaviors of four-step 3-D rectangular braided composites from unit-cell approach. J. Reinf. Plast. Compos. **31**, 233–246 (2012)
78. Wu, L., Sun, B., Gu, B.: Numerical analyses of bending fatigue of four-step 3-D rectangular-braided composite materials from unit cell approach. J. Text. Inst. **106**, 67–79 (2015)
79. Zhang, C., Xu, X.: Finite element analysis of 3D braided composites based on three unit-cells models. Compos. Struct. **98**, 130–142 (2013)
80. Zhang, C., Curiel-Sosa, J.L., Duodu, E.A.: Finite element analysis of the damage mechanism of 3D braided composites under high-velocity impact. J. Mater. Sci. **52**, 4658–4674 (2017)
81. Zhou, H., Zhang, W., Liu, T., Gu, B., Sun, B.: Finite element analyses on transverse impact behaviors of 3-D circular braided composite tubes with different braiding angles. Compos. Part A. **79**, 52–62 (2015)
82. Zhou, H., Hu, D., Zhang, W., Gu, B., Sun, B.: The transverse impact responses of 3-D braided composite I-beam. Compos. Part A. **94**, 158–169 (2017)
83. Zhou, H., Li, C., Zhang, L., Crawford, B., Milani, A.S., Ko, F.K.: Micro-XCT analysis of damage mechanisms in 3D circular braided composite tubes under transverse impact. Compos. Sci. Technol. **155**, 91–99 (2018)
84. Zhou, H., Hu, D., Gu, B., Sun, B.: Transverse impact behavior and finite element analysis of three dimensional braided composite tubes with different braiding layers. Compos. Struct. **168**, 345–359 (2017)
85. Zhou, H., Pan, Z., Gideon, R.K., Gu, B., Sun, B.: Experimental and numerical investigation of the transverse impact damage and deformation of 3-D circular braided composite tubes from meso-structure approach. Compos. Part B. **86**, 243–253 (2016)
86. Zhao, Z., Liu, P., Chen, C., Zhang, C., Li, Y.: Modeling the transverse tensile and compressive failure behavior of triaxially braided composites. Compos. Sci. Technol. **172**, 96–107 (2019)
87. Lyu, Z.-Q., Li, Y.-Y., Wang, Y.-L., Chen, T., Zhang, Y., Guo, C.-B.: Transverse impact responses and failure mechanism of 3D braided composites by mesostructure model. Polym. Compos. **41**(4), 1203–1214 (2020)
88. Li, D., Fang, D., Jiang, N., Xuefeng, Y.: Finite element modeling of mechanical properties of 3D five-directional rectangular braided composites. Compos. Part B. **42**, 1373–1385 (2011)
89. Li, D., Lu, Z., Jiang, N., Fang, D.: High strain rate behavior and failure mechanism of 3-D five-directional carbon/phenolic braided composites under transverse compression. Compos. Part B. **42**, 309–317 (2011)
90. Karbhari, V.M., Falzon, P.J., Herzberg, I.: Energy absorption characteristics of hybrid braided composite tubes. J. Compos. Mater. **31**(12), 1164–1186 (1997)
91. Chiu, C.H., Tsai, K.H., Huang, W.J.: Effects of braiding parameters on energy absorption capability of triaxially braided composite tubes. J. Compos. Mater. **32**(21), 1964–1983 (1998)
92. Karbhari, V.M., Haller, J.E.: Rate and architecture effects on progressive crush of braided tubes. Compos. Struct. **43**, 93–108 (1998)
93. Chiu, C.H., Tsai, K.H., Huang, W.J.: Crush-failure modes of 2D triaxially braided hybrid composite tubes. Compos. Sci. Technol. **59**(11), 1713–1723 (1999)
94. Chiu, C.H., Lu, C.K., Wu, C.M.: Crushing characteristics of 3-D braided composite square tubes. J. Compos. Mater. **31**(22), 2309–2327 (1997)
95. Xiao, X., Mcgregor, C., Vaziri, R., et al.: Progress in braided composite tube crush simulation. Int. J. Impact Eng. **36**(5), 711–719 (2009)

96. Carla, M., Reza, V., Xinran, X.: Finite element modelling of the progressive crushing of braided composite tubes under axial impact. Int. J. Impact Eng. **37**(6), 662–672 (2010)
97. Zhang, P., Gui, L., Fan, Z., et al.: Finite element modeling of the quasi-static axial crushing of braided composite tubes. Comput. Mater. Sci. **73**, 146–153 (2013)
98. Wu, X., Zhang, Q., Zhang, W., et al.: Axial compressive deformation and damage of four-step 3-D circular braided composite tubes under various strain rates. J. Text. Inst. **107**, 1584–1600 (2016)
99. Wu, X., Gu, B., Sun, B.: Comparisons of axial compression behaviors between four-directional and five-directional braided composite tubes under high strain rate loading. J. Compos. Mater. **50**, 3905–3924 (2016). https://doi.org/10.1177/0021998316628619
100. Gideon, R.K., Sun, B., Gu, B.: Mechanical behaviors of 4-step 1 × 1 braided carbon/epoxy 3-D composite tubes under axial compression loading. Polym. Compos. (2015). https://doi.org/10.1002/pc.23519
101. Wu, X.Q., Li, J.L., Cui, Z.X., et al.: Design and manufacture of 3-D integrally braided composite pipes. Fiber Reinf. Plast. Compos. **4**, 21–23 (1998) [In Chinese]
102. Tate, J.S., Kelkar, A.D., Whitcomb, J.D.: Effect of braid angle on fatigue behavior of biaxial braided composites. Int. J. Fatigue. **28**(10), 1239–1247 (2006)
103. Gideon, R.K., Zhou, H., Li, Y., et al.: Quasi-static compression and compression–compression fatigue characteristics of 3D braided carbon/epoxy tube. J. Text. Inst. **107**, 938–948 (2016)
104. Gideon, R.K., Zhou, H., Wu, X., et al.: Finite element analysis of 3D circular braided composites tube damage based on three unit cell models under axial compression loading. Int. J. Damage Mech. **25**(4), 574–607 (2016)
105. Calme, O.: 3D braided composite rings under lateral compression. Compos. Sci. Technol. **65** (1), 95–106 (2005)
106. Hamada, H., Nakai, A., Masui, M.: Mechanical properties for self-reinforced braided composite tubes. J. Thermoplast. Compos. Mater. **12**(3), 164–176 (1999)
107. Karbhari, V.M., Haller, J.E., Falzon, P.K., et al.: Post-impact crush of hybrid braided composite tubes. Int. J. Impact Eng. **22**, 419–433 (1999)
108. Potluri, P., Manan, A., Francke, M., et al.: Flexural and torsional behaviour of biaxial and triaxial braided composite structures. Compos. Struct. **75**(1–4), 377–386 (2006)
109. Harte, A.M., Fleck, N.A.: Deformation and failure mechanisms of braided composite tubes in compression and torsion. Acta Mater. **48**(6), 1259–1271 (2000)
110. Ko, F.K., Hartman, D.: Impact behavior of 2-D and 3-D glass/epoxy composites. In: Materials Science for the Future: 31st International SAMPE Symposium and Exhibition, pp. 1272–1284. SAMPE, Covina (1986)
111. Gong, J., Sankar, B.: Impact properties of 3-D braided graphite/epoxy composites. J. Compos. Mater. **25**(6), 715–731 (1991)
112. Mark, J.F., Kurtis, W.: Characterization and manufacture of braided composites for large commercial aircraft structures. FAA, Ninth DOD(NASA)FAA Conference on Fibrous Composites in Structural Design, Volume 2 (1992)
113. Sun, B., Liu, F., Gu, B.: Influence of the strain rate on the uniaxial tensile behavior of 4-step 3D braided composites. Compos. Part A. **36**(11), 1477–1485 (2005)
114. Zhang, D., Chen, L., Sun, Y., et al.: Transverse tensile damage behaviors of 3-D five-directional braided composites by meso-scale finite element approach. J. Reinf. Plast. Compos. **34**(15), 1202–1220 (2015)
115. Sun, B., Gu, B.: High strain rate behavior of 4-step 3D braided composites under compressive failure. J. Mater. Sci. **42**(7), 2463–2470 (2006)
116. Zhang, F., Wan, Y., Gu, B., et al.: Impact compressive behavior and failure modes of four-step 3-D braided composites-based meso-structure model. Int. J. Damage Mech. **24**(6), 805–827 (2015)

117. Li, Y., Gan, X., Gu, B., et al.: Dynamic responses and damage evolutions of four-step 3-D braided composites subjected to high strain rate punch shear loading. J. Compos. Mater. **50** (12), 1635–1650 (2015)
118. Li, Y., Zhang, W., Gideon, R.K., et al.: Finite element analyses on punch shear behaviors of 3-D braided composites at microstructure level. Int. J. Damage Mech. **26**, 968–988 (2017). https://doi.org/10.1177/1056789516639744
119. Zhang, W., Gu, B., Sun, B.: Transverse impact behaviors of 3D braided composites T-beam at elevated temperatures. J. Compos. Mater. **50**, 3961–3971 (2016)
120. Zhang, W., Pan, Z., Gideon, R.K., et al.: Effects of temperature and strain rate on impact compression behaviors of 3-D carbon fiber/epoxy braided composites. J. Compos. Mater. **49** (7), 771–782 (2014)
121. Li, D., Lu, Z., Jiang, N., et al.: Influence of braiding parameters on the dynamic compressive properties and failure mechanism of 3D braided composites. Mech. Adv. Mater. Struct. **21**(9), 716–721 (2014)
122. Wan, Y., Wang, Y., Gu, B.: Finite element prediction of the impact compressive properties of 3-D braided composites using multi-scale model. Compos. Struct. **128**, 381–394 (2015)
123. Feng, Z.H., Xu, B., Gao, W., et al.: Study on the explosive impact behavior of 3D braided C/E composites. Aerosp. Mater. Technol. **05**, 45–48 (2000) [In Chinese]
124. Shen, H.R.: Impact experiment and analysis of three braided composite materials. J. Inst. Command Technol. **02**, 84–87 (2002) [In Chinese]
125. Gu, B., Xu, J.: Finite element calculation of 4-step 3-dimensional braided composite under ballistic perforation. Compos. Part B. **35**(4), 291–297 (2004)
126. Gu, B.: A refined quasi-microstructure model for finite element analysis of 3-D braided composites under ballistic penetration. J. Compos. Mater. **39**(8), 685–710 (2005)
127. Li, D., Lu, Z., Fang, D.: Longitudinal compressive behavior and failure mechanism of 3-D five-directional carbon/phenolic braided composites at high strain rates. Mater. Sci. Eng. A. **526**(1), 134–139 (2009)
128. Shi, Y., Jinyang, Z., Xiamei, L., et al.: An experimental investigation on the low-velocity impact behavior of 3D five-directional braided composites. Polym. Adv. Technol. **25**(12), 1386–1390 (2014)
129. Zhang, D., Chen, L., Sun, Y., et al.: Meso-scale progressive damage of 3D five-directional braided composites under transverse compression. J. Compos. Mater. **50**, 3345–3361 (2016). https://doi.org/10.1177/0021998315619999
130. Pan, Z., Gu, B., Sun, B.: Experimental investigation of high-strain rate properties of 3-D braided composite material in cryogenic field. Compos. Part B. **77**, 379–390 (2015)
131. Pan, Z., Gu, B., Sun, B.: Thermo-mechanical behaviors of 3-D braided composite material subject to high strain rate compressions under different temperatures. Mech. Adv. Mater. Struct. **23**(4), 385–401 (2016)
132. Zheng, X.X., Zheng, X.T., Gou, L.H.: The research progress on multiscale method for the mechanical analysis of composites. Adv. Mech. **40**(1), 41–56 (2010) [In Chinese]
133. Ji, X.B., Khatri, A.M., Chia, E.S.M., et al.: Multi-scale simulation and finite-element-assisted computation of elastic properties of braided textile reinforced composites. J. Compos. Mater. **48**(8), 931–949 (2014)
134. Zuo, Z.H., Huang, X.D., Rong, J.H., et al.: Multi-scale design of composite materials and structures for maximum natural frequencies. Mater. Des. **51**, 1023–1034 (2013)
135. Jia, X.W., Xia, Z.H., Gu, B.H.: Nonlinear viscoelastic multi-scale repetitive unit cell model of 3D woven composites with damage evolution. Int. J. Solids Struct. **50**, 3539–3554 (2013)
136. Zhang, C., Binienda, W.K.: A meso-scale finite element model for simulating free-edge effect in carbon/epoxy textile composite. Mech. Mater. **76**, 1–19 (2014)
137. Jia, X.W., Sun, B.Z., Gu, B.H.: Ballistic penetration of conically cylindrical steel projectile into 3D orthogonal woven composite: a finite element study at microstructure level. J. Compos. Mater. **45**(9), 965–987 (2011)

138. Hou, Y.Q., Jiang, L.L., Sun, B.Z., et al.: Strain rate effects of tensile behaviors of 3-D orthogonal woven fabric: experimental and finite element analyses. Text. Res. J. **83**(4), 337–354 (2013)
139. Thammandra, V., Potluri, P.: Influence of uniaxial and biaxial tension on meso-scale geometry and strain fields in a woven composite. Compos. Struct. **77**(3), 405–418 (2007)
140. Liu, Y., Chen, Z.F., Zhu, J.X., et al.: Comparison of 3D four-directional and five-directional braided SiO 2f/SiO 2 composites with respect to mechanical properties and fracture behavior. Mater. Sci. Eng. A. **558**, 170–174 (2012)
141. Fang, G.D., Liang, J., Wang, B.L., et al.: Effect of interface properties on mechanical behavior of 3D four-directional braided composites with large braid angle subjected to uniaxial tension. Appl. Compos. Mater. **18**(5), 449–465 (2011)
142. Jiang, L.L., Zeng, T., Yan, S., et al.: Theoretical prediction on the mechanical properties of 3D braided composites using a helix geometry model. Compos. Struct. **100**, 511–516 (2013)
143. Gan, X.H., Yan, J.H., Gu, B.H., et al.: Impact tensile behavior and frequency response of 3D braided composites. Text. Res. J. **82**(3), 280–287 (2012)
144. Xu, K.: A numerical study on the progressive failure of 3D four-directional braided composites. Adv. Mater. Sci. Eng. **2013**, 1–14 (2013)
145. Gu, B.H., Li, Y.L.: Ballistic perforation of conically cylindrical steel projectile into 3-D braided composites. AIAA J. **43**(2), 426–434 (2005)
146. Sun, B.Z., Liu, R.Q., Gu, B.H.: Numerical simulation of three-point bending fatigue of four-step 3-D braided rectangular composite under different stress levels from unit-cell approach. Comput. Mater. Sci. **65**, 239–246 (2012)
147. Sun, B.Z., Zhang, Y., Gu, B.H.: Low-velocity impact response and finite element analysis of four-step 3-D braided composites. Appl. Compos. Mater. **20**(4), 397–413 (2013)
148. Tang, X.D., Whitcomb, J.D., Kelkar, A.D., et al.: Progressive failure analysis of 2× 2 braided composites exhibiting multiscale heterogeneity. Compos. Sci. Technol. **66**(14), 2580–2590 (2006)
149. Yu, X.G., Cui, J.Z.: The prediction on mechanical properties of 4-step braided composites via two-scale method. Compos. Sci. Technol. **67**(3), 471–480 (2007)
150. Flesher, N.D., Chang, F.K., Janapala, N.R.: A dynamic crash model for energy absorption in braided composite materials. Part I: Viscoplastic material model. J. Compos. Mater. **45**(8), 853–865 (2011)
151. Deng, Y., Chen, X.H., Wang, H.: A multi-scale correlating model for predicting the mechanical properties of tri-axial braided composites. J. Reinf. Plast. Compos. **32**(24), 1934–1955 (2013)
152. Cai, Y.M., Sun, H.Y.: Prediction on viscoelastic properties of 3-Dly braided composites by multi-scale model. J. Mater. Sci. **48**(19), 6499–6508 (2013)
153. Lu, Z.X., Zhou, Y., Yang, Z.Y., et al.: Multi-scale finite element analysis of 2.5 D woven fabric composites under on-axis and off-axis tension. Comput. Mater. Sci. **79**, 485–494 (2013)
154. Fang, G.D., Liang, J., Wang, B.L.: Progressive damage and nonlinear analysis of 3D four-directional braided composites under unidirectional tension. Compos. Struct. **89**(1), 126–133 (2009)
155. Li, J.C., Chen, L., Zhang, Y.F., et al.: Microstructure and finite element analysis of 3D five-directional braided composites. J. Reinf. Plast. Compos. **31**(2), 107–115 (2012)
156. Halvorsen, A., Salehi-Khojn, A., Mahinfalah, M., Nakhaei-Jazar, R.: Temperature effects on the impact behavior of fiberglass and fiberglass/Kevlar sandwich composites. Appl. Compos. Mater. **13**, 369–383 (2006)
157. Li, D., Fang, D., Zhang, G., Hu, H.: Effect of temperature on bending properties and failure mechanism of 3-D braided composite. Mater. Des. **41**, 167–170 (2012)
158. Li, D., Zhao, C., Ge, T., Jiang, L., Huang, C., Jiang, N.: Experimental investigation on the compression properties and failure mechanism of 3D braided composites at room and liquid nitrogen temperature. Compos. Part B. **56**, 647–659 (2014)

159. Li, D., Chen, H., Ge, D., Jiang, N., Jiang, L.: Split Hopkinson pressure bar testing of 3D multi-axial warp knitted carbon/epoxy composites. Compos. Part B. **79**, 692–705 (2015)
160. Li, D., Duan, H.-W., Wang, W., Ge, D., Jiang, L., Yao, Q.: Strain rate and temperature effect on mechanical properties and failure of 3D needle-punched carbon/carbon composites under dynamic loading. Compos. Struct. **172**, 229–241 (2017)
161. Li, P., Jia, N., Pei, X., Wan, Z., Li, J., Zheng, Z., et al.: Effects of temperature on bending properties of 3-D and five-directional braided composite. Molecules. **24**, 3977 (2019)
162. Song, L., Li, J.: Effects of heat accelerated aging on tensile strength of three dimensional braided/epoxy resin composites. Polym. Compos. **33**, 1635–1643 (2012)
163. Singh, S., Singh, V., Kumari, S., Udayakumar, A., Bhanu Prasad, V.V.: Thermo-mechanical characterization of carbon fiber reinforced silicon carbide composites having boron nitride as an interphase material. Ceram. Int. **44**, 20755–20761 (2018)
164. Pan, Z., Gu, B., Sun, B.: Longitudinal compressive behaviour of 3D braided composite under various temperatures and strain rates. Appl. Phys. A Mater. Sci. Process. **118**, 1315–1337 (2015)
165. Pan, Z., Zhang, Q., Gu, B., Sun, B.: Effect of temperature and strain rate on biaxial warp-knitted composite. J. Reinf. Plast. Compos. **35**, 295–304 (2016)
166. Pan, Z., Gu, B., Sun, B.: Numerical analyses of thermo-mechanical behaviors of 3-D rectangular braided composite under different temperatures. J. Text. Inst. **106**, 173–186 (2015)
167. Zhang, L.W., Song, Z.G., Qiao, P., Liew, K.M.: Modeling of dynamic responses of CNT-reinforced composite cylindrical shells under impact loads. Comput. Methods Appl. Mech. Eng. **313**, 889–903 (2017)
168. Liu, S., Zhang, J., Shi, B., Wang, L., Gu, B., Sun, B.: Damage and failure mechanism of 3D carbon fiber/epoxy braided composites after thermo-oxidative ageing under transverse impact compression. Compos. Part B. **161**, 677–690 (2019)
169. Zhang, M., Sun, B., Gu, B.: Meso-structure ageing mechanism of 3-D braided composite's compressive behaviors under accelerated thermo-oxidative ageing environment. Mech. Mater. **115**, 47–63 (2017)
170. Zhang, M., Zuo, C., Sun, B., Gu, B.: Thermal ageing degradation mechanisms on compressive behavior of 3-D braided composites in experimental and numerical study. Compos. Struct. **140**, 180–191 (2016)
171. Zhang, M., Sun, B., Gu, B.: Accelerated thermal ageing of epoxy resin and 3-D carbon fiber/epoxy braided composites. Compos. Part A. **85**, 163–171 (2016)
172. Zhang, M., Sun, B., Gu, B.: Experimental and numerical analyses of matrix shrinkage and compressive behavior of 3-D braided composite under thermo-oxidative ageing conditions. Compos. Struct. **204**, 320–332 (2018)
173. Pan, Z., Sun, B., Gu, B.: Thermo-mechanical numerical modeling on impact compressive damage of 3-D braided composite materials under room and low temperatures. Aerosp. Sci. Technol. **54**, 23–40 (2016)
174. Pan, Z., Sun, B., Shim, V.P.W., Gu, B.: Transient heat generation and thermo-mechanical response of epoxy resin under adiabatic impact compressions. Int. J. Heat Mass Transf. **95**, 874–889 (2016)
175. Pan, Z., Wu, X., Wu, L.: Temperature rise caused by adiabatic shear failure in 3D braided composite tube subjected to axial impact compression. J. Compos. Mater. **54**(10), 1305–1326 (2020)
176. Zhang, W., Gu, B., Sun, B.: Thermal-mechanical coupling modeling of 3D braided composite under impact compression loading and high temperature field. Compos. Sci. Technol. **140**, 73–88 (2017)
177. Shariyat, M., Niknami, A.: Layerwise numerical and experimental impact analysis of temperature-dependent transversely flexible composite plates with embedded SMA wires in thermal environments. Compos. Struct. **153**, 692–703 (2016)

178. Liu, S., Zhang, J., Chen, Z., Gu, B., Sun, B.: Modeling the coupling effects of braiding structure and thermo-oxidative aging on the high-speed impact responses of 3D braided composites. Thin-Walled Struct. **150**, 106705 (2020)
179. Liu, S., Shi, B., Siddique, A., Du, Y., Sun, B., Gu, B.: Numerical analyses on thermal stress distribution induced from impact compression in 3D carbon fiber/epoxy braided composite materials. J. Therm. Stresses. **41**, 903–919 (2018)
180. Su, Y.Q., Yao, X.F., Wang, S., Ma, Y.J.: Simultaneous determination of virtual fields and material parameters for thermo-mechanical coupling deformation in orthotropic materials. Mech. Mater. **124**, 33–44 (2018)
181. Ye, J., Wang, Y., Li, Z., Saafi, M., Jia, F., Huang, B., et al.: Failure analysis of fiber-reinforced composites subjected to coupled thermo-mechanical loading. Compos. Struct. **235**, 111756 (2020)

Chapter 2
Thermo-mechanical Coupling Constitutive Equations of Braided Composites

This chapter reports the thermo-mechanical coupled constitutive relationship materials under impact loading. The stiffness matrix is calculated using the bridge model. Elastic-plastic behavior of fiber tows is defined using Hooke's law and Hill's anisotropic plasticity model. The elastic-plastic behavior of the matrix is defined using Hooke's law, J2 isotropic hardening plasticity theory, and von Mises yield criterion. Constitutive equation is constructed with a combination of stiffness damage evolution law, adhesive-cohesive model of interface damage, and thermo-mechanical coupling. The ductile failure and shear failure criteria are introduced to calculate the impact damage.

2.1 Stiffness Matrix of Constituents

The yarns in the braided composite are in the untwisted state, which are actually parallel arranged filaments. After curing of the composite, the yarns are impregnated in the resin matrix, which can be considered as unidirectional lamina (UD), and have transverse isotropic characteristics. The resin is isotropic, and the compliance matrix of the two materials is shown in Eqs. (2.1) and (2.2).

M. Hu, B. Gu, *Impact Damages of Braided Composites*, Engineering Materials,
https://doi.org/10.1007/978-981-16-5703-0_2

$$[S^f] = \begin{bmatrix} 1/E_{11}^f & -\nu_{12}^f/E_{11}^f & -\nu_{12}^f/E_{11}^f & 0 & 0 & 0 \\ -\nu_{12}^f/E_{11}^f & 1/E_{22}^f & -\nu_{23}^f/E_{22}^f & 0 & 0 & 0 \\ -\nu_{12}^f/E_{11}^f & -\nu_{23}^f/E_{22}^f & 1/E_{22}^f & 0 & 0 & 0 \\ 0 & 0 & 0 & 1/G_{23}^f & 0 & 0 \\ 0 & 0 & 0 & 0 & 1/G_{12}^f & 0 \\ 0 & 0 & 0 & 0 & 0 & 1/G_{12}^f \end{bmatrix} \tag{2.1}$$

$$[S_m] = \begin{bmatrix} 1/E_m & -\nu_m/E_m & -\nu_m/E_m & 0 & 0 & 0 \\ -\nu_m/E_m & 1/E_m & -\nu_m/E_m & 0 & 0 & 0 \\ -\nu_m/E_m & -\nu_m/E_m & 1/E_m & 0 & 0 & 0 \\ 0 & 0 & 0 & 1/G_m & 0 & 0 \\ 0 & 0 & 0 & 0 & 1/G_m & 0 \\ 0 & 0 & 0 & 0 & 0 & 1/G_m \end{bmatrix} \tag{2.2}$$

The engineering constants for the fiber tows [1] can be predicted as follows:

$$E_{11} = \delta_f E_{11}^f + \delta_m E_m \tag{2.3}$$

$$E_{22} = E_{33} = E_m / \left[1 - \sqrt{\delta_f}\left(1 - E_m/E_{22}^f\right)\right] \tag{2.4}$$

$$\nu_{12} = \nu_{13} = \delta_f \nu_{12}^f + \delta_m \nu_m \tag{2.5}$$

$$\nu_{23} = \delta_f \nu_{23}^f + \delta_m \nu_m \tag{2.6}$$

$$G_{12} = G_{13} = G_m / \left[1 - \sqrt{\delta_f}\left(1 - G_m/G_{12}^f\right)\right] \tag{2.7}$$

$$G_{23} = G_m / \left[1 - \sqrt{\delta_f}\left(1 - G_m/G_{23}^f\right)\right] \tag{2.8}$$

$$\delta_f + \delta_m = 1 \tag{2.9}$$

where S^f; E_{11}^f, E_{22}^f; ν_{12}^f, ν_{23}^f; and G_{12}^f, G_{23}^f are the compliance matrix, elastic modulus, Poisson's ratio, and shear modulus of the carbon fibers in the axial and radial directions, respectively; S_m, E_m, ν_m, and G_m are the compliance matrix, elastic modulus, Poisson's ratio, and shear modulus of the resin matrix, respectively; E_{11}, E_{22}; ν_{12}, ν_{23}; and G_{12}, G_{23} are the elastic modulus, Poisson's ratio, and shear modulus of the unidirectional lamina in the axial and radial directions, respectively; δ_f and δ_m are the volume fraction occupied by the carbon fibers and resin, respectively. δ_f is also known as the fiber filling factor; it can be obtained from the following equations [2]:

$$\delta_f = \frac{\pi D_f^2}{4S_y} \tag{2.10}$$

$$D_f = \sqrt{\frac{4\lambda}{\pi\rho}} \tag{2.11}$$

where S_y is the yarn cross-sectional area and D_f is the equivalent diameter of carbon fiber tow, which is related to the fiber linear density λ and density ρ.

The global compliance matrix is transformed into an engineering constant for fiber tows impregnated with resin using the bridge model [3]:

$$[S]^y = \left(\delta_f\left[S^f\right] + \delta_m[S^m][A]\right)\left(\delta_f[I] + \delta_m[A]\right)^{-1} \tag{2.12}$$

where $[S^f]$ and $[S^m]$ are the compliance matrices of the fiber and matrix, δ_f and δ_m are the volume fractions of the fibers and matrix in the fiber tow, $[I]$ is the unit matrix, and $[A]$ is the bridging matrix:

$$[A] = \begin{bmatrix} \frac{E_m}{E_{11}^f} & \frac{\left(S_{12}^f - S_{12}^m\right)(a_{11}-a_{22})}{\left(S_{11}^f - S_{11}^m\right)} & \frac{\left(S_{12}^f - S_{12}^m\right)(a_{11}-a_{22})}{\left(S_{11}^f - S_{11}^m\right)} & 0 & 0 & 0 \\ 0 & \frac{1}{2}\left(1+\frac{E_m}{E_{22}^f}\right) & 0 & 0 & 0 & 0 \\ 0 & 0 & \frac{1}{2}\left(1+\frac{E_m}{E_{22}^f}\right) & 0 & 0 & 0 \\ 0 & 0 & 0 & \frac{1}{2}\left(1+\frac{E_m}{E_{22}^f}\right) & 0 & 0 \\ 0 & 0 & 0 & 0 & \frac{1}{2}\left(1+\frac{G_m}{G_{12}^f}\right) & 0 \\ 0 & 0 & 0 & 0 & 0 & \frac{1}{2}\left(1+\frac{G_m}{G_{12}^f}\right) \end{bmatrix} \tag{2.13}$$

where E and G are the modulus of elastic and shear modulus and f and m denote the fiber and matrix, respectively.

The equivalent thermal expansion coefficient [4] of the fiber tow varies with δ_f:

$$\alpha_{11} = \left(\alpha_{11}^f E_{11}^f \delta_f + \alpha_m E_m \delta_m\right) / \left(E_{11}^f \delta_f + E_m \delta_m\right) \tag{2.14}$$

$$\alpha_{22} = \alpha_{33} = \left(1 + \nu_{12}^f\right)\alpha_{22}^f \delta_f + (1 + \nu_m)\alpha_m \delta_m - \alpha_{11}\nu_{12} \tag{2.15}$$

where α_{11} and α_{22} are the equivalent CTE for the longitudinal and transverse directions of the fiber tow impregnated with resin and ν denotes the Poisson's ratio. The mechanical parameters of the fiber tows are entered into the finite element model for calculation in the subsequent content.

The resin matrix is considered to be an isotropic elastic-plastic material, and the specific heat change of the epoxy resin was determined using differential scanning calorimetry (DSC) [5], which is also used as input parameters for the constitutive model.

Braided composites are often divided into three regions from the fiber tow arrangement, corresponding to three typical unit cells (UC), i.e., interior unit cell, surface unit cell, and corner unit cell. To obtain the mechanical properties of the macroscopic homogeneous composites, the transpose matrix is used to calculate the mechanical properties in the global coordinate system. The compliance matrix in the local coordinate system is transformed into the global coordinate system by means of Eqs. (2.16), (2.17a), and (2.17b):

$$[S]_g = \left[T_{ij}\right]_s [S]_l \left[T_{ij}\right]_s^T \tag{2.16}$$

where $[S]_g$ and $[S]_l$ are the compliance matrices in the global and local coordinate systems and $[T_{ij}]_s$ are the transpose matrix:

$$\left[T_{ij}\right]_s = \begin{bmatrix} l_1^2 & l_2^2 & l_3^2 & l_2 l_3 & l_3 l_1 & l_1 l_2 \\ m_1^2 & m_2^2 & m_3^2 & m_2 m_3 & m_3 m_1 & m_1 m_2 \\ n_1^2 & n_2^2 & n_3^2 & n_2 n_3 & n_3 n_1 & n_1 n_2 \\ 2m_1 n_1 & 2m_2 n_2 & 2m_3 n_3 & m_2 n_3 + m_3 n_2 & n_3 m_1 + n_1 m_3 & m_1 n_2 + m_2 n_1 \\ 2l_1 n_1 & 2l_2 n_2 & 2l_3 n_3 & l_2 n_3 + l_3 n_2 & n_3 l_1 + n_1 l_3 & l_1 n_2 + l_2 n_1 \\ 2m_1 l_1 & 2m_2 l_2 & 2m_3 l_3 & l_2 m_3 + l_3 m_2 & m_3 l_1 + m_1 l_3 & l_1 m_2 + l_2 m_1 \end{bmatrix} \tag{2.17a}$$

$$\begin{bmatrix} l_1 & m_1 & n_1 \\ l_2 & m_2 & n_2 \\ l_3 & m_3 & n_3 \end{bmatrix} = \begin{bmatrix} \cos\eta_x \sin\eta_z & \sin\eta_x \sin\eta_z & \cos\eta_z \\ -\sin\eta_x & \cos\eta_x & 0 \\ -\cos\eta_x \cos\eta_z & -\sin\eta_x \cos\eta_z & \sin\eta_z \end{bmatrix} \tag{2.17b}$$

where η_x is the angle between the projection of the yarn in the X-Y plane and the X-axis and η_z is the angle between the straight yarn and the Z-axis.

The stiffness matrix of each unit cell was calculated using the mixing law. The overall stiffness matrix of the macroscopic composite is derived by the volume averaging method:

$$[C] = [S]^{-1} \tag{2.18a}$$

$$[C]_{total} = \sum V_n [C_n]_g \tag{2.18b}$$

where V_n is the volume fraction of each unit cell.

2.2 Elastic-Plastic Constitutive Equations of Constituents

Yarns are considered to be elastic-plastic materials, and yarns are in an elastic state until yielding and obey Hooke's law, expressed by the equation:

$$\{\varepsilon_i\} = [S_{ij}]\{\sigma_j\} \tag{2.19a}$$

$$[S_{ij}] = \begin{bmatrix} [S_{ij}]_\sigma & 0 \\ 0 & [S_{ij}]_\tau \end{bmatrix} \tag{2.19b}$$

where $i, j = 1, 2, 3, 4, 5, 6$ and $[S_{ij}]$ is the symmetric compliance matrix of the five independent mechanical constants of the yarn given by the bridge model. Since yarns are considered to be transversely isotropic unidirectional composites, thus:

$$[S_{ij}]_\sigma = \begin{bmatrix} \frac{1}{E_{11}} & -\frac{\nu_{12}}{E_{11}} & -\frac{\nu_{12}}{E_{11}} \\ -\frac{\nu_{12}}{E_{11}} & \frac{1}{E_{22}} & -\frac{\nu_{23}}{E_{22}} \\ -\frac{\nu_{12}}{E_{11}} & -\frac{\nu_{23}}{E_{22}} & \frac{1}{E_{22}} \end{bmatrix} \tag{2.20a}$$

$$[S_{ij}]_\tau = \begin{bmatrix} \frac{1}{G_{23}} & 0 & 0 \\ 0 & \frac{1}{G_{12}} & 0 \\ 0 & 0 & \frac{1}{G_{12}} \end{bmatrix} \tag{2.20b}$$

where E, G, and ν denote the elastic modulus, shear modulus, and Poisson's ratio, respectively.

When the yielding stage is reached, Hill's anisotropic plasticity model is used to define the plastic behavior of the yarn [6]. The function can be expressed as:

$$f(\sigma) = \sqrt{F(\sigma_{22} - \sigma_{33})^2 + G(\sigma_{33} - \sigma_{11})^2 + H(\sigma_{11} - \sigma_{22})^2 + 2L\sigma_{23}^2 + 2M\sigma_{31}^2 + 2N\sigma_{12}^2} \tag{2.21a}$$

$$F = \frac{\sigma_0^2}{2}\left(\frac{1}{\overline{\sigma}_{22}^2} + \frac{1}{\overline{\sigma}_{33}^2} - \frac{1}{\overline{\sigma}_{11}^2}\right) \tag{2.21b}$$

$$H = \frac{\sigma_0^2}{2}\left(\frac{1}{\overline{\sigma}_{33}^2} + \frac{1}{\overline{\sigma}_{11}^2} - \frac{1}{\overline{\sigma}_{22}^2}\right) \tag{2.21c}$$

$$G = \frac{\sigma_0^2}{2}\left(\frac{1}{\overline{\sigma}_{11}^2} + \frac{1}{\overline{\sigma}_{22}^2} - \frac{1}{\overline{\sigma}_{33}^2}\right) \tag{2.21d}$$

$$L = \frac{3}{2}\left(\frac{\tau_0}{\overline{\tau}_{23}}\right)^2 \tag{2.21e}$$

$$M = \frac{3}{2}\left(\frac{\tau_0}{\overline{\tau}_{13}}\right)^2 \tag{2.21f}$$

$$N = \frac{3}{2}\left(\frac{\tau_0}{\overline{\tau}_{12}}\right)^2 \tag{2.21g}$$

where σ_0 is the user-defined yield stress and $\tau_0 = \sigma_0/\sqrt{3}$. $\overline{\sigma}_{ij}$ is the measured yield stress. The flow law of the yarn is defined as:

$$d\varepsilon_p = d\lambda \frac{\partial f}{\partial \sigma} \tag{2.22}$$

where $d\lambda$ ($d\lambda > 0$) is the hardening parameter.

We consider the resin matrix as an isotropic elastic-plastic material and define the elastic-plastic behavior using Hooke's law (Eqs. 2.19a and 2.19b) and the von Mises yield criterion. The expression of the von Mises yield function is given by:

$$f = \sqrt{\frac{(\sigma_1 - \sigma_2)^2 + (\sigma_2 - \sigma_3)^2 + (\sigma_3 - \sigma_1)^2}{2}} - \sigma_y \tag{2.23}$$

where σ_y is the yield stress. The J2 isotropic hardening plasticity theory was selected to define the plastic behavior of the epoxy resin at yield.

The strain rate effect of resin matrix was based on the Cowper-Symonds model [7]:

$$\dot{\overline{\varepsilon}}^{pl} = D(R-1)^n \tag{2.24}$$

$$R = \sigma/\sigma_0 \tag{2.25}$$

where $\dot{\overline{\varepsilon}}^{pl}$ is the plastic strain rate; D and n are the material constant; R is the yield stress rate; σ and σ_0 are the yield stress at the current strain rate and the yield stress under quasi-static loading, respectively; and $\sigma > \sigma_0$ is satisfied.

2.3 Failure Criteria

Under impact loading, there is significant plastic deformation of the composite prior to damage initiation and evolution. The ductile criterion and shear criterion are used to simulate the failure and damage processes of composites under transverse impact loading. The simplified formulas for the ductile and shear criteria are as follows:

$$\overline{\varepsilon}^{pl}_{ductile}\left(\eta,\dot{\overline{\varepsilon}}^{pl}\right)=\frac{\varepsilon^{+}_{ductile}\,sinh\left[k_{ductile}(\eta^{-}-\eta)\right]+\varepsilon^{-}_{ductile}\,sinh\left[k_{ductile}(\eta-\eta^{+})\right]}{sinh\left[k_{ductile}(\eta^{-}-\eta^{+})\right]} \quad (2.26)$$

$$\overline{\varepsilon}^{pl}_{shear}\left(\tau_{shear},\dot{\overline{\varepsilon}}^{pl}\right)=$$
$$\frac{\varepsilon^{+}_{shear}\,sinh\left[f_{shear}\left(\tau_{shear}-\tau^{-}_{shear}\right)\right]+\varepsilon^{-}_{shear}\,sinh\left[f_{shear}\left(\tau^{+}_{shear}-\tau_{shear}\right)\right]}{sinh\left[f_{shear}\left(\tau^{+}_{shear}-\tau^{-}_{shear}\right)\right]} \quad (2.27)$$

where $\varepsilon^{+}_{ductile}$ and $\varepsilon^{-}_{ductile}$ are the equivalent plastic strains of uniaxial tension and compression at the beginning of ductile damage, ε^{+}_{shear} and ε^{-}_{shear} correspond to the equivalent plastic strains of uniaxial tension and compression at the beginning of shear damage, and $\eta^{+}=1/3$ and $\eta^{-}=-1/3$ are the stress triaxiality in uniaxial tension and compression. In addition, $\tau_{shear}=(1-k_{shear}\eta)/\phi$, where k_{shear} is the material parameter and $\phi=\tau_{max}/\sigma_{eq}$. $k_{ductile}$ and f_{shear} are the material parameters.

When Eqs. (2.28a) and (2.28b) is satisfied, the damage initiation criterion takes effect:

$$\omega_D=\int\frac{d\overline{\varepsilon}^{pl}}{\overline{\varepsilon}^{pl}_D\left(\eta,\dot{\overline{\varepsilon}}^{pl}\right)}=1 \quad (2.28a)$$

$$\omega_S=\int\frac{d\overline{\varepsilon}^{pl}}{\overline{\varepsilon}^{pl}_S\left(\tau_S,\dot{\overline{\varepsilon}}^{pl}\right)}=1 \quad (2.28b)$$

where ω_D and ω_S are state variables and ε_D and ε_S are the equivalent plastic strains at the beginning of ductile damage and shear damage, respectively.

After damage initiation, the material stiffness gradually decreases according to the specified damage evolution law [8]:

$$\sigma=(1-D)\overline{\sigma} \quad (2.29a)$$

$$D=\frac{L\dot{\overline{\varepsilon}}^{pl}}{\overline{u}^{pl}_f}=\frac{\dot{\overline{u}}^{pl}_f}{\overline{u}^{pl}_f} \quad (2.29b)$$

where D is the damage variable scalar and $\overline{\sigma}$ is the final stress tensor calculated in the current increment step. The equivalent plastic displacement at damage is calculated by $\overline{u}_f^{pl} = \frac{2G_f}{\sigma_{y0}}$, and σ_{y0} is the yield stress. $G_f\left(G_f = \int_0^{\varepsilon} \sigma d\varepsilon\right)$ represents the fracture energy per unit area of the crack opening.

The equivalent plastic fracture strain versus stress triaxiality for the braided yarn and epoxy resin was calculated using the above relationship as shown in Figs. 2.1 and 2.2, according to which the braided yarn and epoxy resin were judged to be ductile and shear failure damage.

2.4 Adhesive-Cohesive Model of Interface Damage

Simulation of the damage process of composites under transverse loading also requires consideration of the interfacial damage between the fiber tows and the matrix. Interfacial debonding is a nonnegligible part in the impact damage process of composites. Therefore, cohesive zone model (CZM) based on the surface contact model is used to characterize the interface between the fiber tow and the matrix. While the CZM simulates the debonding process between the yarn and the matrix, the surface contact model is designed to avoid unrealistic interpenetration between the yarns and the matrix after debonding [9]. We assume that the model is initially a linear elastic behavior, followed by damage initiation and evolution. The elastic properties can be written as:

$$\boldsymbol{t} = \begin{Bmatrix} t_n \\ t_s \\ t_t \end{Bmatrix} = \begin{Bmatrix} K_{nn} & K_{ns} & K_{nt} \\ K_{ns} & K_{ss} & K_{st} \\ K_{nt} & K_{st} & K_{tt} \end{Bmatrix} \begin{Bmatrix} \delta_n \\ \delta_s \\ \delta_t \end{Bmatrix} = \boldsymbol{K\delta} \tag{2.30}$$

In this traction-separation law, $\boldsymbol{t}$ denotes the traction vector, consisting of the normal component t_n and two orthogonal shear components t_s and t_t. δ_n, δ_s, and δ_t denote the contact separation vectors.

The maximum nominal stress criterion was chosen to determine the damage initiation for CZM as follows:

$$max\left\{\frac{\langle \tau_n \rangle}{\tau_n^0}, \frac{\tau_s}{\tau_s^0}, \frac{\tau_t}{\tau_t^0}\right\} = 1 \tag{2.31}$$

where τ_n^0, τ_s^0, and τ_t^0 are the peak contact stresses in the normal direction and in both shear directions. The material interface damage initiation occurs when the maximum stress ratio reaches 1.

Once the criterion given in Eq. (2.31) is satisfied, the interface damage extension is simulated by reducing the bond stiffness. The scalar damage variable D (0 for no damage and 1 for complete damage) represents the overall damage to the interface and affects the stress components of the CZM model:

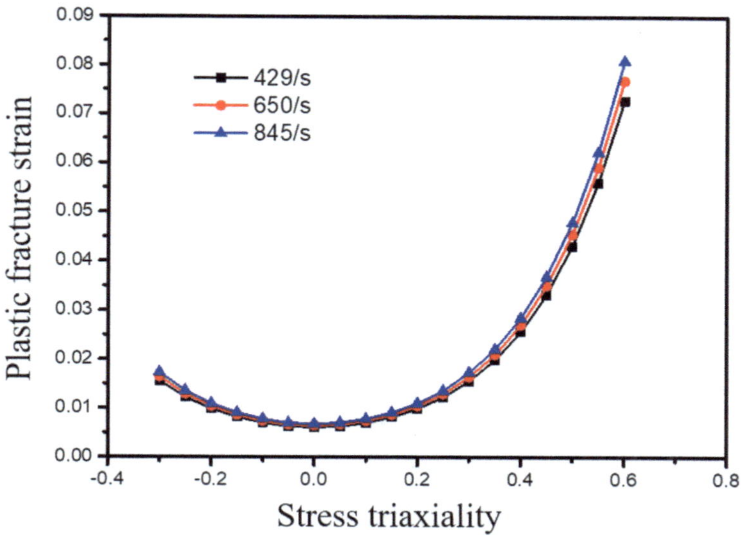

(a)

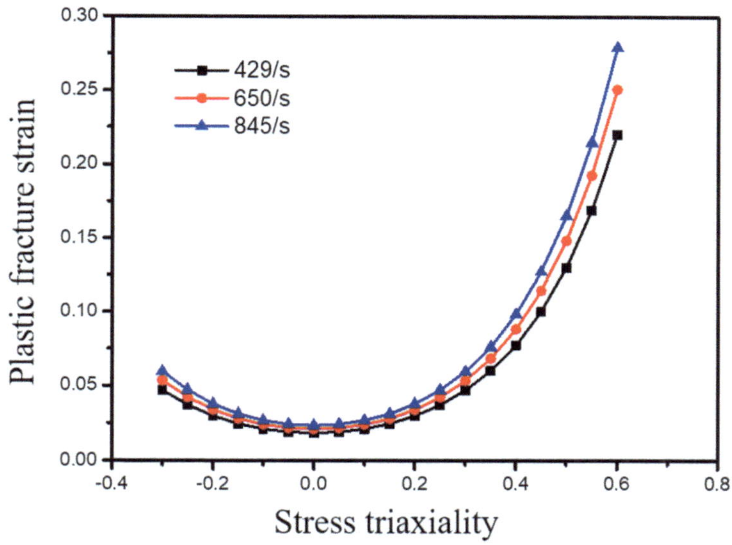

(b)

Fig. 2.1 Relationship between plastic fracture strain and stress triaxiality of braided yarn and epoxy resin in the ductile failure criterion. (**a**) Braided yarn, (**b**) epoxy resin

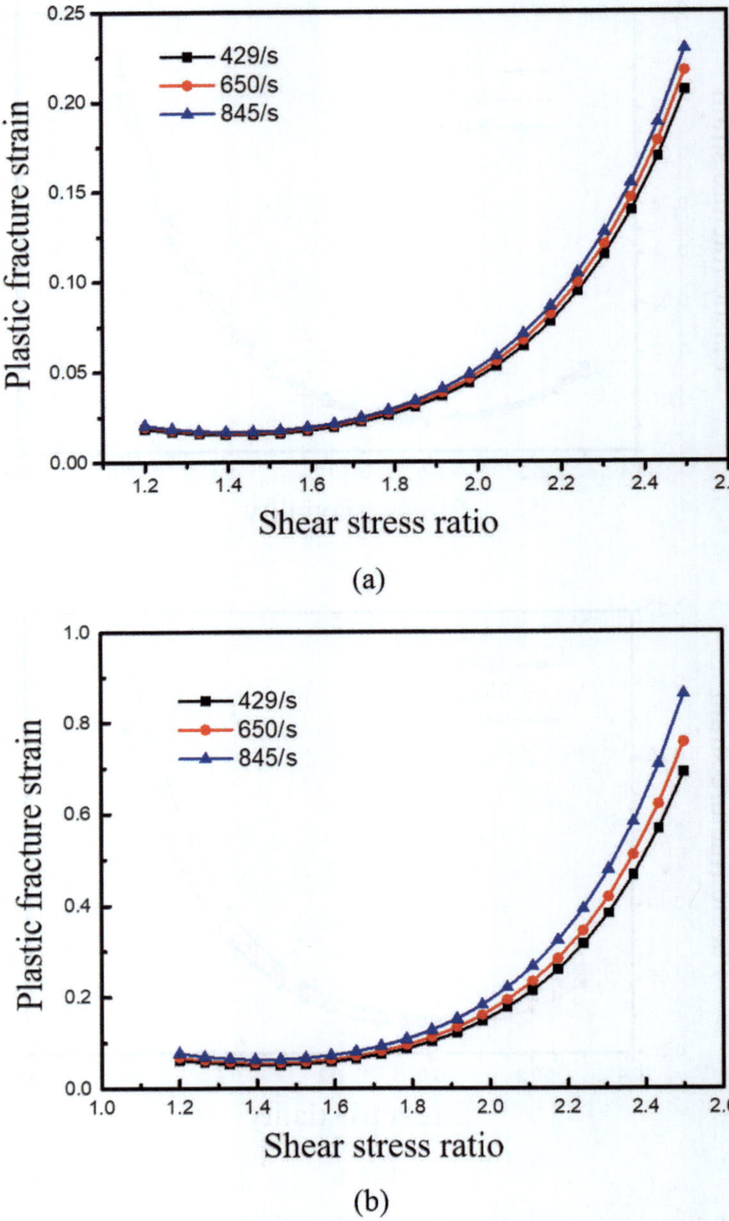

Fig. 2.2 Relationship between plastic fracture strain and shear stress ratio of braided yarn and epoxy resin in the shear failure criterion. (**a**) Braided yarn, (**b**) epoxy resin

$$D = \frac{\delta_m^f(\delta_{max} - \delta_m^0)}{\delta_{max}(\delta_m^f - \delta_m^0)} \tag{2.32a}$$

$$t_n = \begin{cases} (1 - D)\bar{t}_n, & \bar{t}_n \geq 0 \\ \bar{t}_n, & \bar{t}_n < 0 \end{cases} \tag{2.32b}$$

$$t_s = (1 - D)\bar{t}_s \tag{2.32c}$$

$$t_t = (1 - D)\bar{t}_t \tag{2.32d}$$

where δ_m^0, δ_{max}, and δ_m^f denote damage initiation separation, maximum effective separation, and complete failure separation along the interface, respectively, and $\bar{t}_n$, $\bar{t}_s$, and $\bar{t}_t$ are the stress components predicted from the interface constitutive response at the current strain without damage.

The damage extension is defined as an energy-based damage evolution scheme of the Benzeggagh-Kenane form:

$$G_n^C + (G_s^C - G_n^C)\left\{\frac{G_S}{G_T}\right\} = G^C \tag{2.33a}$$

$$G_S = G_s + G_t \tag{2.33b}$$

$$G_T = G_n + G_s \tag{2.33c}$$

where G_s^C and G_n^C are user-defined parameters.

2.5 Thermo-mechanical Coupling Constitutive Equations

Fully coupled thermal stress analysis is required when the stress analysis depends on the temperature distribution, while the temperature distribution depends on the stress state. Adiabatic thermal stress analysis is often used to simulate high-speed loading processes containing large inelastic strains. Due to the temperature characteristics of the material, the temperature rise caused by the deformation of composite is an important factor. During the impact process, the plastic work generated by the inelastic deformation of the material is thermally dissipated, resulting in a significant temperature rise, which in turn changes the material properties. When the flow stress is reduced by the temperature rise, the temperature rise causes thermal expansion and local deformation.

In this case, both thermal and stress solutions must be obtained simultaneously. We specify the inelastic heat fraction η in the fully thermo-mechanical coupled analysis that represents the heat generated by the dissipation energy during inelastic deformation and damage. In adiabatic analysis, the plastic strain produces a heat flux per unit volume of:

$$r^{pl} = \eta \boldsymbol{\sigma} : \dot{\varepsilon}^{pl} \tag{2.34}$$

where r^{pl} is the heat flux added to the heat energy balance, η is the inelastic heat fraction, and σ and $\dot{\varepsilon}^{pl}$ are the stress and plastic strain rates, respectively. The plastic strain increment is expressed by the flow potential as:

$$\dot{\boldsymbol{\varepsilon}}^{pl} = \dot{\varepsilon}^{pl} \boldsymbol{n} \tag{2.35}$$

where $\boldsymbol{n}$ is the direction of flow and ε^{pl} is the plastic strain scalar metric.

The heat equation to be solved at each integration point is:

$$\rho c(\theta) \dot{\theta} = r^{pl} \tag{2.36}$$

where ρ is the density of the material and $c(\theta)$ is the specific heat.

For elastic-plastic materials, the relationship between the mechanical field and the heat generated in the deformation part of the heat energy equation that is related to the rate is given by:

$$k \nabla^2 T - \alpha(3\lambda + 2\mu) T \dot{\varepsilon}^{el} + \eta \sigma : \dot{\varepsilon}^{pl} = \rho C \dot{T} \tag{2.37}$$

where k, T, α, and C denote the thermal conductivity, absolute temperature, thermal expansion coefficient, and specific heat, respectively; λ and μ are the Lamé elastic constants; and ρ, σ, $\dot{\varepsilon}^{el}$, and $\dot{\varepsilon}^{pl}$ are the density, stress component, elastic strain component, and plastic strain component, respectively. Under high strain rate loading, the thermal conduction term ($k\nabla^2 T$) and the thermoelastic term ($\alpha(3\lambda + 2\mu) T \dot{\varepsilon}^{el}$) can be neglected. Thus, Eq. (2.37) can be written as:

$$\eta \sigma : \dot{\varepsilon}^{pl} = \rho C \dot{T} \tag{2.38}$$

The above equation provides the coupling between the temperature rise in the material and the dissipation process that generates heat during highly transient deformation. Substituting Eqs. (2.34) into (2.38), the heat flux can be written as:

$$r^{pl} = \rho C \dot{T} \tag{2.39}$$

Then, we use the inverse Eulerian format to integrate the plastic strain, so r^{pl} at the end of the increment can be approximated as:

$$r^{pl} = \frac{1}{2\Delta t} \eta \Delta \varepsilon^{pl} \boldsymbol{n} : (\boldsymbol{\sigma} + \boldsymbol{\sigma}_t) \tag{2.40}$$

From Eqs. (2.39) and (2.40), it follows that:

$$\Delta T = \Delta t \frac{r^{pl}}{\rho C} = \frac{\Delta t}{\rho C} \cdot \frac{1}{2\Delta t} \eta \Delta \varepsilon^{pl} \boldsymbol{n} : (\boldsymbol{\sigma} + \boldsymbol{\sigma}_t) = \frac{\eta \Delta \varepsilon^{pl}}{2\rho C} \cdot \boldsymbol{n} : (\boldsymbol{\sigma} + \boldsymbol{\sigma}_t) \tag{2.41}$$

Thus, the temperature at the end of each incremental step can be calculated from Eq. (2.41).

The difference between CTE of the fiber tow and the matrix prevent the composite from expanding freely under the effect of the temperature field. This results in thermal strain and thermal stress, expressed as:

$$\varepsilon^{th} = \alpha(\theta, f_\eta)(\theta - \theta^0) - \alpha\left(\theta^I, f_\eta^I\right)(\theta^I - \theta^0) \tag{2.42}$$

$$\sigma^{th} = -C(T)\varepsilon^{th} \tag{2.43}$$

where θ, θ^0, and θ^I represent the current temperature, the reference temperature, and the initial temperature, respectively, f_η and f_η^I are the current and initial values of the field variables, and $C(T)$ is the stiffness matrix of the material at temperature T.

References

1. Christos, C.C.: Mechanics of composite materials_ past, present and future. J. Compos. Technol. Res. **11**(1), 3–14 (1989)
2. Chen, L., Tao, X., Choy, C.: A structural analysis of 3-D braids. J. China Text. Univ. (Eng Ed). **14**(3), 8–13 (1997)
3. Huang, Z.M.: The mechanical properties of composites reinforced with woven and braided fabrics. Compos. Sci. Technol. **60**, 479–498 (2000)
4. Farooqi, J.K., Sheikh, M.A.: Finite element modelling of thermal transport in ceramic matrix composites. Comput. Mater. Sci. **37**, 361–373 (2006)
5. Liu, S., Zhang, J., Shi, B., Wang, L., Gu, B., Sun, B.: Damage and failure mechanism of 3D carbon fiber/epoxy braided composites after thermo-oxidative ageing under transverse impact compression. Compos. Part B. **161**, 677–690 (2019)
6. Hill, R., Orowan, E.: A theory of the yielding and plastic flow of anisotropic metals. Proc. R. Soc. Lond. Ser. A Math. Phys. Sci. **193**, 281–297 (1948)
7. Pankow, M., Waas, A.M., Yen, C.F., et al.: Modeling the response, strength and degradation of 3D woven composites subjected to high rate loading. Compos. Struct. **94**(5), 1590–1604 (2012)
8. Hu, M., Zhang, J., Sun, B., Gu, B.: Finite element modeling of multiple transverse impact damage behaviors of 3-D braided composite beams at microstructure level. Int. J. Mech. Sci. **148**, 730–744 (2018)
9. Ahmadian, H., Yang, M., Soghrati, S.: Effect of resin-rich zones on the failure response of carbon fiber reinforced polymers. Int. J. Solids Struct. **188–189**, 74–87 (2020)

Chapter 3
Multiscale Geometric Model of 3-D Braided Composites

This chapter focuses on the modeling methods for 3-D braided composite beam and tube, namely, the full-scale meso-structure models and unit cell/meso-structure hybrid structural models, and the development of transverse impact models and compression impact models that meet test conditions.

3.1 Full-Size Geometric Model of Beam

In real 3-D braided composites, thousands of fibers in each fiber tow are impregnated by the surrounding uniformly distributed resin, i.e., each braided yarn is a transverse isotropic material. Figure 3.1 shows the representative unit cell (RUC) model of a fiber tow impregnated in matrix in a 3-D braided composite preform.

Figure 3.2 shows a full-scale meso-structure geometry model of the 3-D braided composite. Based on scanning electron microscope (SEM) image analysis, the cross-sectional shape of the braided yarn can be idealized as a regular hexagon, as shown in Fig. 3.3. The braided composite consists of matrix and preform, namely, epoxy resin and braided yarns (interior yarns, surface yarns, and corner yarns). The geometric model is locally enlarged to show the different topologies for braiding angles of 20°, 30°, and 40°, as shown in Fig. 3.2b. In the geometric model, the composite beam has a rectangular shape. The matrix structure is generated by the Boolean operator [1], namely, the complementary set of braided preforms in the rectangular composite beam is epoxy resin.

Figure 3.4 shows mesh scheme. A linear hexahedral element is used for the mesh of the braided yarn. Since continuous thermal diffusion analysis requires fully coupled temperature-displacement analysis, the most straightforward approach is to use coupled temperature-displacement elements throughout the adiabatic analysis. Therefore, the four-node linear tetrahedral element with coupled temperature-displacement is used to mesh the epoxy resin. Since carbon fiber is a temperature-insensitive material, we do not consider the adiabatic temperature rise.

M. Hu, B. Gu, *Impact Damages of Braided Composites*, Engineering Materials,
https://doi.org/10.1007/978-981-16-5703-0_3

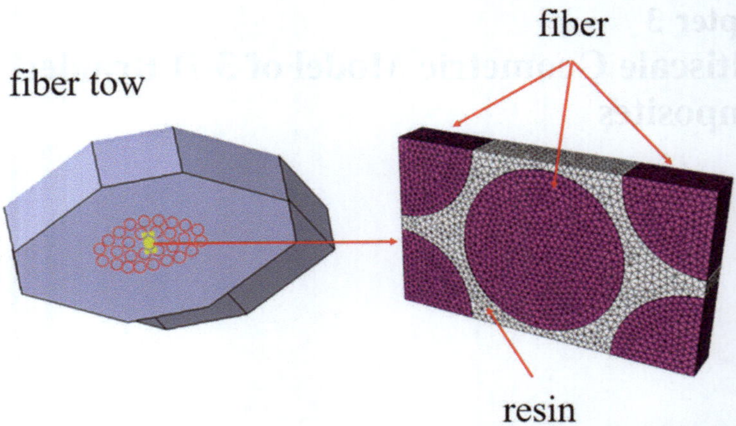

Fig. 3.1 Representative unit cell (RUC) model of fiber tow impregnated in matrix

3.2 Full-Size Geometric Model of Tube

A geometric modeling method suitable for 3-D circular braided meso-structure is proposed. Firstly, the yarn motion cycle structure is obtained according to the four-step motion law, and the whole fabric structure is obtained through the array splicing cycle structure.

As shown in Fig. 3.5, the geometric modeling process of 3-D circular braided preform is described in detail with the example of yarn arrangement 3 × 32 ($m \times n$). As shown in Fig. 3.5a, all yarns with consistent movement patterns are marked with the same color according to the braiding process, and it can be found that the whole disc can be obtained by splicing four different colored yarns through a circular array. As shown in Fig. 3.5b, the representative area is divided into three parts: the outer surface yarn layer, the core yarn layer, and the interior surface yarn layer. The circumferential angle occupied by both the interior and outer surface yarn layers is 2θ ($\theta = 2\pi/n$), and they are composed of two yarns of different alignment, each of which moves one braided length (h) along the braiding direction and the θ angle along the circumferential direction. The core yarn layer occupies the circumferential angle of 3θ ($m\theta$) and is composed of four complete monochromatic yarns of different orientations (whereas the conventional unit cell model contains all yarns in this unit cell [2]), each yarn moving half a braided length ($h/2$) in the braiding direction and the 2θ angle ($(m - 1)\theta$) in the circumferential direction. As shown in Fig. 3.5c, the three regions are arrayed at the 2θ angle in the circumferential direction and at one braided length h in the braiding direction to obtain the final fabric structure. Figure 3.5d shows the path of single yarn movement in the fabric, alternating inside and outside in the circumferential direction and spiraling along the braiding direction.

This 3-D braided structural preform modeling method can be implemented by changing the braiding parameters for various structural geometry modeling. As

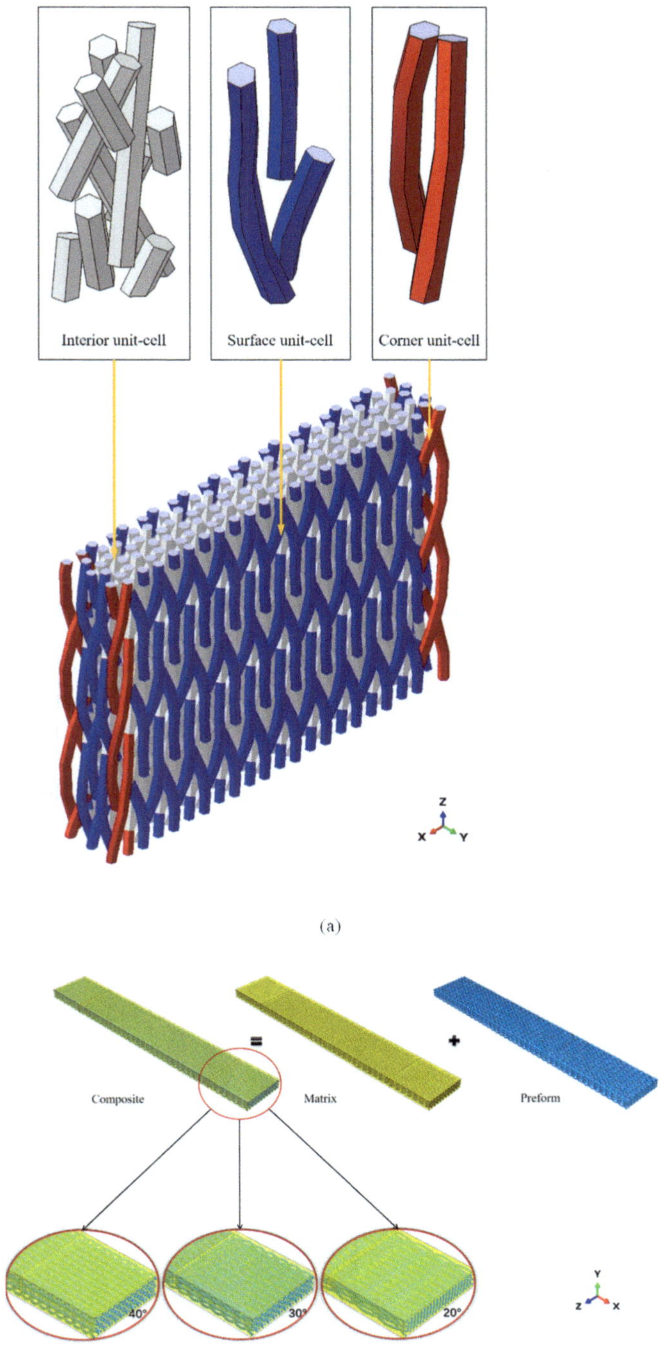

Fig. 3.2 Geometric model of the 3-D braided composite. (**a**) 3-D braided preforms, (**b**) braided composite beams

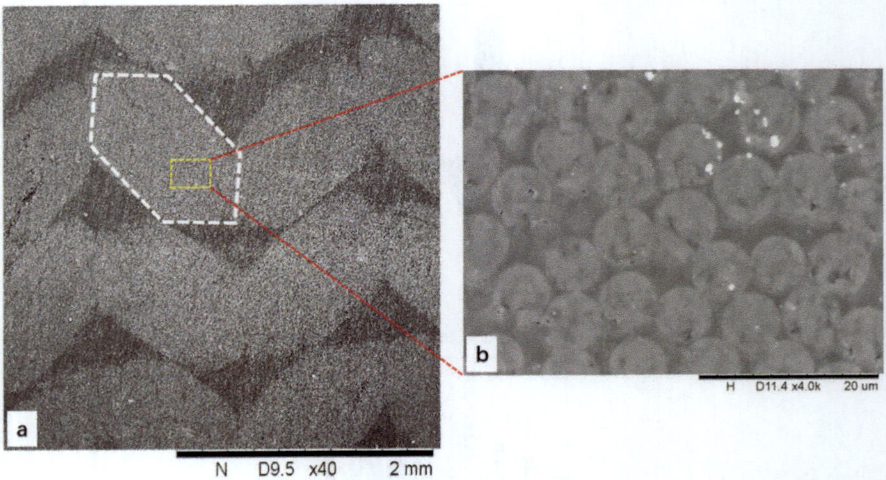

Fig. 3.3 Scanning electron microscope (SEM) photographs of cross-sections. (**a**) 3-D braided composite cross-section, (**b**) single carbon fiber and epoxy resin matrix

shown in Table 3.1, the modeling process can be achieved by varying the θ ($\theta = 2\pi/n$), braided length h, and number m of braided layers. In this study, there are six different braided composite tubes, including different numbers of layers, different braiding angles, and with or without axial yarns. The above methods are used to establish the geometric models of meso-structure of various structural preforms, as shown in Fig. 3.6.

The 3-D braided composite tubes were mainly simulated with establishing a homogenized unit cell model [3, 4]. This method has two disadvantages: (1) the unit cell model is established based on the braided structural parameters, while the change of material behaviors cannot be shown in the unit cell model, and (2) the unit cell model homogenizes the components and cannot predict the contributions of each constituent. The FEA model at meso-structure level can visualize the load deformation and damage of the yarn and matrix during the impact process.

As shown in Fig. 3.7, the geometric model of the 3-D braided composite tube consists of resin and 3-D braided preform. From the cross-section of the composite tube, it can be found that the cross-sectional shape of the yarn is a flat ellipse [5]. In order to reduce the surface in the model and improve the quality of meshing, the yarn uses a flat octagonal cross-section.

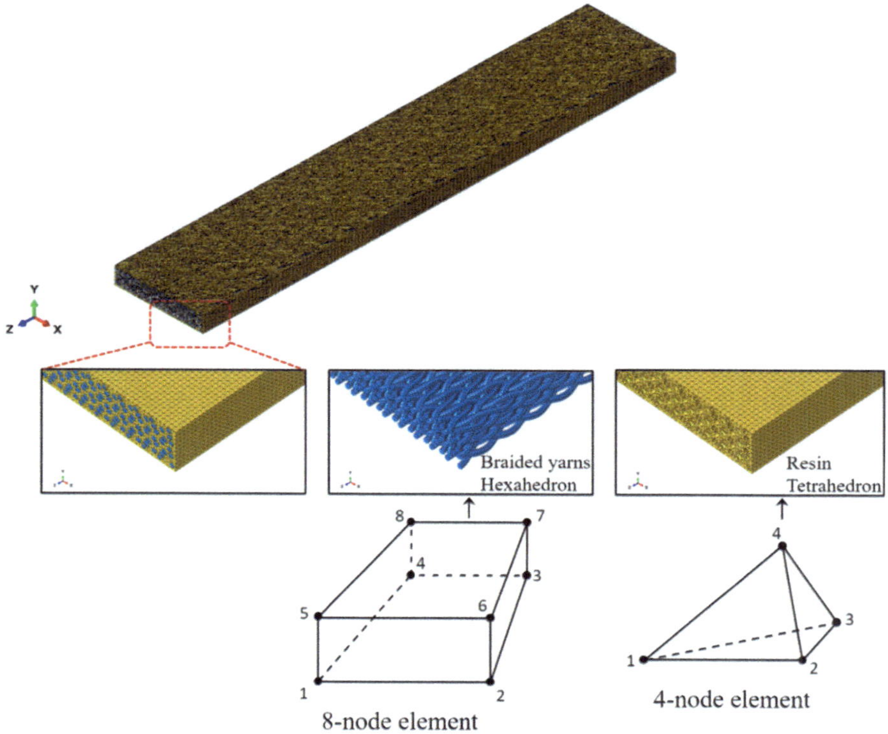

Fig. 3.4 Mesh scheme

3.3 Unit Cell/Microstructure Hybrid Model

Figure 3.8a shows a multiscale modeling scheme has proposed to analyze impact damages of 3-D four-directional braided composites. The fiber tows are made of thousands of carbon fibers impregnated with resin, and micro-scale unit cells are established based on the impregnated yarn morphology. The results of the mechanical parameters calculated from the unit cell model are input into the meso-scale unit cell model. The 3-D braided composite preforms are composed of three kinds of unit cells, interior unit cell, surface unit cell, and corner unit cell, as the periodic components to transversely and longitudinally integrate the mesoscopic scale model and establish the 3-D braided composite meso-structure and macro-structure hybrid model. The impact region is a meso-structure model based on the real geometric topology of the composite assembled by yarns and epoxy resin, and other regions are equivalent macroscopic homogeneous geometric models. The meso-structure geometric model of the 3-D four-directional braided composite consists of interior yarns, corner yarns, surface yarns, and resin, while the 3-D five-directional braided composite also has axial yarns. Due to the extruded state

Fig. 3.5 Geometric model of 3-D circular braided preform. (**a**) Yarn movement path, (**b**) cycle structure, (**c**) cycle structure splicing combination of 3-D braided preforms, (**d**) single fiber towpath in the fabric

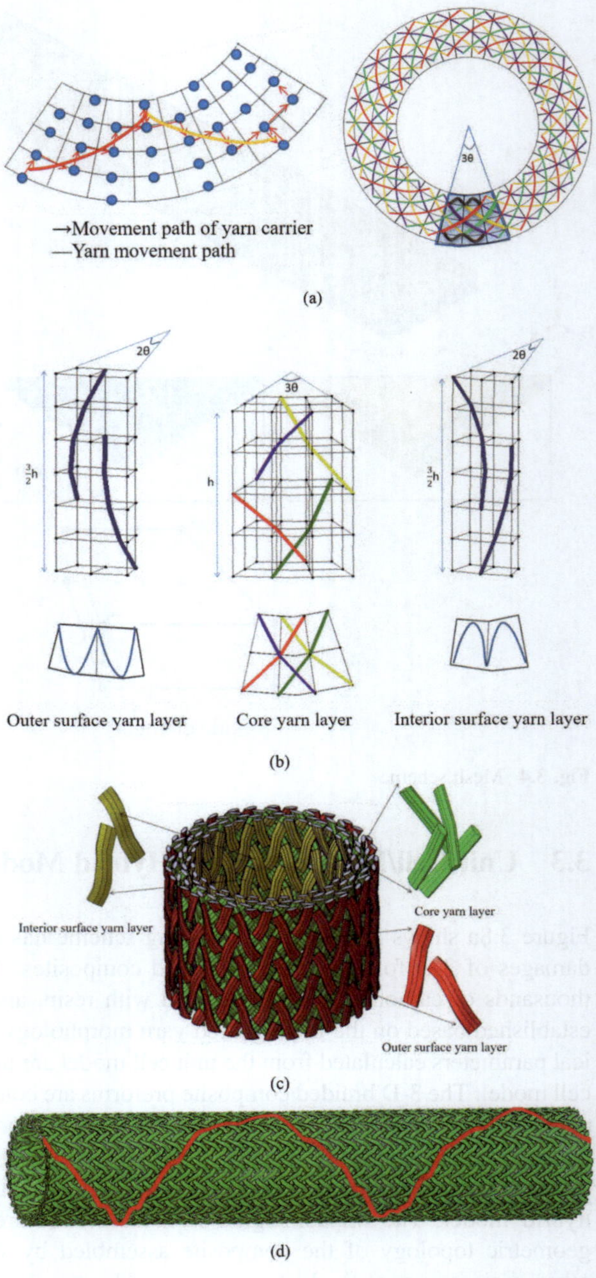

Table 3.1 Geometric modeling parameters of 3-D circular braided preforms

Parameters	Core yarn	Inner side yarns	Outer side yarns
Yarn number	4	2	2
Layer angle	$m\theta$	2θ	2θ
Yarn angle	$(m-1)\theta$	θ	θ
Layer height	h	$3/2 \cdot h$	$3/2 \cdot h$
Yarn height	$1/2 \cdot h$	h	h
Circular array angle	2θ	2θ	2θ
Axial array height	h	h	h

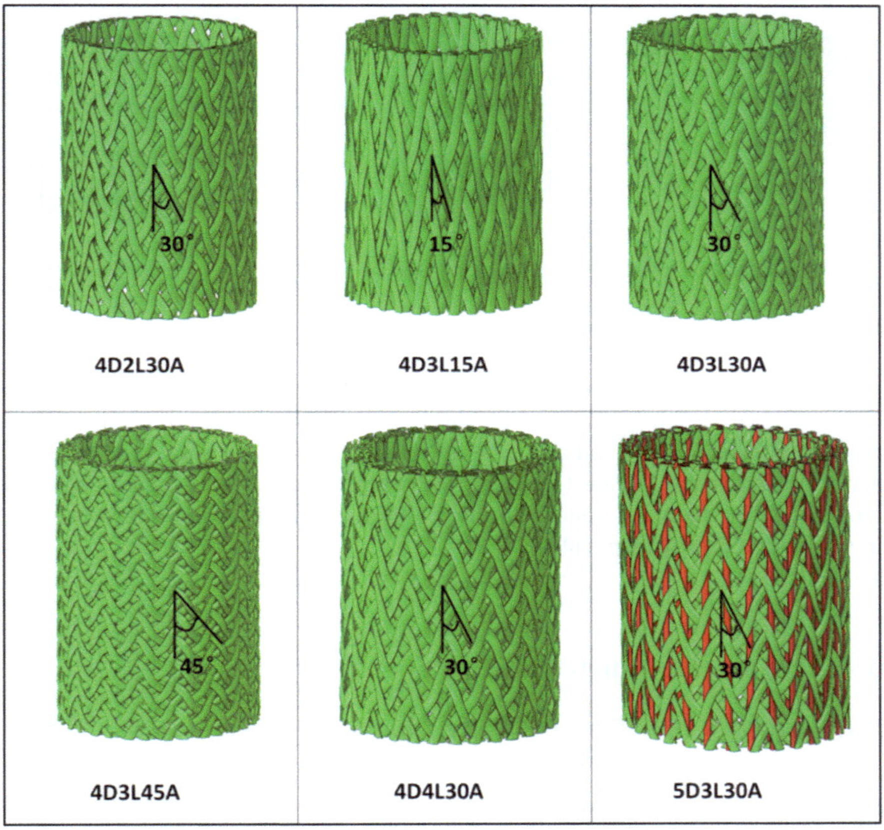

Fig. 3.6 Digital diagram of 3-D circular braided tubes

of the yarns in the braided composites, the cross-sectional shapes of the braided and axial yarns can be idealized as regular hexagonal and quadrilateral shapes [6].

Figure 3.8b shows the braided structure of the preform and the meshing scheme of the model. The four-node linear tetrahedral element is used for the yarn, while the six-node linear triangular prism element is used for the macroscopic part. As

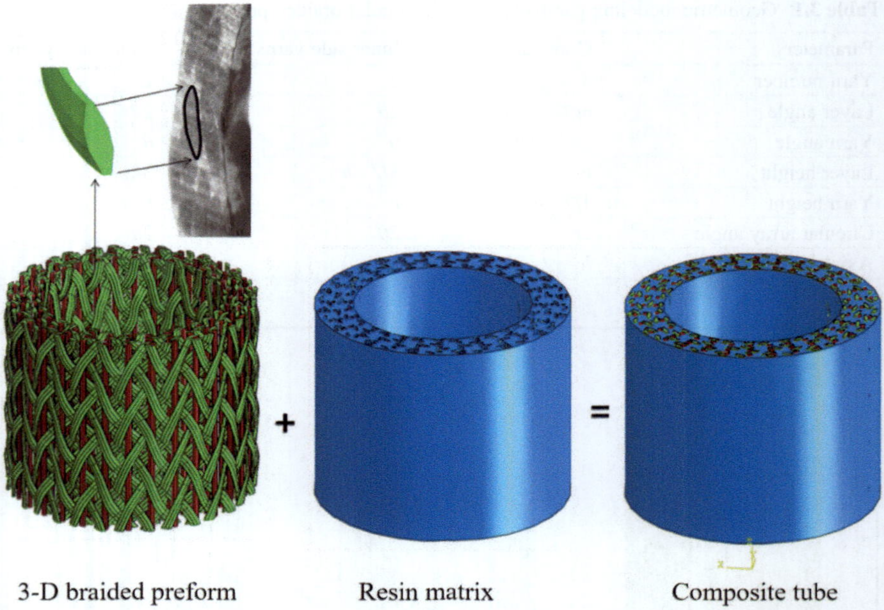

Fig. 3.7 3-D braided composite tube

mentioned in Sect. 3.1, we used a coupled temperature-displacement four-node linear tetrahedral element for meshing the epoxy resin during the adiabatic analysis without considering the adiabatic temperature rise of the yarn. The co-nodal approach is used to connect the macrostructure model with the meso-structure model, which makes the transfer of thermo-mechanical properties have continuity in the future finite element calculation.

3.4 Transverse Impact Model

3.4.1 Full-Scale Meso-structure Composite Beam

A full-scale meso-structure transverse impact model based on the 3-D braided composite microstructure was developed using the computer-aided engineering (CAE) software CATIA V5R20 and the commercial finite element analysis (FEA) software ABAQUS. The ABAQUS package was run on a Linux platform. Figure 3.9 shows the boundary conditions and mesh scheme of the transverse impact test model. The ends of the composite beam are fixedly displaced and rotated (25 mm each), and the incident bar is only displaced along the impact direction. Stress waves are input at the end face of the incident bar, and different impact pressures are converted into corresponding stress-time curves as shown in Fig. 3.10. Considering

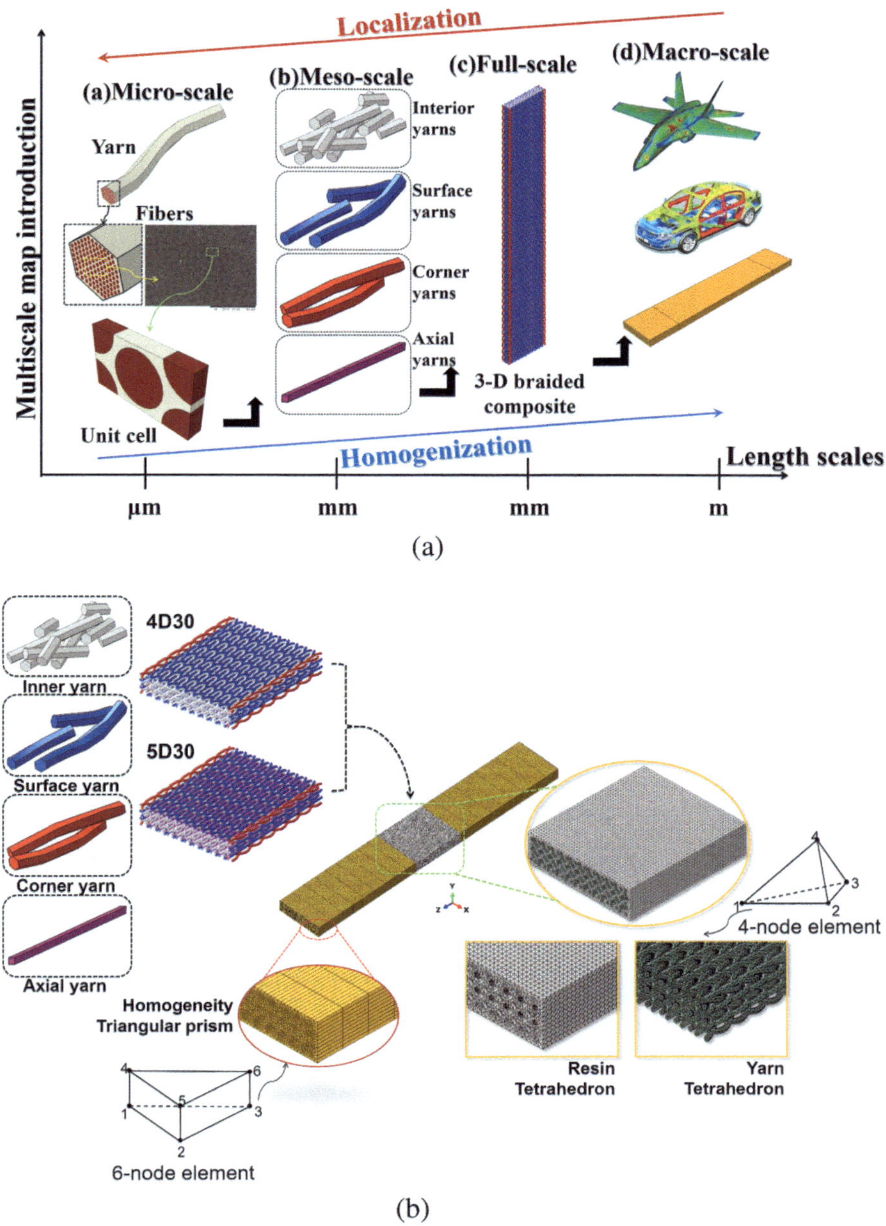

Fig. 3.8 Multiscale modeling of 3-D braided composite. (**a**) Multiscale structure levels, (**b**) braided preform and mesh scheme

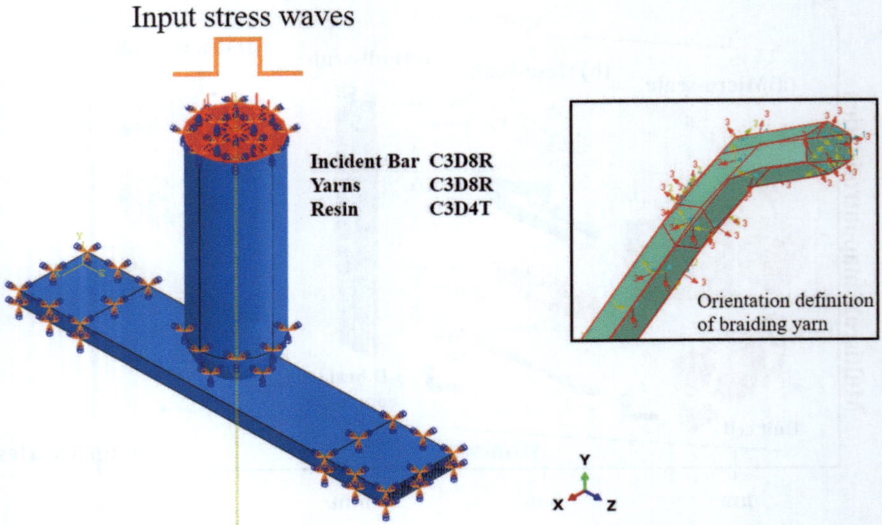

Fig. 3.9 Boundary conditions and mesh scheme and yarn orientation definition in transverse impact test model

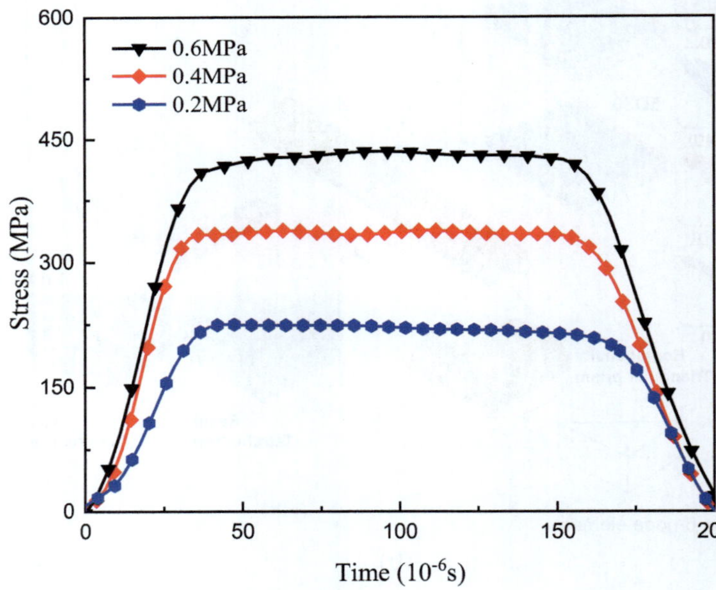

Fig. 3.10 Input stress waves (0.2, 0.4, and 0.6 MPa)

Table 3.2 Number of meshes and volume fractions for the full-scale meso-structure transverse impact model[a]

Braided structure		4D20	4D30	4D40
Number of elements		1,549,790	1,469,715	1,718,432
V_y	FEA	38.3	40.0	42.4
V_f	FEA	26.8	28.0	29.7
	EXP	27.7	28.6	30.8

[a] V_y denotes yarn volume fraction (%), V_f denotes fiber volume fraction (%)

that the fiber tow is made of thousands of carbon fibers impregnated with resin, the yarns representing the fiber tow can be considered as transverse isotropic unidirectional composites. The orientation of the braided yarn is defined as shown in Fig. 3.9. The number of meshes and volume fractions of the model are summarized in Table 3.2. The adiabatic temperature fields are set for the composite at 25, 90, 150, and 210 °C.

The computation time required to complete the simulation depends on the mesh density. For the eight-node solid element, the critical time step Δt_e is [7]ε:

$$\Delta t_e = \frac{L_e}{Q + \left(Q^2 + C^2\right)^{1/2}} \tag{3.1}$$

where Q is a function of the bulk viscosity coefficient C_0 and C_1:

$$Q = \begin{cases} C_1 c + C_0 L_e |\dot{\varepsilon}_{kk}| & \text{for } \dot{\varepsilon}_{kk} < 0 \\ 0, & \text{or } \dot{\varepsilon}_{kk} \geq 0 \end{cases} \tag{3.2}$$

where L_e is the characteristic length and $L_e = \nu_e/A_{e\ max}$. ν_e is the volume of the element. $A_{e\ max}$ is the area of the largest side. c is the adiabatic sound velocity of material.

For a material with m elements, the time step in the next calculation step Δt is:

$$\Delta t^{n+1} = \alpha \cdot min\left(\Delta t_{e1}, \Delta t_{e2}, \cdots \Delta t_{em}\right) \tag{3.3}$$

The scale factor α in Eq. (3.3) is 0.9 or the lesser value of stability.

From Eq. (3.3), it can be seen that the higher the mesh density, the longer the time required for calculation and the closer the calculation results are to the actual situation. The element size varies according to the area, and the contact between the incident bar and the composite is meshed at a finer size.

(a)

(b)

Fig. 3.11 Meshing scheme of 3-D braided composite tube

3.4.2 Meso-structure of the Braided Composite Tube

Finite element preprocessing is conducted with the commercial FEA software ABAQUS/Explicit.

The analysis step type is dynamic explicit, and the time set for this analysis step is based on the propagation time of the stress wave in the incident bar:

$$T = 2m \cdot L_0 / C_0 \tag{3.4}$$

where T is the total time of stress wave propagation, m is the number of impacts, L_0 is the length of the incident bar, and C_0 is the propagation velocity of the stress wave in the incident bar. The total time required for the four impact processes is 0.0025 s. The finite element result output includes field output and history output. The field output variable is the variable describing the change with spatial position, including a large number of element or node calculation results. The output frequency is low, and the stress, strain, and displacement are set in the field output, while the history output mainly describes the variable changing with time and outputs part of the node or element calculation results.

As shown in Fig. 3.11, different meshing types are used for the 3-D braided composite tube preform and the matrix, respectively. For the 3-D braided composite tube preform, it is composed of several independent braided yarns with regular

Table 3.3 Number of elements of composite tubes with different braided structures

	Preform		Matrix		Composite
Structure	Mesh type	Number of elements	Mesh type	Number of elements	Total number of elements
4D2L30A	C3D8R	66,735	C3D4	1,283,038	1,349,773
4D3L15A	C3D8R	81,408	C3D4	1,622,345	1,703,753
4D3L30A	C3D8R	103,842	C3D4	1,456,835	1,560,677
4D3L45A	C3D8R	145,920	C3D4	2,083,819	2,229,739
4D4L30A	C3D8R	131,136	C3D4	2,344,428	2,475,564
5D3L30A	C3D8R	223,104	C3D4	2,654,476	2,877,580

octagonal cross section. The hexahedral mesh type (C3D8R) with high accuracy and fast calculation speed is adopted, as shown in Fig. 3.11a. In contrast, for the 3-D braided composite tube matrix, an independent whole matrix is interspersed with complex yarn path holes inside. This complex structure is difficult to be divided by a regular hexahedral mesh type, so a relatively low accuracy tetrahedral element type (C3D4) is used, as shown in Fig. 3.11b. The number of elements for the various full-scale meso-structure composite models in this study ranged from 1.3 to 2.9 million, as shown in Table 3.3. The number of elements for matrix is much larger than that of corresponding preform. Due to the large number of elements, this full-scale finite element model of the meso-structure is computationally expensive, and the computation time increases with the increase of the number of elements.

Figure 3.12 shows the model of the SHPB test system for transverse impact of 3-D braided composite tube. Both the incident bar and the 3-D braided composite tube are adopted in full-scale model. Since the braided structure is asymmetric, the 1/2 or 1/4 model cannot be used to shorten the computation time. The incident bar only defines the displacement degrees of freedom along the impact direction, and the other directions are fully constrained. The holding area of the composite tube is set as the fully constrained degree of freedom. In the test, the stress wave signals are generated by the impact bar impacting the incident bar and recorded by the strain gauge. In the finite element model, the stress wave signals obtained from the test are loaded on the flat end face of the incident bar to obtain one-dimensional compressive stress wave. The contact between the bar and the composite adopts general contact (hard contact without considering friction), the hemispherical end face of the incident bar is set as the main surface, and the contact surface between the composite and the bar is set as the slave surface, so as to ensure that the slave surface does not penetrate into the main surface. In order to smoothly propagate the stresses between the interface nodes, the mesh interfaces of the composite preform and the matrix are connected by co-nodes with a perfectly bonding constraint (tie) type.

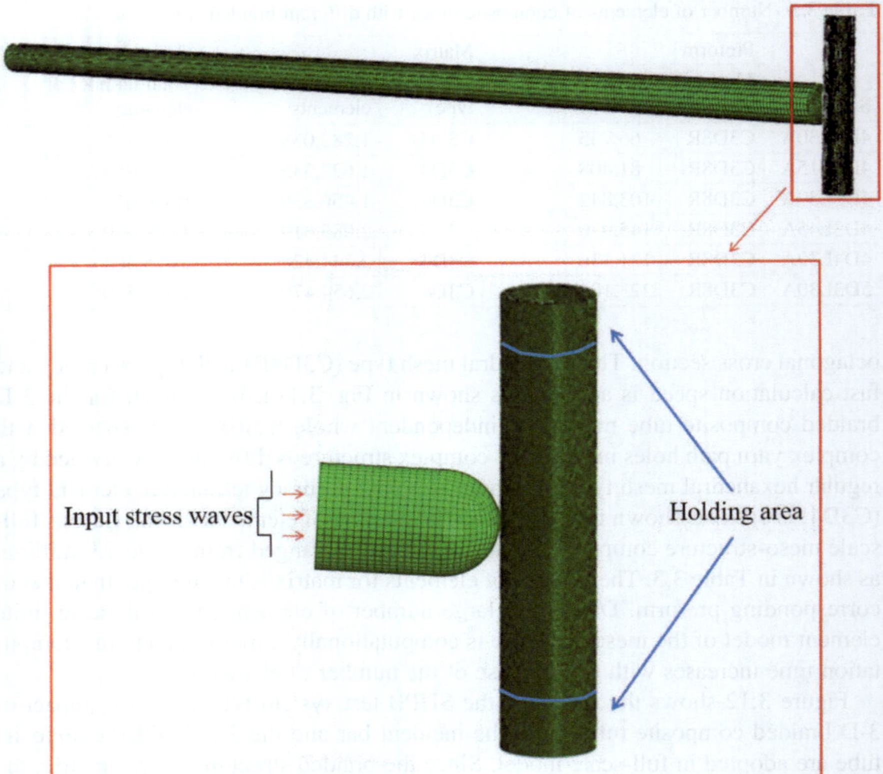

Fig. 3.12 Transverse impact model with SHPB apparatus

3.4.3 Hybrid Unit Cell/Meso-structure Composite Beam

The computer-aided engineering (CAE) software named CATIA V5R20 and the commercial FEA software ABAQUS were used to establish the transverse impact model of hybrid unit cell/meso-structure based on the microstructure of 3-D braided composites. As before, the ABAQUS package was also run on a Linux platform. The finite element meshing software named HYPERMESH was used to optimize the mesh elements and combine the mesoscopic and macroscopic parts by co-nodes. 460,002 and 739,771 elements were used for the 4D30 and 5D30 transverse impact test models, which greatly reduce the number of elements and improve the computational efficiency compared to the full-scale meso-structure model. The boundary conditions and mesh types of the transverse impact test model are shown in Fig. 3.13. The homogeneity and yarn are considered as transversely isotropic unidirectional composites, the fiber direction is one direction of the yarn, and the Z-axis direction is one direction of the homogeneity. Similarly, the composite beam

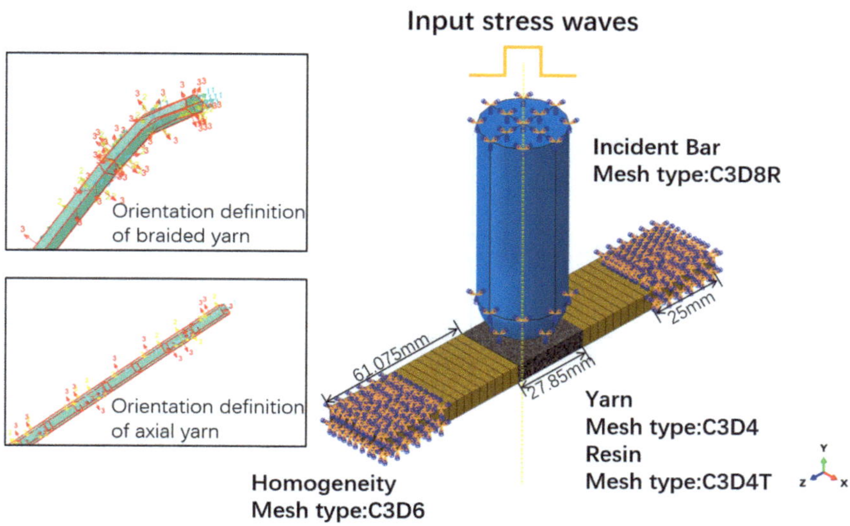

Fig. 3.13 Boundary conditions, mesh scheme, and yarn orientation definition for the transverse impact test model

is completely stationary at 25 mm at each end, and the incident bar is displaced only along the impact direction. The stress wave is set in the form of stress-time curve on the end face of the incident bar, as shown in Fig. 3.10.

3.5 Axial Impact Model

A finite element model of the 3-D braided composite tube under quasi-static and impact compression loading is established as shown in Fig. 3.14. The preprocessing of finite element calculation includes analysis step setting, contact condition setting, constraints and loading conditions, and mesh element division.

3.5.1 Analysis Step Settings

Two analysis steps are set up for this model: the initial analysis step and the compressive load analysis step. The quasi-static compression test takes tens of seconds or minutes, so the ABAQUS/Standard solver module is used to calculate the quasi-static axial compression process. In the simulation, the analysis step is set to "Static General," and the compression load analysis step length is 100s. The dynamic impact compression test is extremely short, so the dynamic ABAQUS/ Explicit solver module is used to simulate the dynamic impact compression process

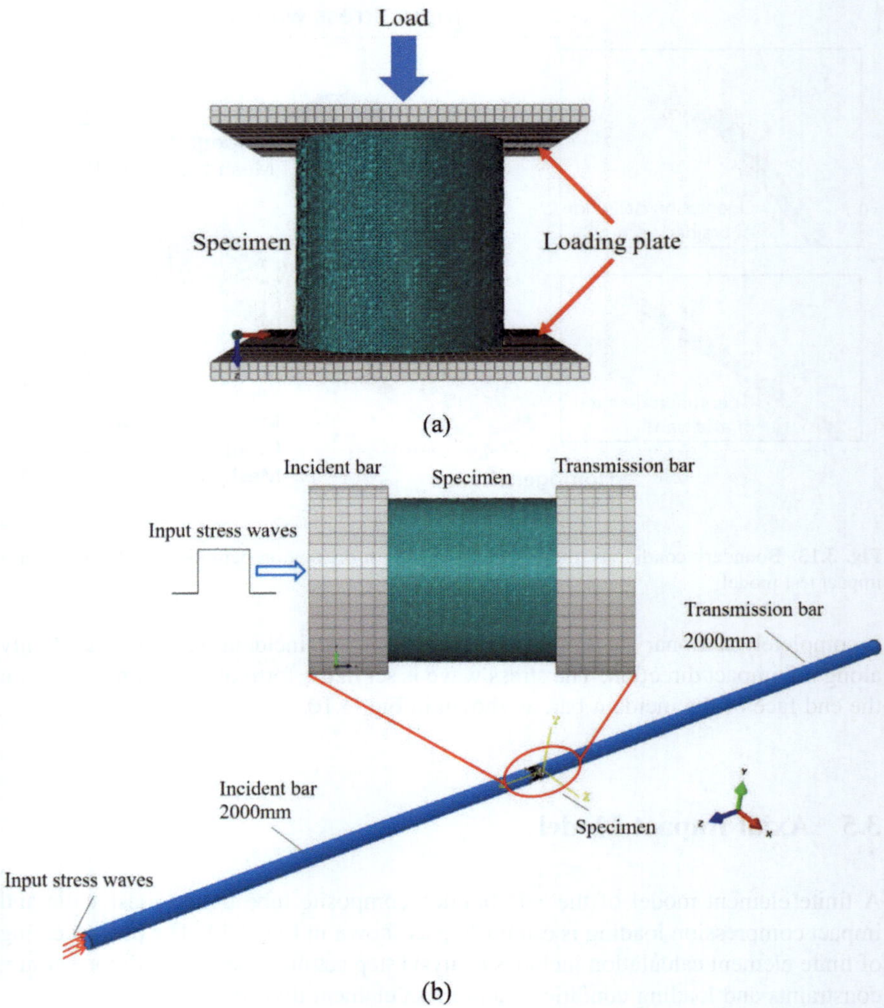

Fig. 3.14 FEA model of 3-D braided composite tube under quasi-static and impact compression. (**a**) Quasi-static axial compression, (**b**) axial impact compression

of the composite tube. The analysis step is set to dynamic explicit, and the impact compression load analysis step length is 300 μs.

In order to obtain the stress-strain curve and the distribution state of stress and strain at different moments of the composite during compression damage in the simulation, field variables and history variables need to be set in the analysis step. The field variables characterize the distribution of the variables in space, and the history variables characterize the change of the variables with time. Thus, the outputs of the field variables are selected as displacement, strain, and stress, and the outputs of the history variables are selected as the force of the contact surface between the incident bar and the composite tube and the displacement of the incident bar and the transmission bar.

3.5.2 Contact Condition Setting

According to the FEA situation, the quasi-static compression model uses the general contact in the ABAQUS/Standard module to define the frictionless contact between the loading plate and each part of the specimen. The braided yarn and the matrix surface are set as "Master Surface" and "Slave Surface," respectively. The braided yarn is defined as "self-contact," and the tangential friction coefficient is 0.15. In order to ensure smooth propagation of stress waves between the braided yarn and the matrix, the interface meshes between the braided yarn and the matrix are defined as a co-node connection, which is supposed to be a perfect adhesion, and is defined by "tie contact." The dynamic impact compression model uses the "general contact" in the ABAQUS/Explicit module to define the contact between the Hopkinson pressure bar and each part of the specimen and the contact between the braided yarns. The definition of contact between braided yarn and matrix is the same as that of quasi-static model.

3.5.3 Constraints and Loading Conditions

In the quasi-static axial compression test, the upper and lower loading plates are not deformed before and after the test. The lower loading plate remains stationary without displacement or rotation degrees of freedom, and the upper loading plate moves downward at a constant velocity of 2 mm/min. Therefore, the lower loading plate is set as a rigid body in the model, and its rotation degrees of freedom and translation degrees of freedom are constrained, and the same velocity load is applied to the upper loading plate along the axial direction.

In the dynamic impact compression test, the SHPB device belongs to uniaxial compression. The displacement degrees of freedom of the incident bar and the transmission bar along the axial direction are not constrained in the model, while the rotation degrees of freedom and the translational degrees of freedom in the vertical axis are constrained. The specimen placed between the two bars is in a completely free state. In the impact compression test, the stress waves generated by the impact of the incident bar and the incident rod are recorded by the strain gauges and output to the computer in the form of voltage-time curves. Therefore, the experimentally measured electrical signals of the input wave at each strain rate could be transformed into stress waves.

Fig. 3.15 Mesh of 3-D braided composite tube

Table 3.4 Number of elements for each component of the quasi-static and impact compression test system

Type of material	Preform	Matrix	Loading plate	SHPB	Number of elements
4D3L25A	98,317	52,7433	12,800	73,216	698,966
4D3L35A	11,8620	55,4914	12,800	73,216	751,768
4D3L45A	12,9770	59,8169	12,800	73,216	801,155
4D2L35A	69,528	55,6492	12,800	73,216	619,112
4D4L35A	13,9770	81,0021	12,800	73,216	1,023,007
5D2L35A	14,2800	81,7754	12,800	73,216	967,880

3.5.4 Mesh Element Division

The preform of the 3-D braided composite tube is meshed with the eight-node hexahedral second reduction integral (C3D8R) element type, and the matrix is meshed with the tetrahedral (C3D4) element type due to the complexity of the matrix shape. The loading plates, incident bar, and transmission bar are also meshed by the regular hexahedral C3D8R element. In the low-temperature compression simulation, the element type of preform of composite tube is C3D8T, and the element type of matrix is C3D4T. Figure 3.15 shows the mesh scheme of the preform and matrix of the 3-D four-directional braided composite tube as an example. Table 3.4 lists the mesh division quantities of the preform, matrix, and the loading plates and Hopkinson pressure bar in the test system for five different structures of 3-D four-directional braided composite tubes and 3-D five-directional braided composite tubes.

References

1. Zhang, M., Sun, B., Gu, B.: Meso-structure ageing mechanism of 3-D braided composite's compressive behaviors under accelerated thermo-oxidative ageing environment. Mech. Mater. **115**, 47–63 (2017)
2. Zhao Yong-li: Analysis of structures and elastic properties of 3D tubular braided composites. J. Tianjin Polytech. Univ. **29**(3), 15–18 (2010) [In Chinese]
3. Hwan, C.L., Tsai, K.H., Chen, W.L., et al.: Predicting the elastic moduli of 3-D (four-step) braided tubes using a spatial spring model. J. Compos. Mater. **47**(8), 991–1000 (2012)
4. Gideon, R.K., Zhou, H., Wu, X., et al.: Finite element analysis of 3D circular braided composites tube damage based on three unit cell models under axial compression loading. Int. J. Damage Mech. **25**(4), 574–607 (2016)
5. Ohtani, A., Nakai, A.: Effect of internal structure on mechanical properties of braided composite tubes. In: 16th International Conference on Composite Materials, pp. 1–5 (2007)
6. Shi, B., Zhang, M., Liu, S., Sun, B., Gu, B.: Multi-scale ageing mechanisms of 3D four directional and five directional braided composites' impact fracture behaviors under thermo-oxidative environment. Int. J. Mech. Sci. **155**, 50–65 (2019)
7. Hallquist, J.O.: LS-DYNA3D Theoretical Manual, pp. 19.1–19.2. Livermore Software Technology Corporation, Livermore (1993)

Chapter 4
Transverse Impact of Braided Beams

This chapter focuses on the preparation of 3-D braided composite beams and the analysis of the thermo-mechanical properties of epoxy resin using dynamic mechanical analysis tests and Fourier transform infrared spectroscopy. The principle of SHPB transverse impact is introduced, and the heating device is designed to test the transverse impact properties of the specimens at different temperatures and different impact pressures. The finite element results are compared and analyzed with the test results to verify the validity of the model. The effects of ambient temperature, loading rate, and braided structure on the transverse impact behavior of the composite beams are mainly investigated by test methods.

4.1 Preparation of 3-D Braided Carbon Fiber/Epoxy Resin Composite Beams

The carbon fiber is T700-12K from Toray® of Japan. Due to the preferred orientation of graphite microcrystalline structure along the fiber axis, it has high modulus and high strength in the fiber axis direction. Carbon fiber has low density, no creep, high specific modulus, and high specific strength, is soft, and can be processed into various fabrics. Carbon fiber filament tows can be used as preform to composite with ceramics, resins, and metals to manufacture advanced composites. Carbon fiber-reinforced epoxy resin composites have the highest specific strength and specific modulus among existing engineering materials. Therefore, it is widely used in various fields of aerospace, defense, and civil industries. In addition, carbon fiber also has good low-temperature resistance, does not brittle at liquid nitrogen temperature, and is widely used in the field of thermo-mechanical coupling protection. Table 4.1 lists the specification of carbon fiber tow.

The epoxy resin (types JC-02A and JC-02B) modified anhydride curing agent produced by Changshu Jiafa Company in China. The epoxy resin has low viscosity,

M. Hu, B. Gu, *Impact Damages of Braided Composites*, Engineering Materials,
https://doi.org/10.1007/978-981-16-5703-0_4

Table 4.1 Specifications of carbon fiber tow

Fiber tow specifications	Fiber density (g/cm^3)	Fiber tow linear density (Tex)
12k	1.8	800

Table 4.2 Specification of epoxy resin

Label	Ingredients	Room temperature viscosity (MPa·s)	Epoxy value (eq/100 g)	Density (g/cm^3)	Probation period 25 °C (h)	Mixed weight ratio
JC-02A	Bisphenol A epoxy	1000–3000	0.50–0.53	1.12–1.14	8 ~ 10	100:85
JC-02B	Modified anhydride	30–50	/	1.20–1.25		

slow gelation speed, and long application period, which can ensure that the resin can enter the preform better and maintain a faster impregnation speed. At the same time, it can ensure that the viscosity of the resin does not change significantly during the impregnation process. The specifications are shown in Table 4.2.

The spatial arrangement of the yarns has a great influence on the mechanical properties and volume fraction of the fiber-reinforced composites. 3-D braided carbon fiber/epoxy resin composite beams are made of carbon fiber tows as preform, and 3-D braided preforms are braided by 1 × 1 four-step square braiding technique, as shown in Fig. 4.1. The braided yarn arrangement for the 3-D four-directional/five-directional braided preforms is 21 × 5, and the axial yarn arrangement for the 3-D five-directional braided preforms is 20 × 5. Three kinds of braiding angles are selected for the 3-D four-directional braided preforms, which are 20°, 30°, and 40°, respectively, denoted as 4D20, 4D30, and 4D40. The braiding angle for the 3-D five-directional braided preforms is 30°, denoted as 5D30. The composites were prepared with vacuum-assisted resin transfer molding (VARTM) method. Table 4.3 shows the mechanical parameters of carbon fiber and epoxy resin. Table 4.4 shows the elastic-plastic parameters of epoxy resin at room temperature. Table 4.5 lists the mechanical parameters of epoxy resin under different strain rates and temperatures. Figure 4.2 shows the 3-D braided composite beam preforms and the specimens with surface braiding angles of 20°, 30°, and 40°, respectively. The specifications are shown in Table 4.6. The fiber volume fraction was determined by the muffle combustion method. The fiber filling coefficient (δ_f) of the braided fiber tows was 0.7, which measured on SEM photographs (see Fig. 3.3). The thermal expansion coefficient, thermal conductivity, and specific heat capacity of carbon fibers and epoxy resin are given in Table 4.3. Thus, the thermo-mechanical properties of the carbon fiber tows are calculated with Eqs. (2.12), (2.13), (2.14), and (2.15) (see Tables 4.7 and 4.8). Table 4.9 shows the calculated elastic parameters of the braided yarns impregnated by the resin at different temperatures and strain rates. The volume fractions of the macroscopic material were calculated to be 67.80%, 30.51%, and 1.69% for the interior, surface, and corner unit cells, respectively. The mechanical properties of the macroscopic homogeneity are shown in Table 4.10.

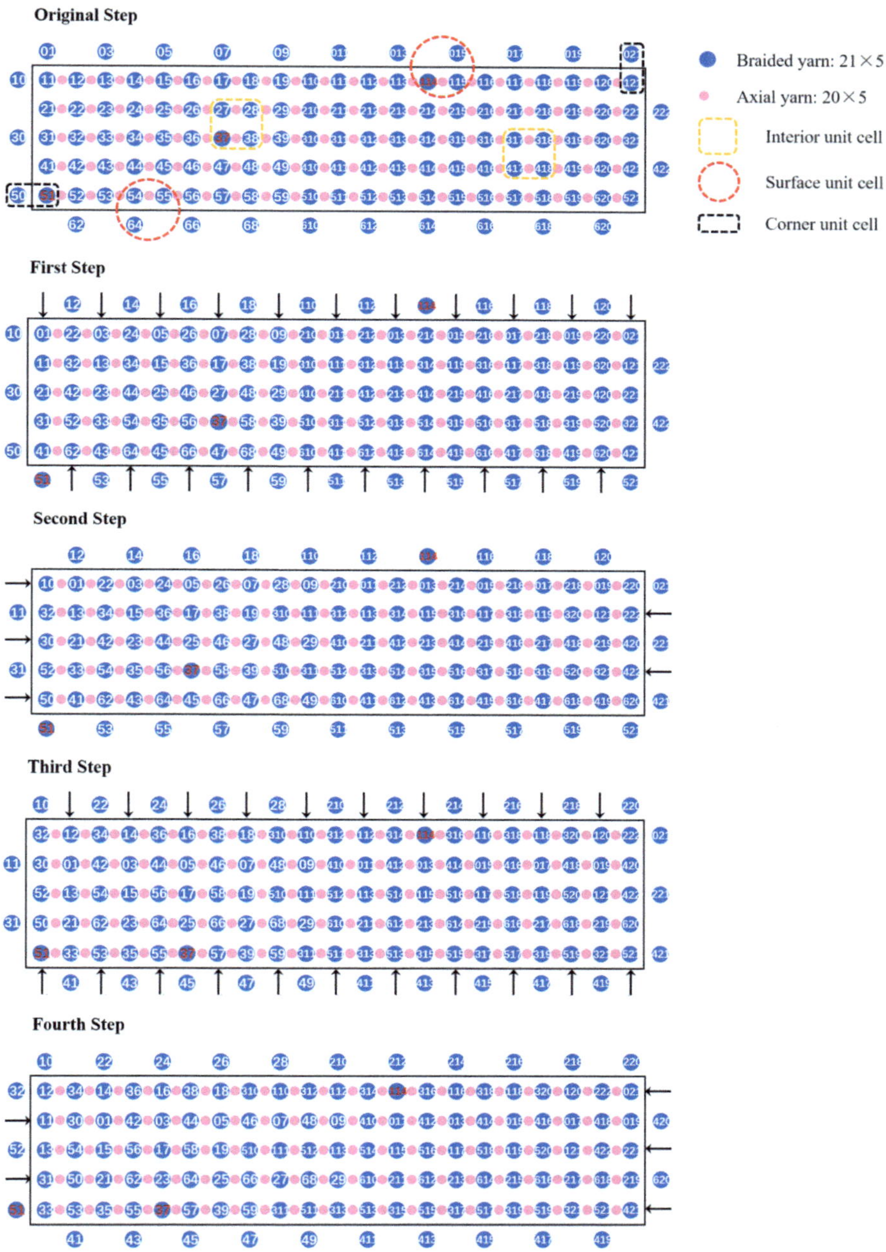

Fig. 4.1 Schematic diagram of the braiding process of the 3-D braided preform

Table 4.3 Mechanical and thermal parameters of carbon fibers and epoxy resin[a]

	E_{11} (GPa)	$E_{22} = E_{33}$ (GPa)	$G_{12} = G_{13}$ (GPa)	G_{23} (GPa)	$\nu_{12} = \nu_{13}$	ν_{23}	ρ (g/cm^3)	α_{11} ($\times 10^{-6}$/K)	$\alpha_{22} = \alpha_{33}$ ($\times 10^{-6}$/K)	k (W/m·K)	C (kJ/kg. K)	T (°C)
Carbon fiber	230	14	9	5	0.25	0.3	1.78	−0.38	12	9.38	0.75	25
Epoxy resin	2.40		0.90		0.33		1.13	67.00		0.5	0.94	25
	1.93		0.73					71.68				90
	0.02		0.01					80.85				150
	0.02		0.01					108.50				210

[a]E denotes elastic modulus, G denotes shear modulus, ν denotes Poisson's ratio, ρ denotes density, α denotes thermal expansion coefficient, T denotes temperature

Table 4.4 Elastic-plastic parameters of epoxy resin

Properties	Parameters
Density ρ (g/cm^3)	1.13
Elastic modulus E (GPa)	2.4
Shear modulus G (GPa)	0.89
Compression strength XC (MPa)	102
Tensile strength XT (MPa)	80
Shear strength SS (MPa)	113
Yield strength δY (MPa)	69
Poisson's ratio	0.3
Elongation at break (%)	5.9

Table 4.5 Mechanical parameters of epoxy resin at different temperatures and strain rates

Strain rate (s^{-1})	Temperature (°C)	E_m (MPa)	G_m (MPa)	ν_m
	23	2435	795	0.33
480	−50	2546	830	
	−100	2667	865	
	23	2642	846	
715	−50	2753	875	
	−100	2865	898	
	23	2861	891	
930	−50	2960	924	
	−100	3116	949	

4.2 Thermo-mechanical Properties of Resin

The thermo-mechanical properties of the epoxy resin casting were tested with Perkin Elmer 8000 dynamic mechanical analysis (DMA). The mechanical behavior of the specimen was studied by applying sinusoidal periodic stress wave under the temperature controlled by program. The functional relationships of storage modulus, loss modulus, and loss factor with time, temperature, and force frequency were measured. It is widely used in the fields of thermoplastic and thermoset plastics and composites. The storage modulus reflects the elasticity of the specimen, which is the scale of whether the specimen can fully recover the deformation. The loss modulus reflects the viscous property of the specimen, which is the heat loss of the specimen during the deformation. The loss factor is the ratio of the loss modulus to the storage modulus, which reflects the vibration absorption.

A temperature scan was made from 25 to 210 °C at a heating rate of 5 °C/min in single cantilever beam bending mode. The size of epoxy resin specimen was 35.6 × 13.1 × 3.5 mm. As shown in Fig. 4.3a, the storage modulus started to drop abruptly from around 105 °C until it leveled off around 125 °C, indicating that the glass transition temperature (T_g) was between 105 and 125 °C. As shown in Fig. 4.4, the onset of thermal degradation is about 300 °C, and the temperature of complete thermal degradation is close to 600 °C according to the thermogravimetric

Fig. 4.2 3-D braided composite beam preparation. (**a**) Specimen, (**b**) VARTM curing

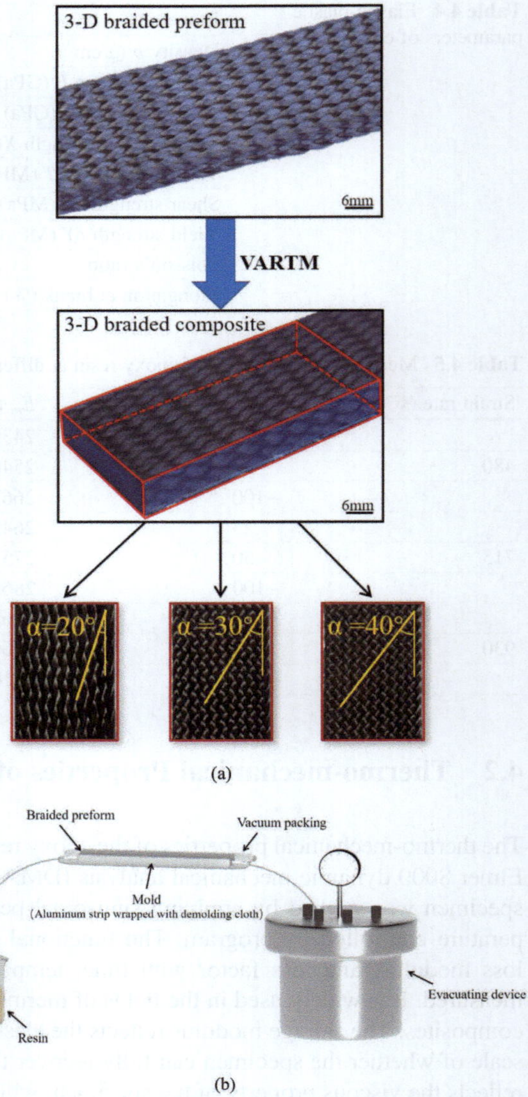

(TG) analysis results. The chemical composition of the epoxy resin changes considerably after thermal degradation, so we chose room temperature (25 °C), 90, 150, and 210 °C as the test ambient temperatures.

In addition, DMA tests in time scan mode were made at 90 and 210 °C to test the stability of the epoxy resin, as shown in Fig. 4.3b. The results show that the storage modulus of epoxy resin is basically unchanged. This stability proves that the effect of heating time on the test results is negligible. Combined with the

Table 4.6 Geometrical parameters of 3-D braided composite

Braided structure	Average thickness (mm)	Average width (mm)	Average length (mm)	Fiber volume fraction (%)	Linear density (g/mm)
4D20	5.53	27.44	150.0	27.65	1.45
4D30	5.87	27.10	150.0	28.63	1.46
4D40	6.28	25.81	150.0	30.84	1.48
5D30	6.27	30.27	150.0	41.22	1.49

thermogravimetric analysis, it is clear that the resin undergoes severe chemical changes only when the temperature is higher than 300 °C.

A temperature scan was made from room temperature to −140 °C at a heating rate of 3 °C/min in single cantilever beam bending mode. The size of epoxy resin specimen was 15.8 × 5.5 × 2.8 mm. As shown in Fig. 4.5, the epoxy resin hardened at low temperatures, the storage modulus increased as the temperature decreased, while the loss modulus rose to a maximum as the temperature dropped to −100 °C and then decreased as the temperature decreased.

4.3 Testing Setup

A modified SHPB apparatus was used to test the transverse impact behavior of 3-D braided composites at different ambient temperatures and impact velocities. The i-Speed 716 high-speed camera (IX Cameras Ltd., UK) was applied to record the damage process of the specimens during the impact at a frequency of 5×10^4 frames/sec with a maximum resolution of 2048 × 1536 pixels. The impact pressures of 0.2, 0.4, and 0.6 MPa were selected for the test. Figure 4.6 shows the schematic diagram of the SHPB and the transverse impact test setup, which consists of a gas gun, impact bar, incidence bar, and signal acquisition device. The impact bar and the incidence bar are both made of 45Gr steel, with a diameter of 30 mm and lengths of 400 mm and 1800 mm, respectively. The strain gauges are attached at a distance of 600 mm from the left end of the incidence bar to collect the stress wave signals. Both ends of the specimen were fixed to the device by a fixture with a test length of 100 mm. The test was repeated for three times under the same conditions, and the average value was taken.

An elastic stress wave is generated when the impact bar strikes the incident bar. It propagates from the incident bar to the contact surface with the specimen. Part of the stress wave passes through the specimen, generating the first impact pulse. The other portion is reflected back to the incident bar. When the reflected stress wave reaches another free surface of the incident bar, the attenuated stress wave is reflected again and propagates to the contact surface between the incident bar and the specimen, which is divided into two parts again. This process continues repeatedly until the impact energy is completely absorbed and dissipated. The elastic wave is reflected

Table 4.7 Mechanical parameters of carbon fiber tows at different temperatures

	E_{11} (GPa)	$E_{22} = E_{33}$ (GPa)	$G_{12} = G_{13}$ (GPa)	G_{23} (GPa)	$\nu_{12} = \nu_{13}$	ν_{23}	α_{11} ($\times 10^{-6}$/K)	$\alpha_{22} = \alpha_{33}$ ($\times 10^{-6}$/K)	T (°C)
Carbon fiber tow	161.58	6.93	3.15	2.57	0.27	0.31	−0.12	39.13	90
	161.01	0.13	0.05	0.06			−0.38	42.86	150
	161.01	0.12	0.04	0.04			−0.38	53.89	210

Table 4.8 Thermal expansion coefficient and thermal conductivity of yarns in the braided composites

Strain rate (s^{-1})	T (°C)	α_{11} ($\times10^{-6}$/K)	α_{22} ($\times10^{-6}$/K)	k_{11} (W/m·K)	k_{22} (W/m·K)
	23	0.2256	28.5614	6.2898	0.7457
480	−50	0.2469	28.5561		
	−100	0.2668	28.5511		
	23	0.2594	28.5529		
715	−50	0.2789	28.5481		
	−100	0.2951	28.5441		
	23	0.2883	28.5457		
930	−50	0.3013	28.5425		
	−100	0.3264	28.5362		

Table 4.9 Mechanical parameters of braided yarns

Strain rate (s^{-1})	T (°C)	E_{11} (GPa)	E_{22} (GPa)	E_{33} (GPa)	ν_{12}	ν_{23}	ν_{13}	G_{12} (GPa)	G_{13} (GPa)	G_{23} (GPa)
	23	133.65	7.05	7.05	0.25	0.3	0.25	4.30	4.30	2.88
480	−50	134.95	7.35	7.35				4.42	4.42	2.99
	−100	136.96	7.57	7.57				4.53	4.53	3.10
	23	137.25	7.53	7.53				4.41	4.41	2.98
715	−50	138.76	7.81	7.81				4.52	4.52	3.11
	−100	140.91	8.16	8.16				4.61	4.61	3.23
	23	142.15	8.22	8.22				4.55	4.55	3.48
930	−50	144.24	8.58	8.58				4.62	4.62	3.57
	−100	146.38	8.97	8.97				4.71	4.71	3.69

Table 4.10 Effective mechanical properties of the braided composites

	E_{11} (GPa)	$E_{22} = E_{33}$ (GPa)	$G_{12} = G_{13}$ (GPa)	G_{23} (GPa)	$\nu_{12} = \nu_{13}$	ν_{23}
4D30	53.76	8.32	12.39	5.74	0.67	0.47
5D30	67.57	8.40	10.53	5.14	0.61	0.48

several times, causing multiple impacts on the specimen. In this test, the final morphology of the specimen can be represented at the end of the fourth impact cycle, so we selected the first four impact cycles. The stress waves recorded by the ultra-dynamic strain gauge were converted into voltage-time curves as shown in Fig. 4.7. The amplitude of stress wave decreases gradually due to the energy absorption of the braided composite. The impact load P, displacement μ, energy absorption W, and wave velocity C_0 can be calculated by the following equations:

$$P(t) = EA[\varepsilon_I(t) + \varepsilon_R(t)] \tag{4.1}$$

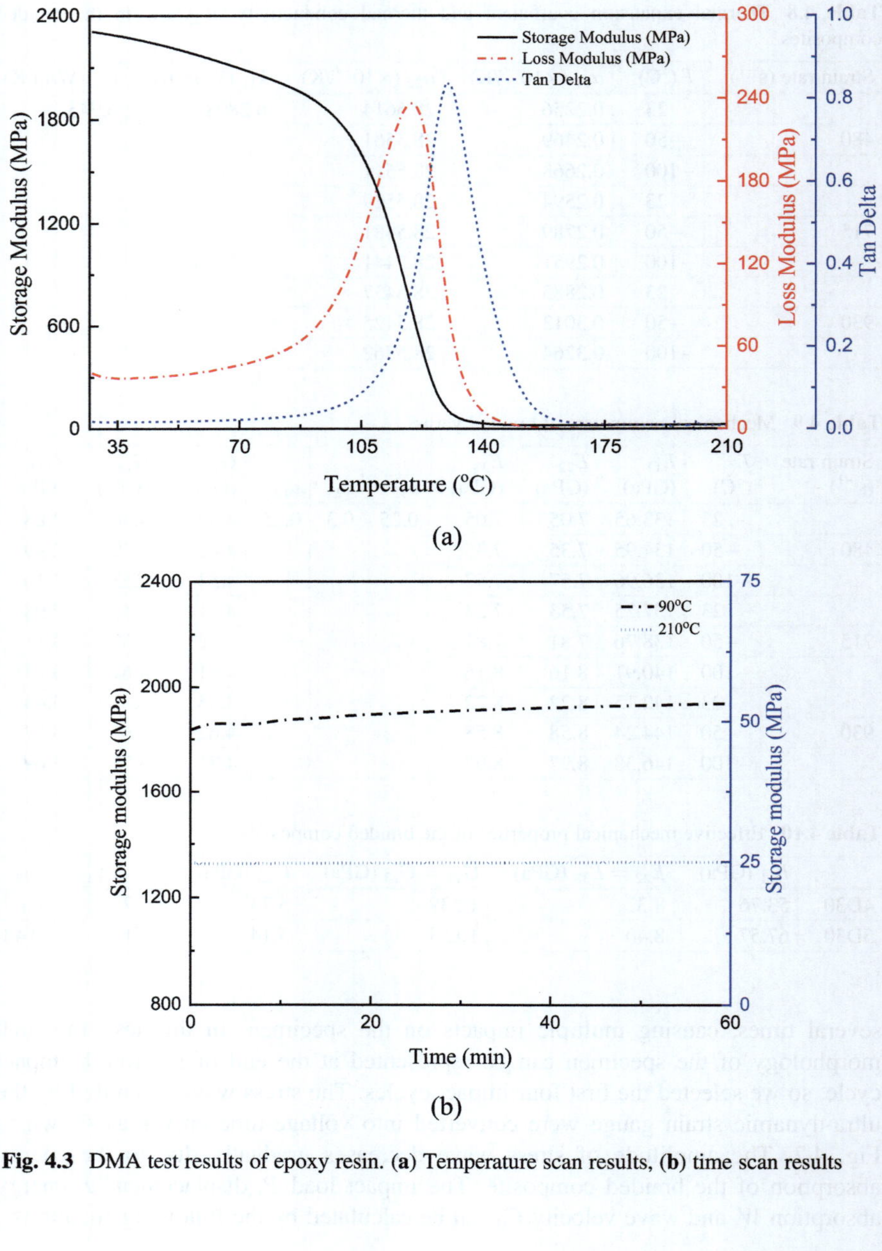

Fig. 4.3 DMA test results of epoxy resin. (**a**) Temperature scan results, (**b**) time scan results

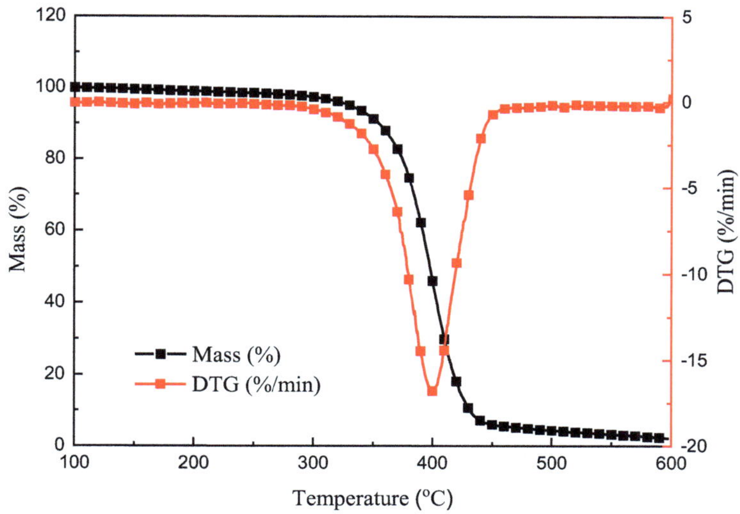

Fig. 4.4 Thermogravimetric analysis of epoxy resin

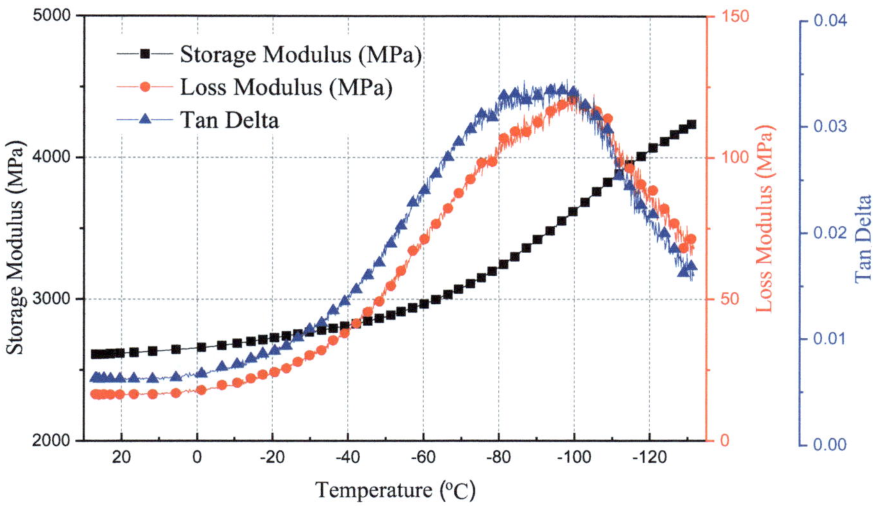

Fig. 4.5 Thermo-mechanical response of epoxy resin during cooling from room temperature to −130 °C

$$\mu(t) = C_0 \int_0^t [\varepsilon_I(t) - \varepsilon_R(t)]dt \tag{4.2}$$

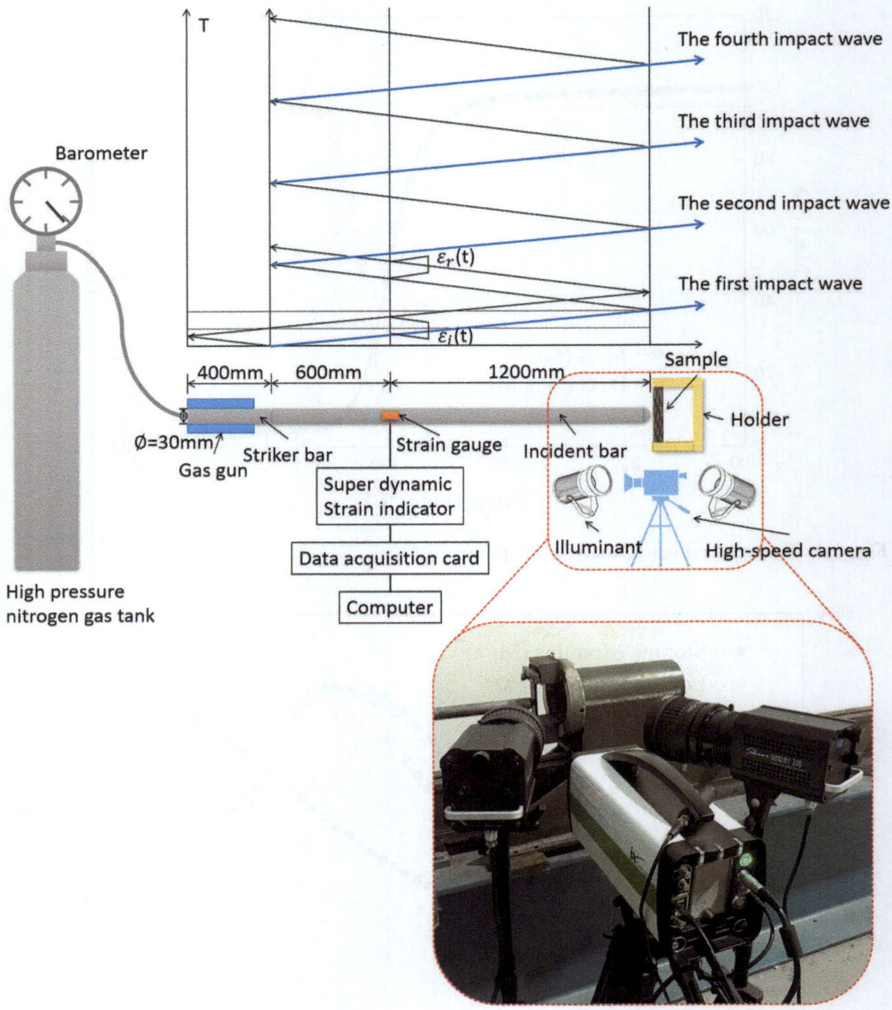

Fig. 4.6 Test principle and transverse impact test setup of SHPB apparatus

$$W = \int_0^t \frac{1}{2} P(t)\mu(t)dt = \frac{1}{2} EAC_0 \int_0^t \left[\varepsilon_I^2(t) - \varepsilon_R^2(t)\right] dt \tag{4.3}$$

$$C_0 = \sqrt{E/\rho} \tag{4.4}$$

where $\varepsilon_I(t)$ and $\varepsilon_R(t)$ denote the incident and reflected pulse signals recorded by the strain gauge, respectively, A denotes the cross-sectional area of the bar, and E and ρ denote the elastic modulus and density of the bar, respectively.

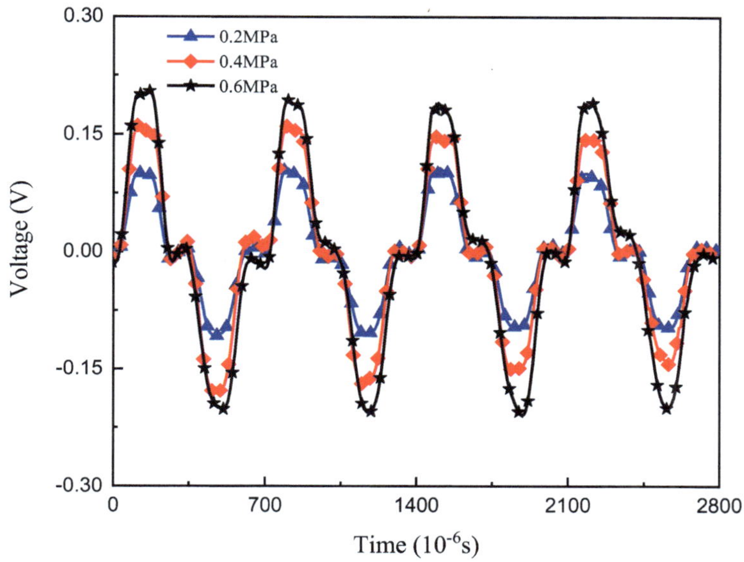

Fig. 4.7 Voltage-time curves for the first four impact cycles

A simple and effective heating device is developed for application in the SHPB apparatus as shown in Fig. 4.8. The device consists of a heating box and a control unit. The heating box has four different layers of components which are heat transmission layer, metal resistance heating layer, ceramic insulation layer, and stainless steel case. The principle of the control unit is that the solenoid valve closes automatically when the temperature of the specimen as measured by the thermocouple reaches the set temperature. Note that during the heating process, the incident bar is moved out of the heating box to avoid the influence of the temperature gradient field inside the bar on the stress wave propagation. According to GB/T 9979-2005 test standard from China, after keeping the specimen temperature at the set temperature for 15 ~ 20 min, taking the incident bar to contact with the specimen immediately for making transverse impact test to avoid the loss of specimen temperature.

4.4 Effects of Braiding Angles and Impact Velocities

The behavior of 3-D multidirectional braided composites with multiple transverse impacts is numerically simulated using a thermo-mechanical coupling model in FEA. The relationship among the damage distribution inside the specimen and braided structure, loading rate, and ambient temperature is explored by combining with μ-CT and comparing the failure behavior and failure modes of the specimen under different conditions.

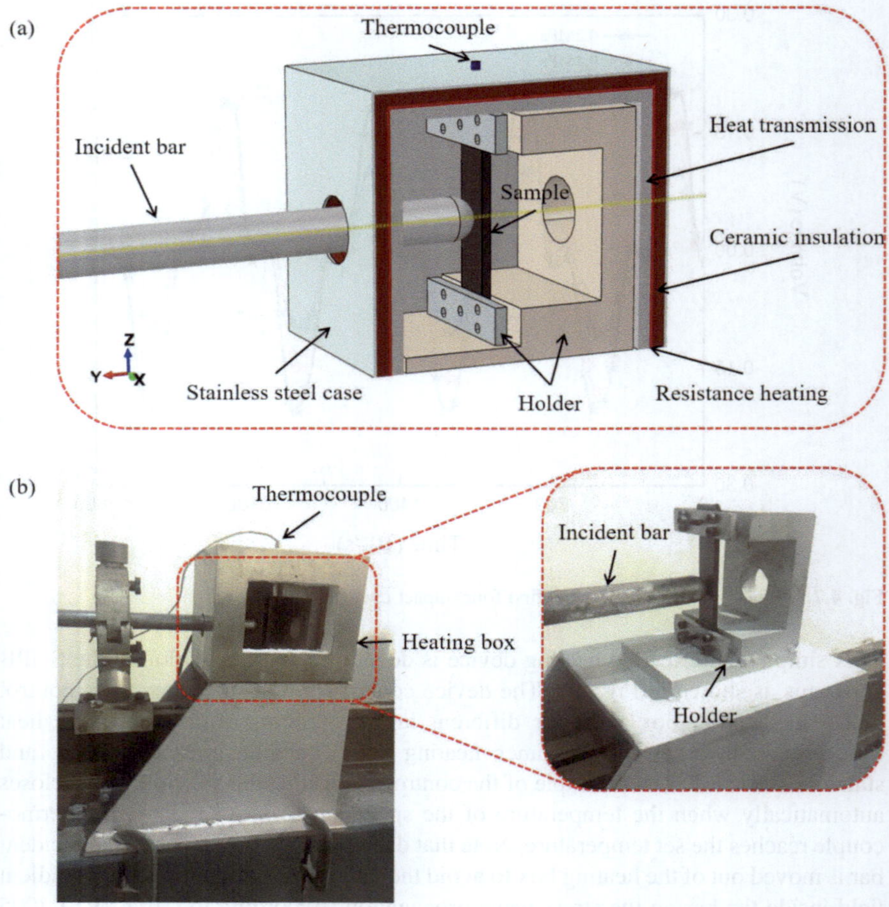

Fig. 4.8 High-temperature device. (**a**) Sketch diagrams, (**b**) photographs

4.4.1 Test and Finite Element Results

Figures 4.9 and 4.10 show the load-displacement histories of the 3-D four-directional braided composites with braiding angles of 20°, 30°, and 40° at 0.2 MPa, 0.4 MPa, and 0.6 MPa, respectively. The results show that:

1. The peak load decreases gradually as the impact cycles increase. This is because when the stress wave reaches the interface between the incident bar and the specimen, part of the stress wave passes through the specimen, part of the impact energy is dissipated, and another part is reflected back to the incident bar, resulting in energy reduction of the next impact cycle, and thus the load peak is gradually reduced.

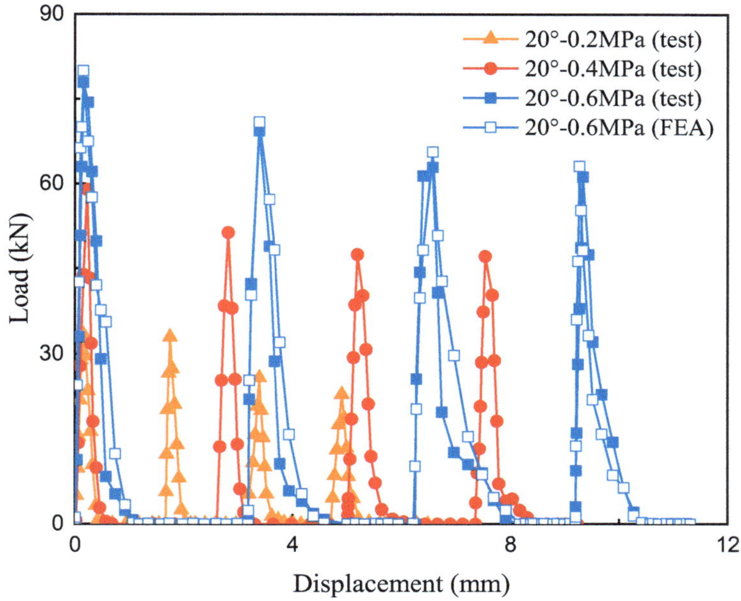

Fig. 4.9 Load-displacement curves for 3-D four-directional braided composites with the braiding angle of 20° under different impact pressures

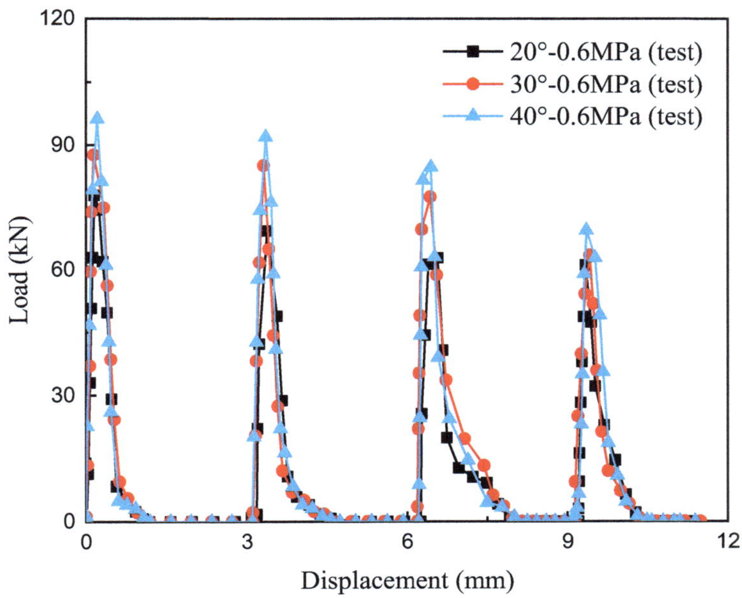

Fig. 4.10 Load-displacement curves of 3-D four-directional braided composites with different braiding angles at 0.6 MPa

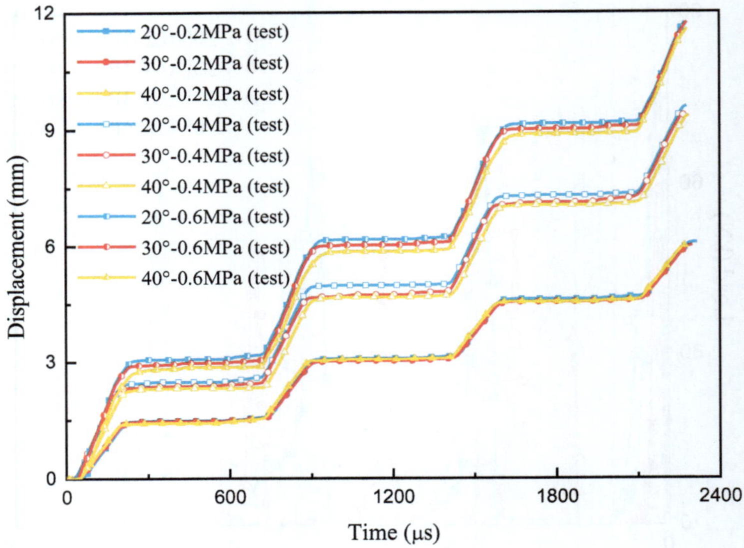

Fig. 4.11 Displacement-time curves

2. By comparison, it is found that with the increase of impact pressure, the load peak and displacement increase, and the deformation of the composite increases.
3. With the increase of the braiding angle, the peak load increases, and the displacement decreases.

The fiber tows of composites with large braiding angles are interwoven more closely, thus increasing the thickness and volume fraction of the composite, so the stiffness and resistance to deformation of the composite increase, and the displacement decreases slightly. As the impact pressure increases, the displacement decreases more significantly. The test results and the FEA results match in terms of peak loads and displacements, indicating that the meso-structure model is able to predict the transverse impact results. Detailed analyses will be presented in Chaps. 8 and 10.

Figure 4.11 shows the displacement histories of 3-D four-directional braided composites with braiding angles of 20°, 30°, and 40° at 0.2 MPa, 0.4 MPa, and 0.6 MPa, respectively. All curves show the same tendency, i.e., the displacement increases with time and remains constant as the stress wave is reflected into the incident bar. It is clear from the comparison that the displacement increases as the impact pressure increases. As the braiding angle increases, the displacement decreases slightly. This is due to the fact that the higher the impact pressure, the higher the impact energy carried, resulting in greater deformation of the composite.

Figure 4.12 shows the energy absorption of 3-D four-directional braided composites with different braiding angles under different impact pressures. The energy absorption can be calculated from the load-displacement curve integral or Eq. (4.3).

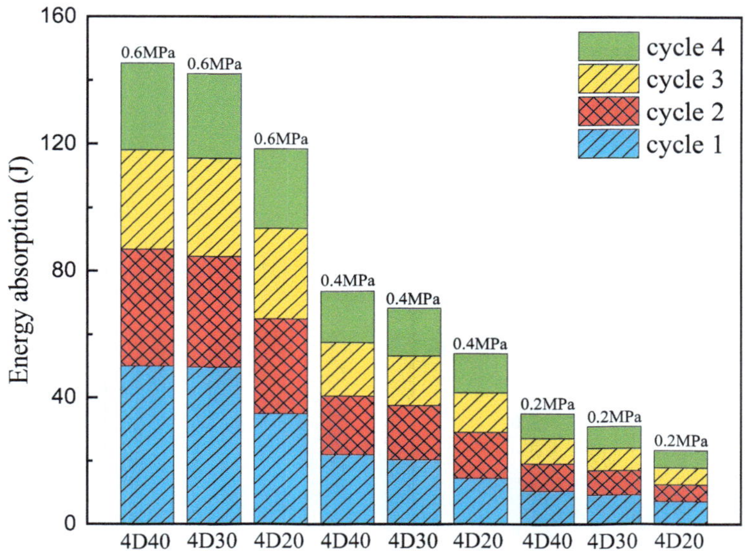

Fig. 4.12 Energy absorption of 3-D four-directional braided composites with different braiding angles under different impact pressures

It is clearly seen that the energy absorption is highest for 0.6 MPa, and the energy absorption for large braiding angles is higher than that for small braiding angles. The energy absorption of the first cycle is the largest and decreases with the increase of impact cycles. With the increase of braiding angle, it can be seen from the geometric model that the number of interlacing points per unit volume of yarn increases. This makes that the thickness and fiber volume fraction of the composite increase and also the stiffness and the energy absorption threshold increase. The effect of impact pressure is more significant than the effect of braiding angle. The test results are consistent with the displacement curves of the finite element method, which indicates that the finite element model is able to predict the impact response of 3-D braided composite beams.

4.4.2 Damage Morphologies and Failure Modes

Figures 4.13 and 4.14 combine the high-speed photographs with FEA to observe the deformation process of the specimen. The top half is the finite element simulations, and the bottom half is the high-speed photographs. The FEA results are in good agreement with the test results, which verifies the reasonableness of the finite element calculation. The results show that as the impact pressure increases and the braiding angle decreases, the more severe the damage of the composite is and the larger the transverse displacement is. The pictures show the damage morphologies of

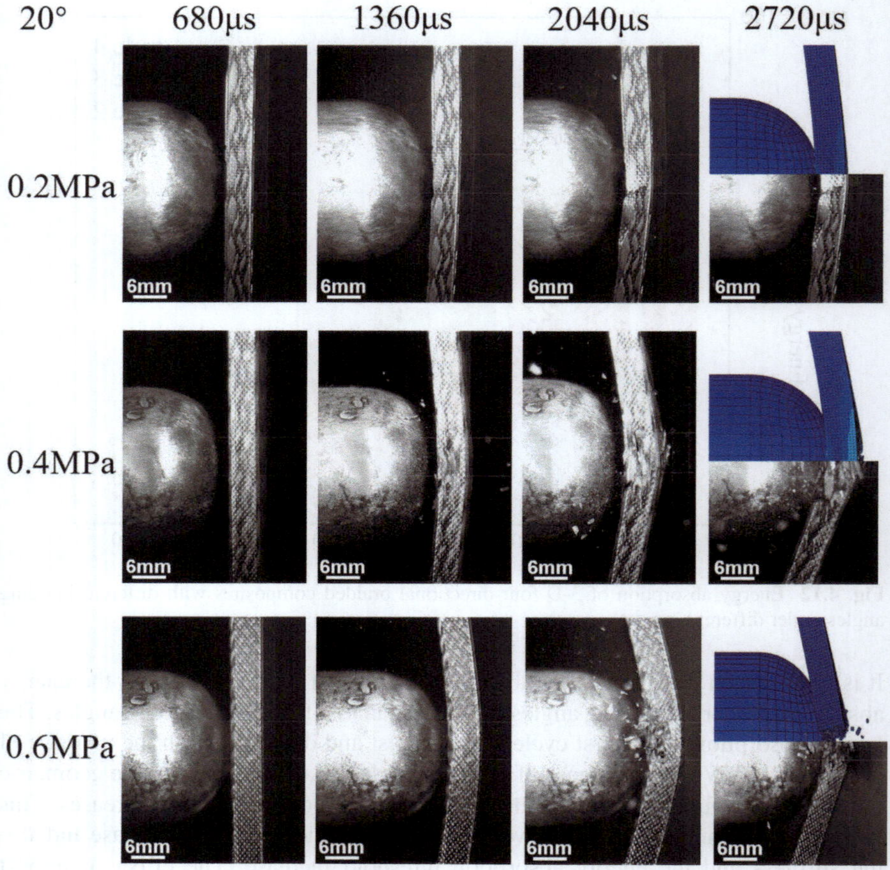

Fig. 4.13 Comparison of deformation and fracture of 3-D four-directional braided composites under different impact pressures in test and FEA (0.2, 0.4, and 0.6 MPa)

the impact surface of the 3-D four-directional braided composite beams with different braiding angles under various impact pressures, and comparing the finite element results with the test results. With the same braiding angle, the impact damage area increases as the impact pressure increases from 0.2 to 0.6 MPa. This is due to the fact that higher impact pressure carries higher impact energy, which leads to more severe damage. At the impact pressure of 0.6 MPa, the impact damage area gradually decreases as the braiding angle increases. This is due to the compact structure of the composite preform with large braiding angle, which increases the thickness and fiber volume fraction of the composite and in turn increases the deformation resistance of the composite. The main failure modes are matrix cracking, fiber breakage at low impact pressure, yarn fracture at high impact pressure, and interfacial debonding at large braiding angles.

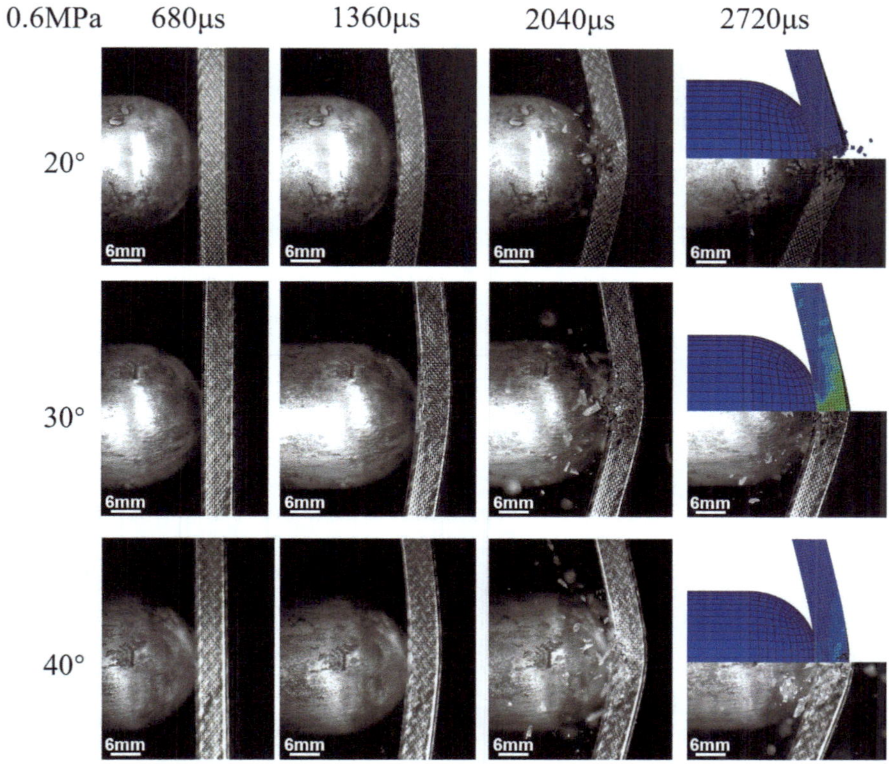

Fig. 4.14 Comparison of deformation and fracture of 3-D four-directional braided composites with different braiding angles in test and FEA (20°, 30°, 40°)

4.5 Effects of Braiding Angles and Temperatures

4.5.1 Impact Compression Behavior of Epoxy Resins at Different Temperatures

Figure 4.15a shows the stress-strain curves of the epoxy resin castings at different ambient temperatures at the impact pressure of 0.6 MPa at ambient temperatures ranging from 25 to 210 °C. The elastic-plastic properties of the epoxy resin were found to be closely related to temperature, and the obtained elastic-plastic properties of the epoxy resin at different temperatures were input into the FEA software package ABAQUS. A high-speed camera was also used to record the damage process of the epoxy resin casting at different temperatures at 0.6 MPa, as shown in Fig. 4.15b. The results show that the epoxy resin starts cracking from 130 μs at 25 °C, and the cracking onset is delayed when the temperature increases. Under the same impact pressure, the epoxy resin can withstand greater deformation without

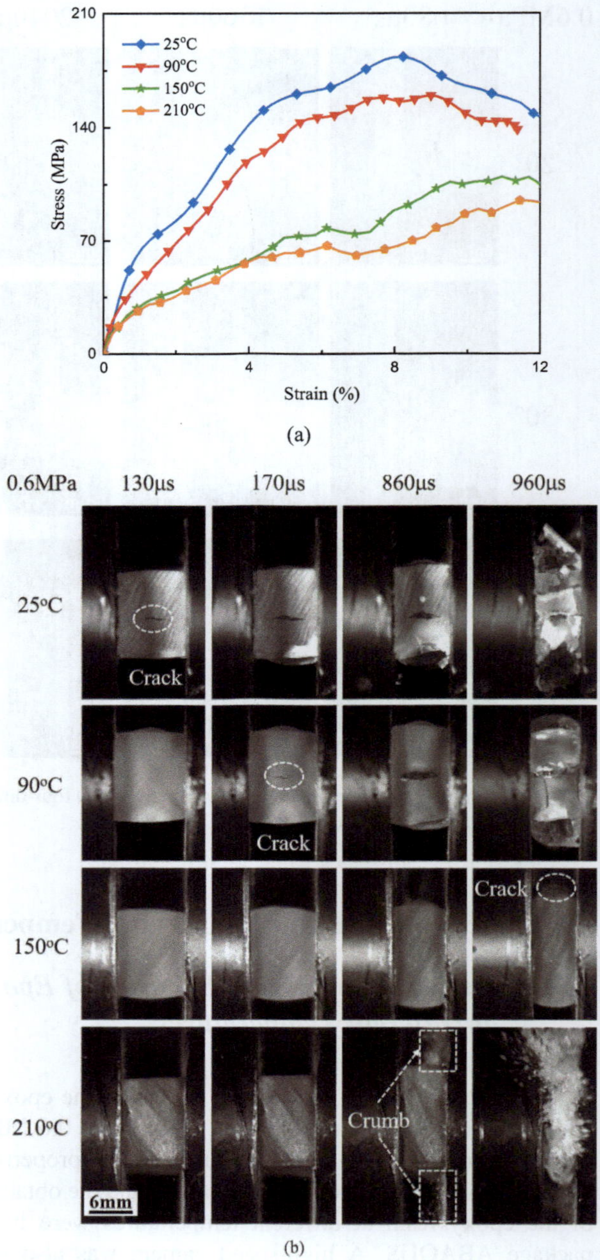

Fig. 4.15 Impact compression test of epoxy resin. (**a**) Stress-strain curve of epoxy resin at 0.6 MPa, (**b**) damage morphologies of epoxy resin at different temperatures during impact

cracking after temperature increase. The epoxy resin cracked from 170 μs at 90 °C, 960 μs at 150 °C, and 860 μs at 210 °C. At 210 °C, the epoxy resin underwent catastrophic damage at 960 μs because the high temperature of 210 °C was much higher than the glass transition temperature. This caused severe changes in both physical and chemical properties of the epoxy resin, resulting in an earlier onset of cracking of the epoxy resin relative to 150 °C and a greater degree of damage than 150 °C. The study of the impact compression properties of epoxy resin castings at different temperatures is necessary to discover the transverse impact behavior of braided composite beams under the same conditions.

4.5.2 Test and Finite Element Results

High-speed transverse impact tests were performed on 3-D four-directional braided composite beams with braiding angles of 20°, 30°, and 40° at ambient temperatures of 25 °C, 90 °C, 150 °C, and 210 °C and impact pressure of 0.6 MPa, respectively. Figure 4.16 shows the load-displacement histories obtained from the test and the finite element model for the first four impact cycles. In the process of elastic stress wave propagation, part of the energy is absorbed by the composite, and the other part of the energy is converted into heat energy. The energy is gradually dissipated during the impact cycle. Therefore, with the continuation of the impact cycle, the peak load gradually decreases. We can find that the peak load decreases with increase in ambient temperature while the displacement increases slightly. This is related to the weakened behavior of the braided composite at high temperatures. As the temperature increases, the physical properties of the epoxy resin weaken and become softer. As shown in Fig. 4.3a, the storage modulus of the epoxy resin gradually decreases from 2400 MPa at room temperature to 19 MPa at 210 °C, while the storage modulus of the carbon fiber tows in the braided composite gradually decreases from 161.72 to 161.01 GPa.

In addition, the significant difference between the CTE of the epoxy resin and carbon fiber is another reason for the decrease in the peak load. As the braiding angle increases, the peak load increases and the displacement decreases. The preform of the composite with braiding angle of 40° maintains structural integrity during the impact process. The results show that the stiffness, deformation resistance, and damage resistance of the composites with large braiding angle are enhanced due to the compact interlacing of yarns and the higher thickness and fiber volume fraction of the specimens. It is concluded that the braiding angle has a significant effect on the structural integrity of the braided composites.

Figure 4.17 shows the displacement histories for the first four impact cycles obtained from the test and finite element model, which can reflect the deformation process of 3-D braided composite beam under transverse impact. All curves show the same trend, i.e., the displacement increases linearly with time at the beginning and remains constant as the elastic stress wave is reflected into the incident bar. Although the displacement increases slightly with increasing temperature and decreasing braiding angle, the effect of impact pressure is more significant. The test results are in good agreement with the finite element analysis results on the

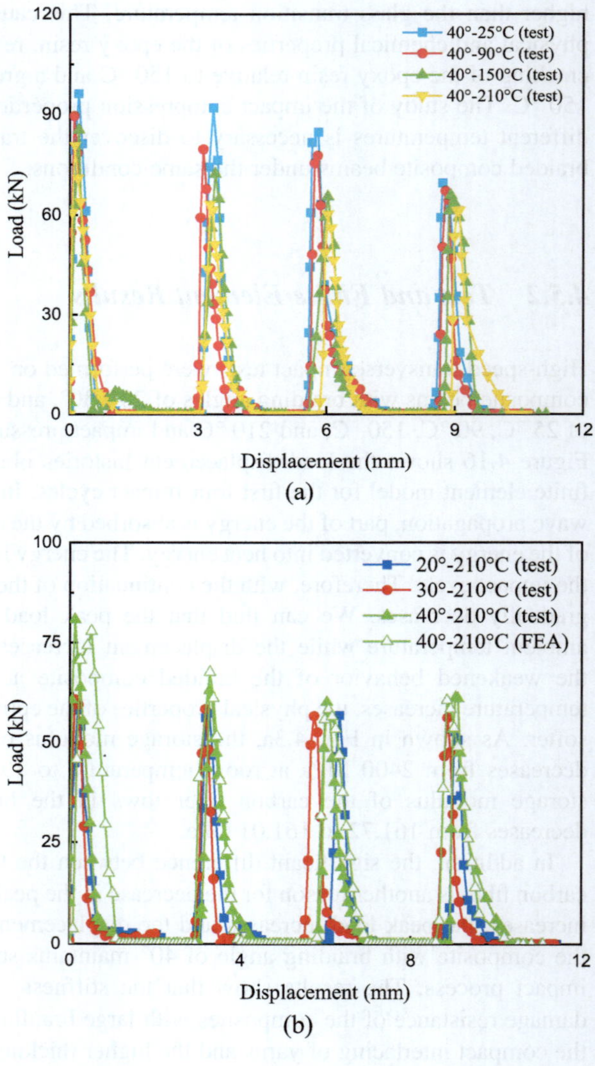

Fig. 4.16 Load-displacement curves at 0.6 MPa. (**a**) Test results at different temperatures, (**b**) test and FEA results for composites with different braiding angles

displacement-time curve. The results indicate that the finite element model under the temperature field can predict the deformation of the braided composite beam well.

Energy absorption is an important parameter to evaluate the ability of composites to withstand impact. From Fig. 4.18, it can be seen that the energy absorption of the composite decreases as the ambient temperature increases and the braiding angle decreases. It is clear that the energy absorption is maximum for the first impact cycle and decreases as the impact cycle continues. This is due to the fact that part of the energy is absorbed by the material and converted into deformation and heat energy, while the other part is absorbed by the incident bar and gradually dissipated.

Fig. 4.17 Displacement-time curves at 0.6 MPa. (**a**) Test results at different temperatures, (**b**) test results for composites with different braiding angles

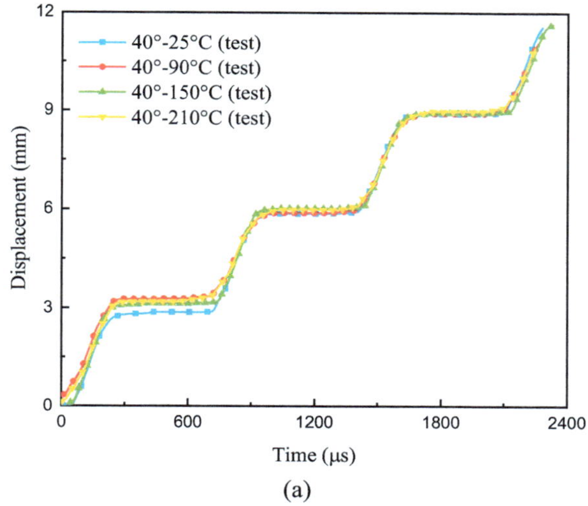

(a)

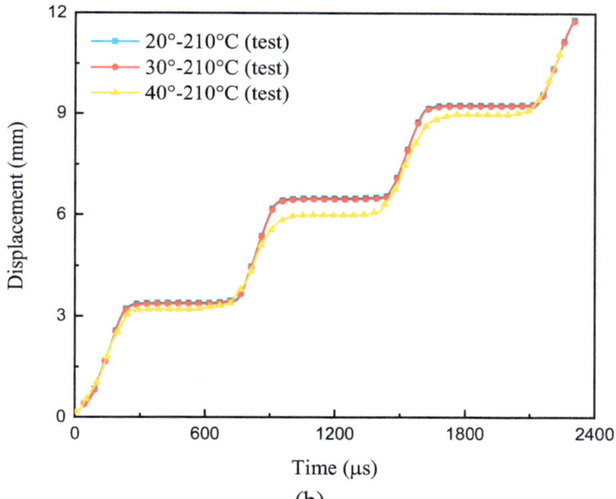

(b)

The deformation and damage processes of 3-D four-directional braided composites with different braiding angles at various temperatures are given in Figs. 4.19 and 4.20. The test results are in good agreement with the FEA results and validate the finite element model. From Fig. 4.19, it can be seen that the deformation of the specimen gradually increases with time. At room temperature, the epoxy resin cracked or even crushed in the third impact cycle, but the resin hardly cracked when the ambient temperature increased. This indicates that the weakened behavior of the epoxy resin at high temperature affects the failure modes of the composite. While at 210 °C, as shown in Fig. 4.20, the resin of the 3-D four-directional braided composite with the braiding angle of 20° shattered into powder and fiber tow fracture was evident. This is mainly attributed to the loose structure of the composites with

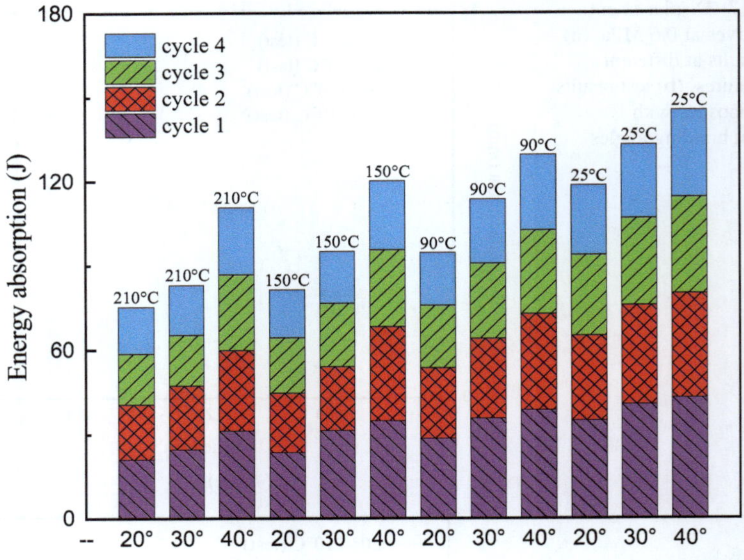

Fig. 4.18 Energy absorption of 3-D four-directional braided composites at different temperatures

small braiding angle, the weakening and debonding of the interface between fiber and resin at high temperature, and the severe weakening of the physical properties of epoxy resin at high temperature.

4.5.3 Coupling Effects of Temperature and Braiding Angle

Figure 4.21 shows the visual understanding of the coupling effect of braiding angle and temperature on the transverse impact behavior. The gradient colors from dark to light represent different values of mechanical behavior. Extracting the peak load from the load-displacement curves, as shown in Fig. 4.21a, it is evident that the peak load decreases as the ambient temperature increases and the braiding angle decreases. Thus, the increase in temperature has a negative effect on the peak load, while the braiding angle has a positive effect. Figure 4.21b extracts the displacement at the end of the fourth impact cycle. The results show that the displacement increases slightly as the ambient temperature increases and the braiding angle decreases. This implies that ambient temperature and braiding angle are not the main factors affecting the displacement. Figure 4.21c shows the total energy absorption for the first four impact cycles. The trend of energy absorption is similar to that of the load, i.e., the energy absorption decreases with the increase of ambient temperature and the decrease of braiding angle.

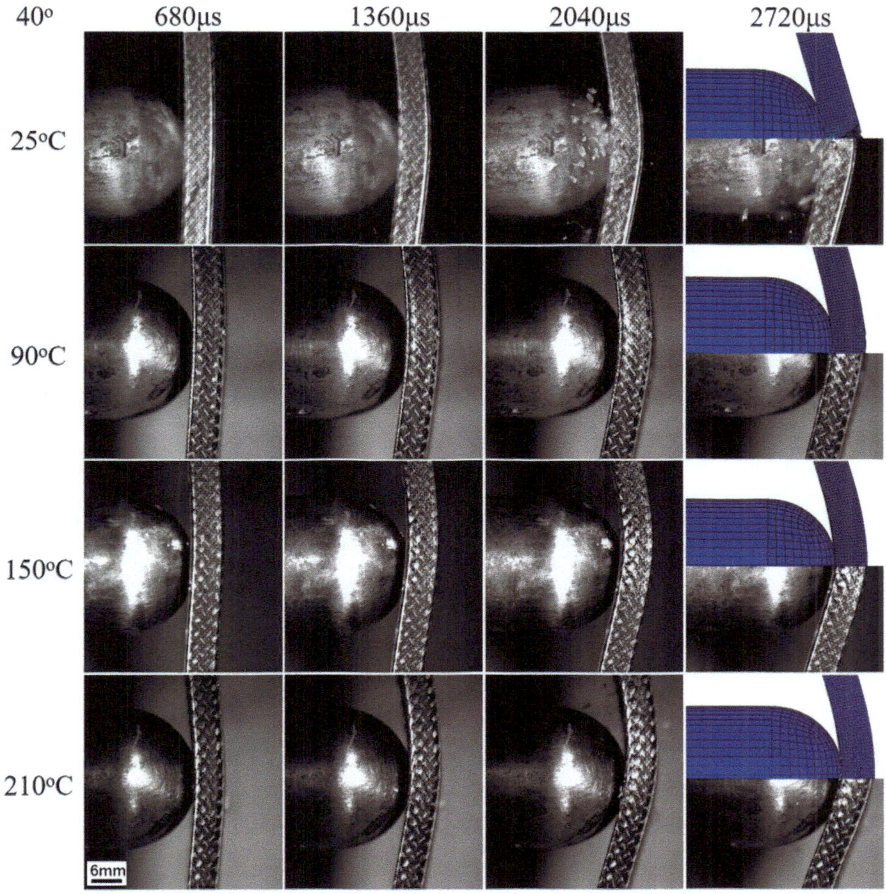

Fig. 4.19 Deformation and damage of 3-D four-directional braided composites with the braiding angle of 40° at different temperatures

4.5.4 *Damage Morphologies and Failure Modes*

The macroscopic view of the final damage morphologies of 3-D four-directional braided composite beams is shown in Fig. 4.22. All composite beams show significant bending and fracture in the impact direction when viewed from the side. The impact damage is mainly located at the impact and clamping positions. As the braiding angle and the ambient temperature decreases, the failure mode of the composites is more severe, and the transverse displacement is greater. For composite beams with the braiding angles of 20° and 30°, the resin cracked along the braiding angle direction rather than the vertical cracking extension commonly occurred in isotropic materials [1], while the resin of composite beams with the braiding angles of 40° peeled off directly. At 90 and 150 °C, some specimens fractured at the

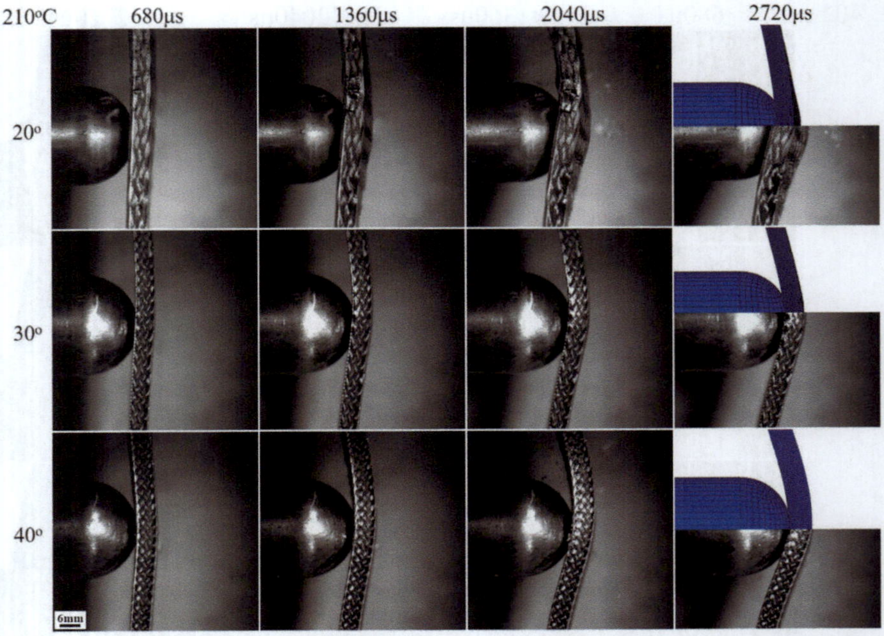

Fig. 4.20 Deformation and damage of 3-D four-directional braided composites with different braiding angles at 210 °C

clamping position. The possible reason is that the impact position and clamping position of the composite beam are in the maximum stress concentration at high temperature and the clamping position could not withstand large deformation, resulting in preferential shear fracture at this position. The final damage morphology of the 20° braiding angle sample at 210 °C is attributed to the degraded behaviors of the epoxy at the environment which is higher than the glass transition temperature. From the front side of the specimen, it was observed that the local indentation on the impact surface became clearer and the impact area increased as the ambient temperature increased. At the same time, the resin cracking and fiber breakage were reduced. This phenomenon reflects the toughness of the epoxy resin tends to rubbery behavior above T_g when it is not completely degraded [2]. The composite with the braiding angle of 20° failed completely at all ambient temperatures. At 25 and 90 °C below T_g, the dominant failure mode of the composite beams was resin brittle cracking and yarn fracture, and the crack propagation path is along the braiding angle. At 150 and 210 °C above T_g, fiber breakage, matrix ductile cracking, and interface debonding between yarn and resin were observed. This is from the weakening of the resin properties at high temperatures and the dominant role of the braided preform in bearing the load.

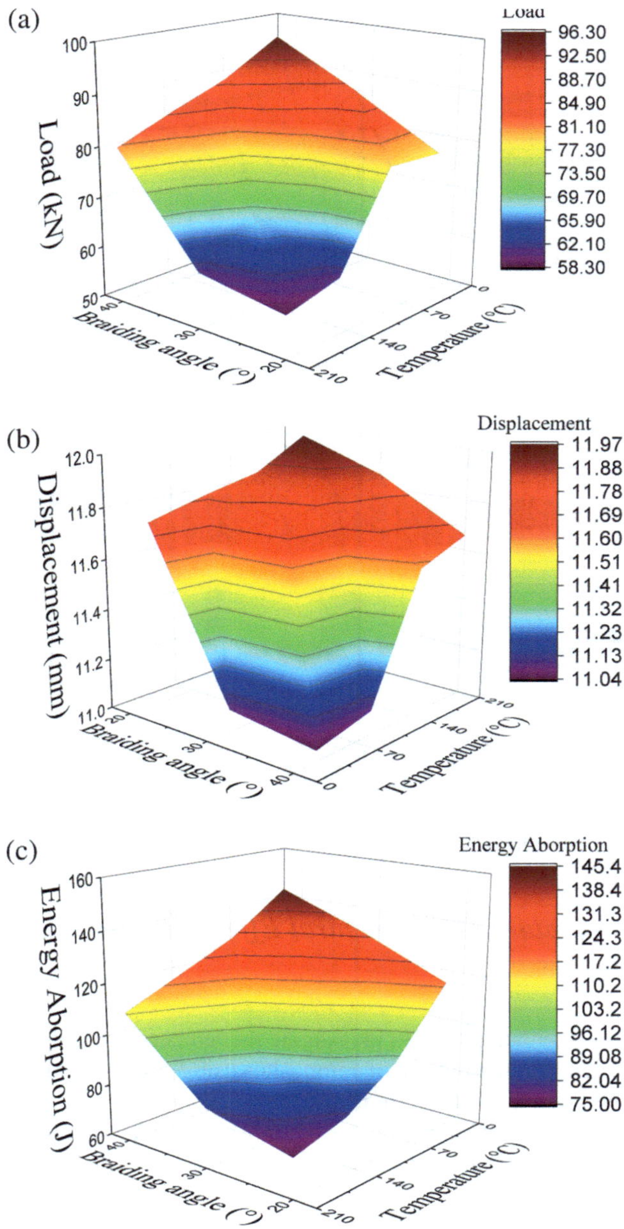

Fig. 4.21 Coupling effect of braiding angle and temperature in transverse impacts. (**a**) Load, (**b**) displacement, (**c**) energy absorption

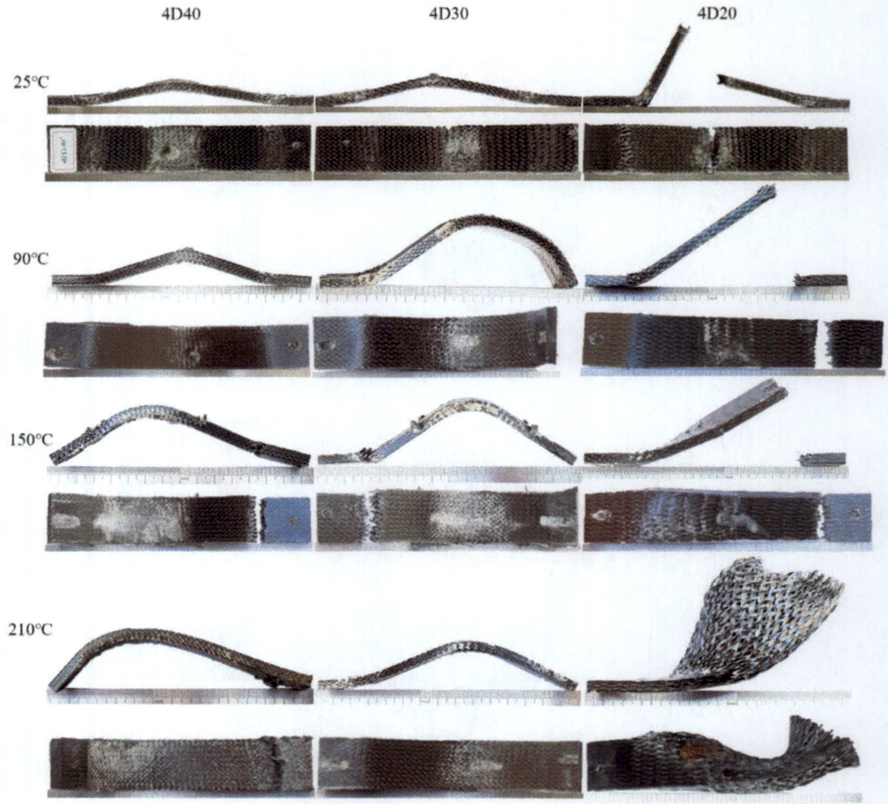

Fig. 4.22 Final damage morphology of 3-D four-directional braided composite

4.5.5 Temperature and Structural Effects on Impact Damages

We selected representative tests to investigate the temperature and structural effects on the impact damage distribution of composites using μ-CT combined with the unit cell/meso-structure hybrid structure model.

The nanoVoxel 4000 X-ray 3D μ-CT equipment from Tianjin Sanying Precision Instruments Co., Ltd. (China) was used to scan the internal structure of the composite specimens after impact.

Figure 4.23 shows the load-displacement curves with the impact deformation process under the first four impact cycles. Figure 4.26a shows the load-displacement curves corresponding to (b), where the finite element results match the test results. Figure 4.26b shows the impact deformation process recorded by high-speed photography. Figure 4.26c shows the load-displacement curves for 3-D four-directional braided composites with the braiding angle of 40° at 25 and 150 °C. Figure 4.26d

Fig. 4.23 Load-displacement curves for the first four impact cycles. (**a**) Comparison of test and FEA results for 40°/25 °C/0.2 MPa specimen, (**b**) impact deformation process of 40°/25 °C/0.2 MPa specimen, (**c**) comparison of test results for different ambient temperatures and impact pressures, (**d**) comparison of test results for different braided structures

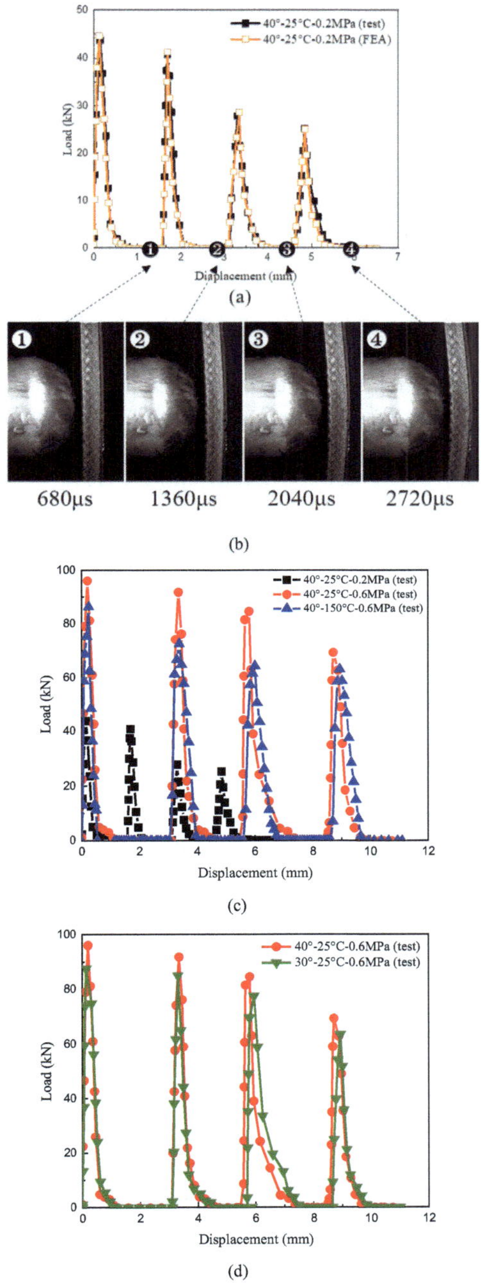

shows the load-displacement curves for the 3-D four-directional braided composites with the braiding angle of 30° and 40° at 25 °C. The peak load increases with increasing braiding angle, decreases with increasing temperature, and increases significantly with increasing impact pressure. The peak load decreases as the impact cycle increases because the energy is gradually converted into kinetic and heat energy.

Figure 4.24 shows the macroscopic damage morphology of the specimen after impact as well as the μ-CT cross-section. Figure 4.24a shows the test results of the 4D30 specimen at 0.6 MPa at room temperature compared to the FEA results. The crushing resin failure on the impact surface and fracture of the fiber tows on the back side due to tensile stress can be observed from the macroscopic image. The braiding angle was reduced and the damage to the specimen was more severe. The cross-sectional μ-CT images showed that there were debonding and splitting between the fiber tows near the impact surface, cracks in the resin-rich area, interface cracking between resin and fiber tows, and fracture failure of fiber tows on the back side of the impact. The main failure modes are fiber tow fracture and interface cracking and debonding. The FEA results present resin cracking and yarn misalignment slip due to extrusion, which affects the original braided structure and weakens the impact resistance. Compared with the specimen with large braiding angle, the energy absorption is reduced (Fig. 10.12), which further reduces the adiabatic temperature rise.

Figure 4.24b shows the damage morphology of 4D30 specimen at different cross-sections after 0.6 MPa impact at room temperature. Five cross-sections are taken from the impact point to the edge of the damage area, with adjacent cross-sections spaced 2.2 mm apart. Different cross-sections exhibit different damage morphologies. From cross-section A-A to C-C, the deformation is larger and larger, and the damage is more and more serious. In the cross-section C-C via the impact point, the interfacial cracking is distributed throughout the cross-section. The debonding and splitting among fiber tows near the impact surface are severe. The fiber tows on the back are fractured. The cross-section B-B is far from the impact point, the fiber tows on the back are not completely failed, and some fibers are broken. The fiber tows of the cross-section B-B are split near the impact surface, and the upper part is seriously damaged than the lower part. The resin cracking delaminates the surface fiber tows from the internal fiber tows. The interface cracking of the cross-section A-A is mainly distributed in the surface fiber tow layer.

Figure 4.25 shows the damage morphologies of the same meso-structure under different impact pressures and temperatures [3].

40°/25 °C/0.2 MPa: There are many macroscopic cracks, that is, the resin on the back of the impact cracks along the direction of braided yarns. Many microcracks were observed inside the specimen from μ-CT images, mainly concentrated between the surface braided yarns and the internal braided yarns near the impact surface, as well as interfacial cracking in the impact direction. The main failure mode was interfacial cracking. Due to the reduction in impact pressure, the energy absorption was reduced to 34.9 J. The degree of interfacial damage and the adiabatic

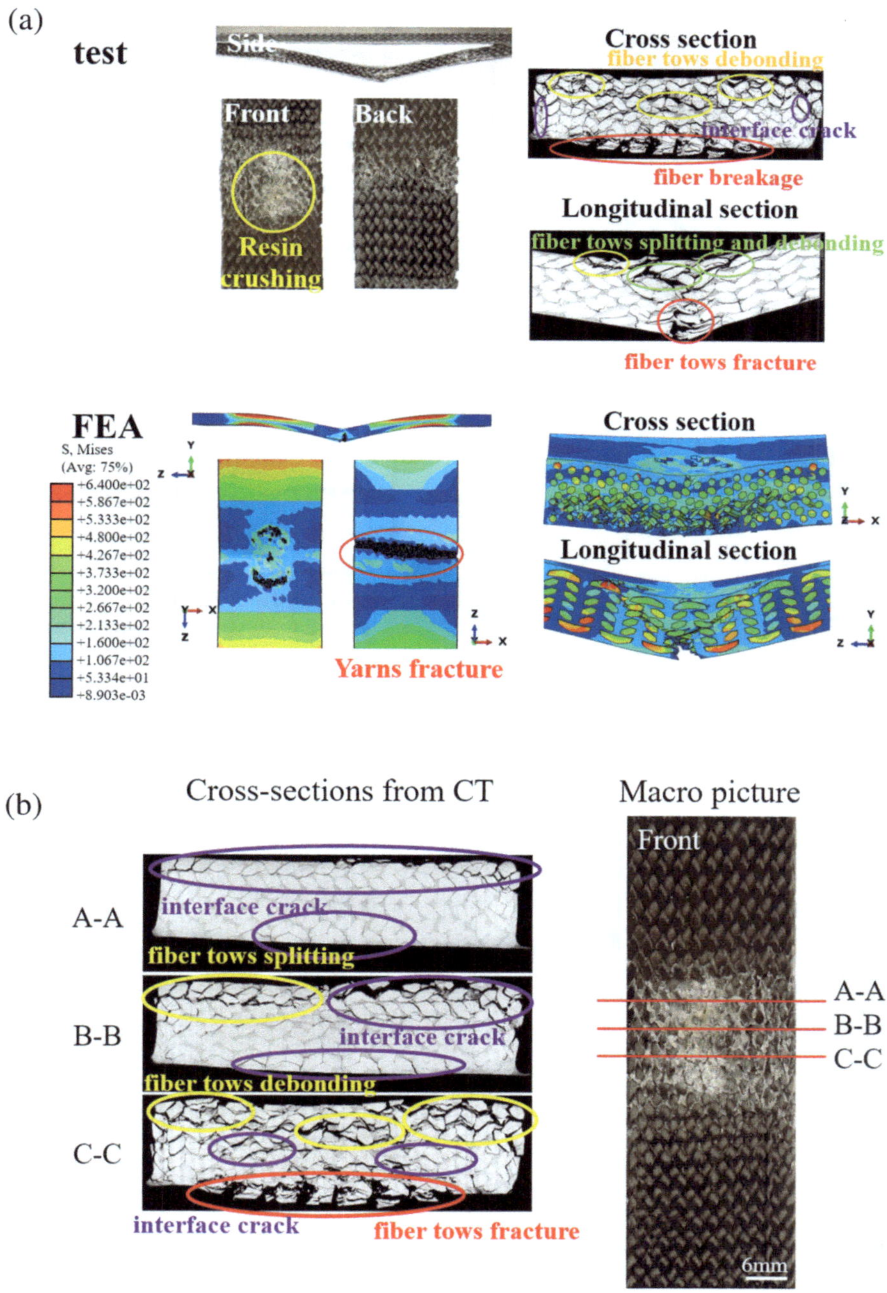

Fig. 4.24 Macroscopic damage morphology and μ-CT cross-section of 3-D four-directional braided composite beam after impact. (**a**) Comparison of test and FEA results for 30°/25 °C/0.6 MPa specimens, (**b**) damage morphologies at different cross-sections

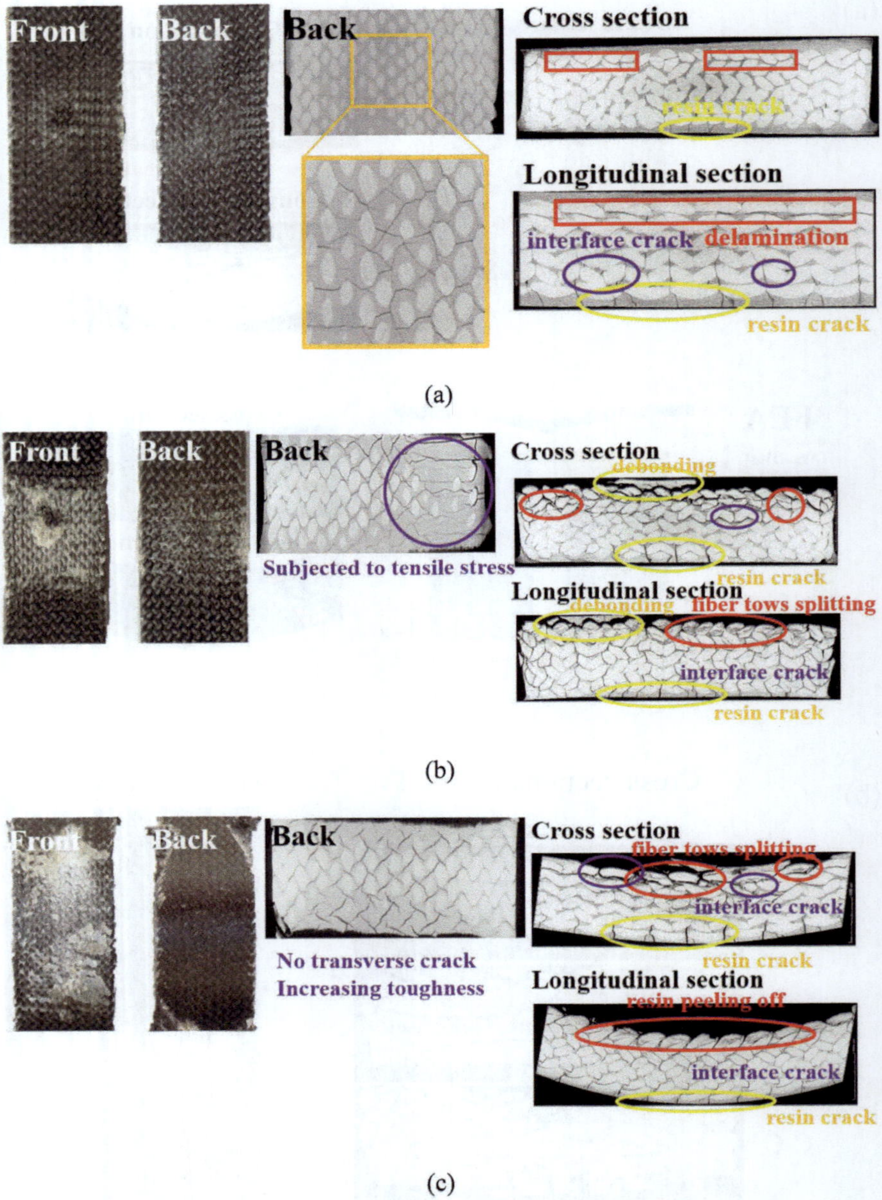

Fig. 4.25 Macroscopic damage morphologies of the braiding angle of 40° after impact and μ-CT cross-section through the impact point. (**a**) 40°/25 °C/0.2 MPa, (**b**) 40°/25 °C/0.6 MPa, (**c**) 40°/150 °C/0.6 MPa

temperature rise were also reduced, indicating that the impact pressure had the most significant effect on the impact damage of the material.

40°/25 °C/0.6 MPa: Resin debonding on the impact surface, transverse cracks on the back surface due to the tensile stress suffered, and cracks along the direction of the braided yarn were observed in the macroscopic images. The damage area of the specimen increased with increasing impact pressure. It was observed from the μ-CT images that the fiber tows were split as well as the interfacial cracking near the impact point. Compared with the specimen at 150 °C, the toughness of the resin was much lower. The surface resin had brittle crack failure, and the brittleness behaviors were obvious. The main failure modes were resin debonding and interfacial cracking. It is found in Fig. 10.12 that the main interfacial damage regions are located in the resin cracking area and the impact tolerance, energy absorption, load peak, and adiabatic temperature rise increase.

40°/150 °C/0.6 MPa: The toughness increased with the increase of the ambient temperature. No transverse cracks were observed in the resin on the back surface in the macroscopic images. The resin on the impact surface is debonding and peeling off, and the braided preform structure is complete. The μ-CT images showed that the surface fiber tows were split and the toughness of the resin was obvious. With the increase of temperature, the interfacial bonding strength decreased, and the interfacial cracking is uniformly distributed throughout the cross section. The main failure modes are resin debonding and peeling off and interfacial cracking. Figure 10.12 reveals that the interfacial damage is spread over almost the entire cross-section. The μ-CT results show a slight decrease in energy absorption compared to room temperature, resulting in a relative decrease in temperature rise as well.

4.6 Effects of Braided Structures

As shown in Fig. 4.26a, b, it can be seen from the load-displacement curves that the impact pressure and the addition of the axial yarns have a positive effect on the peak load. The addition of axial yarns has a negative effect on displacement, while the increase in impact pressure has a positive effect on displacement. The peak load decreases as the number of impact cycles increases (see previous studies for more details (see Sects. 4.4 and 4.5)). The impact loading rate has a more significant effect on the load and displacement than the structure of the preform. The FEA results match the load-displacement curves obtained from the tests and the deformation process of the composite obtained from high-speed photography, as shown in Fig. 4.26c, indicating the validity of the finite element model.

As can be seen from Fig. 4.27a, energy absorption increases with the increase of impact pressure and the addition of the axial yarn. Again, the impact loading rate has a more significant effect on energy absorption than the structure of preform. Figure 4.27b shows the specific energy absorption (energy absorption per unit volume) obtained from the finite element results for evaluating the contribution of different components to energy absorption. It is noteworthy that for all the samples,

Fig. 4.26 Load-displacement curves for the first four impact cycles. **(a)** Comparison of test results of different braided structure specimens at different impact pressures. **(b)** Comparison of test results of 4D30 specimens at different impact pressures. **(c)** Comparison of impact deformation process and test results of 5D30 specimens at 0.4 MPa with finite element results

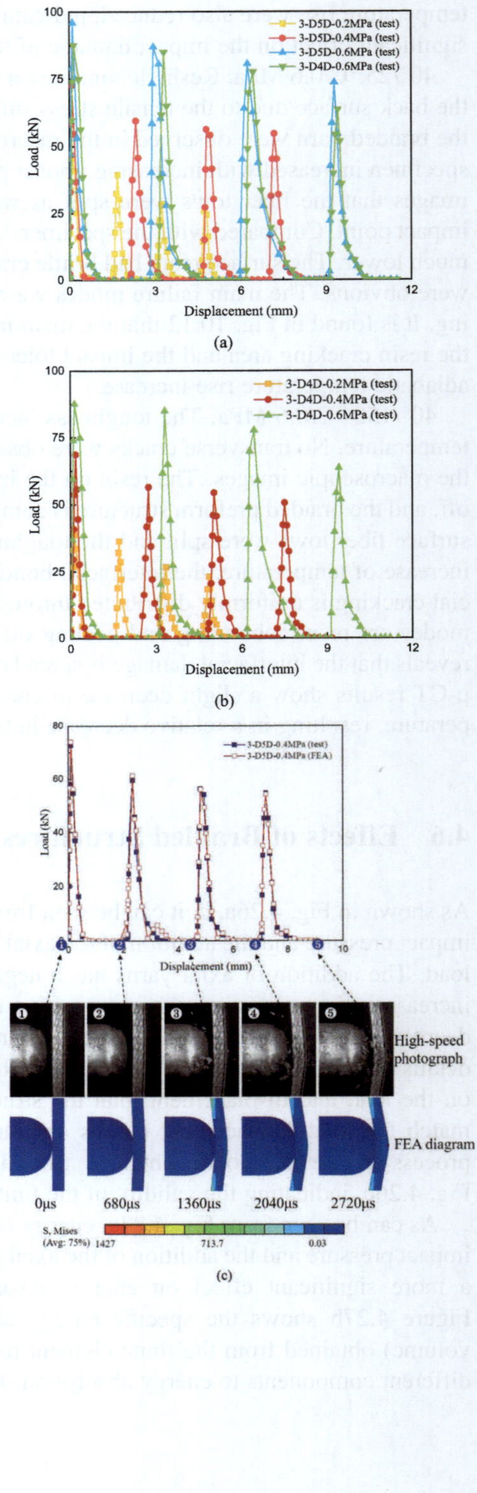

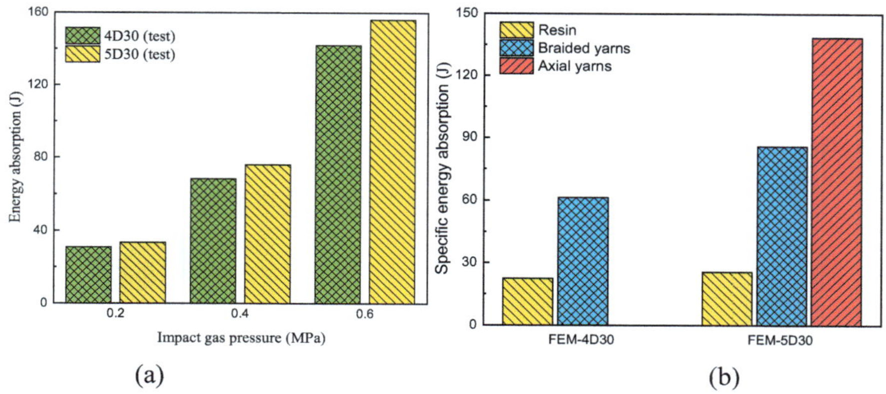

Fig. 4.27 Energy absorption of the composite and specific energy absorption of each constituent

the braided preform is the main energy absorption component during the impact. More importantly, although the volume fraction of the axial yarn is much lower than that of the braided yarn, the specific energy absorption of the axial yarn is significantly greater than that of the braided yarn. This shows that the addition of the axial yarn improves the transverse deformation resistance of the composites and the braided preform plays an important role in the deformation resistance of the composites.

The energy conservation law implied by the first law of thermodynamics states that the time rate of kinetic energy and internal energy for a fixed body is equal to the sum of the rate of work done by the surface and body forces. This can be expressed as:

$$\frac{d}{dt}\int_V\left(\frac{1}{2}\rho\boldsymbol{v}\cdot\boldsymbol{v}+\rho U\right)dV=\int_S\boldsymbol{v}\cdot\boldsymbol{t}dS+\int_V\boldsymbol{f}\cdot\boldsymbol{v}dV \tag{4.5}$$

where ρ is the current mass density, $\boldsymbol{v}$ is the velocity field vector, U is the internal energy per unit mass, $\boldsymbol{t}$ is the surface traction vector, and $\boldsymbol{f}$ is the body force vector. If we introduce the strain decomposition, $\dot{\boldsymbol{\varepsilon}}=\dot{\boldsymbol{\varepsilon}}^{el}+\dot{\boldsymbol{\varepsilon}}^{pl}$ (where $\dot{\boldsymbol{\varepsilon}}^{el}$ and $\dot{\boldsymbol{\varepsilon}}^{pl}$ are elastic and plastic strain rates), the internal energy, E_I, can be expressed as:

$$\begin{aligned}E_I&=\int_0^t\left(\int_V\boldsymbol{\sigma}^C:\dot{\boldsymbol{\varepsilon}}dV\right)d\tau\\&=\int_0^t\left(\int_V\boldsymbol{\sigma}^C:\dot{\boldsymbol{\varepsilon}}^{el}dV\right)d\tau+\int_0^t\left(\int_V\boldsymbol{\sigma}^C:\dot{\boldsymbol{\varepsilon}}^{pl}dV\right)d\tau=E_S+E_P\end{aligned} \tag{4.6}$$

where E_S is the applied elastic strain energy and E_P is the energy dissipated by plasticity.

Figure 4.28 illustrates the final damage morphology of the 3-D four-directional/five-directional braided composites in three directions [4]. Under the impact gas

Fig. 4.28 Final damage morphologies of 3-D four-directional/five-directional braided composites. (**a**) 0.2 MPa, (**b**) 0.4 MPa, (**c**) 0.6 MPa

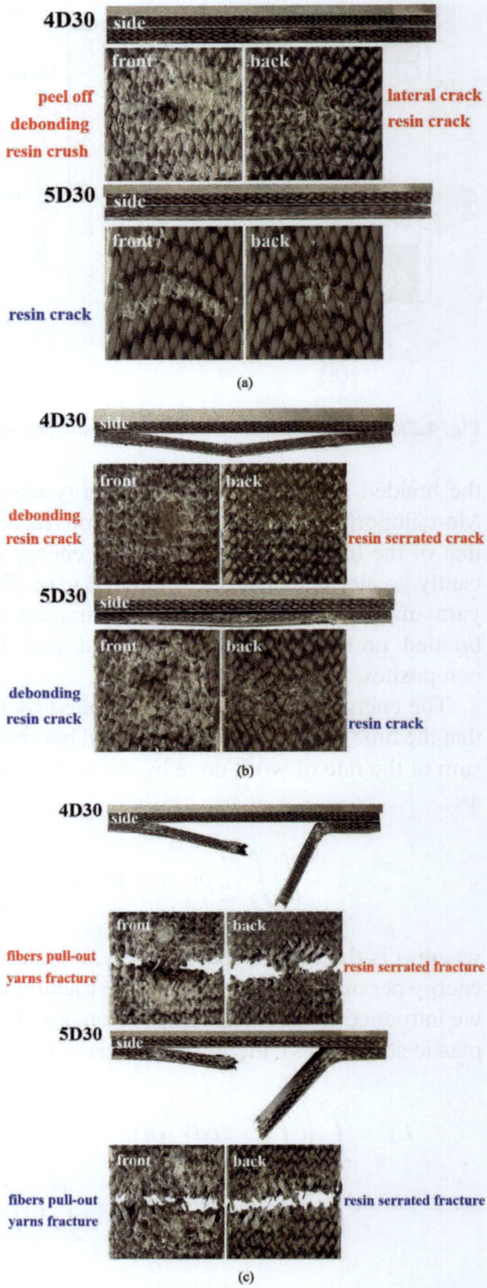

pressure of 0.4 MPa, the difference of final morphology between 4D30 and 5D30 composites is the most significant. The transverse displacement of 4D30 composites without axial yarns is much larger than that of 5D30 composites. The addition of the axial yarns improves the deformation resistance of the composites. Resin cracking and debonding occurred on the impact surfaces of both composites. The resin on the back of 4D30 composites cracked zigzagly, i.e., along the direction of braiding angle. The 5D30 composites cracked in all directions. It can be seen that the failure mode of the composite was changed by the braided preform architecture. Due to the decrease of impact energy of 0.2 MPa, the braided preform of the composites maintained structural integrity, while the resin debonded, cracked, and peeled off. Meanwhile, there were lateral cracks on the back of 4D30 composite due to tensile stress. The composites failed completely under the impact pressure of 0.6 MPa. Due to the joint action of tensile stress and shear stress, the yarns fractured, and the fibers pulled out. The damage morphology on composites heavily depended on the micro-scale architecture.

References

1. Hu Meiqi, Liu Shengkai, Zhang Junjie, Wang Lei, Sun Baozhong, Gu Bohong: Multiple transverse impact damage behaviors of 3-D-braided composite beams under room and high temperatures. Int. J. Damage Mech. **29**(5), 715–747 (2020)
2. Hu Meiqi, Zhang Junjie, Sun Baozhong, Gu Bohong: Finite element modeling of multiple transverse impact damage behaviors of 3-D braided composite beams at microstructure level. Int. J. Mech. Sci. **148**, 730–744 (2018)
3. Hu Meiqi, Sun Baozhong, Gu Bohong*: Temperature and structure effects on the impact damage distribution of 3D braided composites. Zhongguo Kexue Jishu Kexue/Scientia Sinica Technologica. **51**(1), 108–118 (2021)
4. Hu Meiqi, Sun Baozhong, Gu Bohong: Microstructure modeling multiple transverse impact damages of 3-D braided composite based on thermo-mechanical coupling approach. Compos. Part B. **214**, 108741 (2021)

Chapter 5
Transverse Impact of Braided Tubes

This chapter focuses on the effects of impact velocity and braided structure on the transverse impact behaviors of 3-D braided composite tubes. The test results of load-displacement curves, deformation, and damages under transverse impact are presented. The inner damages are photographed with μ-CT to show the fiber tow breakages and crack distributions.

5.1 Preparation of 3-D Braided Carbon Fiber/Epoxy Resin Composite Tubes

The 3-D braided structure of the circular preform is braided using a 1×1 four-step circular braiding technique. Figure 5.1a shows the 3-D circular braiding yarn arrangement diagram, where the main yarn carriers are arranged in the circumferential and radial directions to form the main yarn array. The number of braided yarn carriers along the radial direction represents the number of braided layers m, and the number of braided yarn carriers in the circumferential direction is the number of braided columns n. The side yarns are arranged at intervals inside and outside the main yarns, and the axial yarn carriers are arranged at intervals with the main braided yarn carriers.

The four-step circular braiding process [1] is shown in Fig. 5.1b, where each braiding cycle consists of four movements: the first step, radial movement, with adjacent columns of braided yarn carriers moving in opposite directions; the second step, circumferential movement, with adjacent layers of carriers moving in opposite directions; the third step, radial movement, with each carrier moving in the opposite direction to the first step; and the fourth step, circumferential movement, with each carrier moving in the opposite direction to the second step. During the entire braiding cycle, the axial yarns are not involved in the braiding. The braiding angle is controlled by changing the beating-up force, the greater the force, the greater the braiding angle.

M. Hu, B. Gu, *Impact Damages of Braided Composites*, Engineering Materials,
https://doi.org/10.1007/978-981-16-5703-0_5

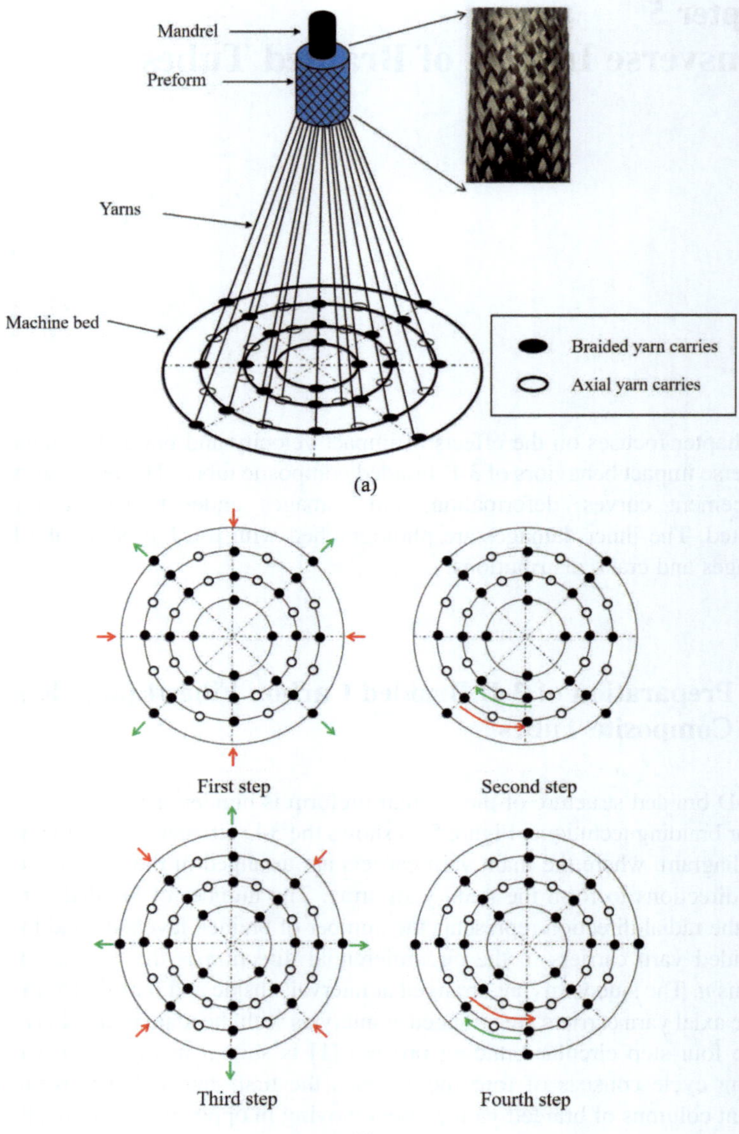

Fig. 5.1 3-D circular braided preform braiding. (**a**) Yarn arrangement diagram, (**b**) four-step movement process

Table 5.1 Braiding parameters for different tubes

Type of structure	$m \times n$	Braiding angle (°)	Axial yarn
4D2L30A	2 × 32	30	–
4D3L15A	3 × 32	15	–
4D3L30A	3 × 32	30	–
4D3L45A	3 × 32	45	–
4D4L30A	4 × 32	30	–
5D3L30A	3 × 32	30	√

Note: *4D* without axial yarn, *5D* with axial yarn, *L* number of braiding layers (2L, two layers), *A* braiding angle (30A, 30°), *m* number of layers of yarn arrangement, *n* number of columns of yarn arrangement

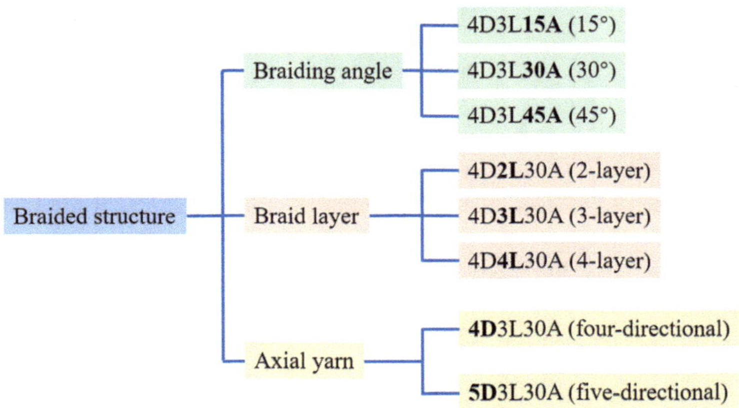

Fig. 5.2 Numbering of different braiding parameters

3-D circular braided preform can be defined by the number of layers and columns of the main body yarns, and for 3-D four-directional (referred to as 3D4D) $m \times n$ circular braided preform, the total number N of braided yarns can be expressed as:

$$N = n(m + 1) \tag{5.1}$$

For 3-D five-directional (referred to as 3D5D) circular braided preform containing axial yarns, the total number N of braided yarns can be expressed as:

$$N = n(2m + 1) \tag{5.2}$$

In order to investigate the effect of different braiding structure parameters on the transverse impact behaviors of 3-D braided composite tubes, six different braiding structures were designed, as shown in Table 5.1. The braiding parameters include braiding angle (15°, 30°, and 45°), number of braiding layers (two, three, and four layers), and axial yarn (four-directional and five-directional). Figure 5.2 shows

Table 5.2 Structural parameters of 3D braided preforms

Braided structure	Braiding angle (°)	h (mm)	*m*	*n*	*N*
3D4D	25	8	3	32	128
3D4D	35	5	3	32	128
3D4D	45	3	3	32	128
3D4D	35	5	2	32	96
3D4D	35	5	4	32	160
3D5D	35	5	2	32	160

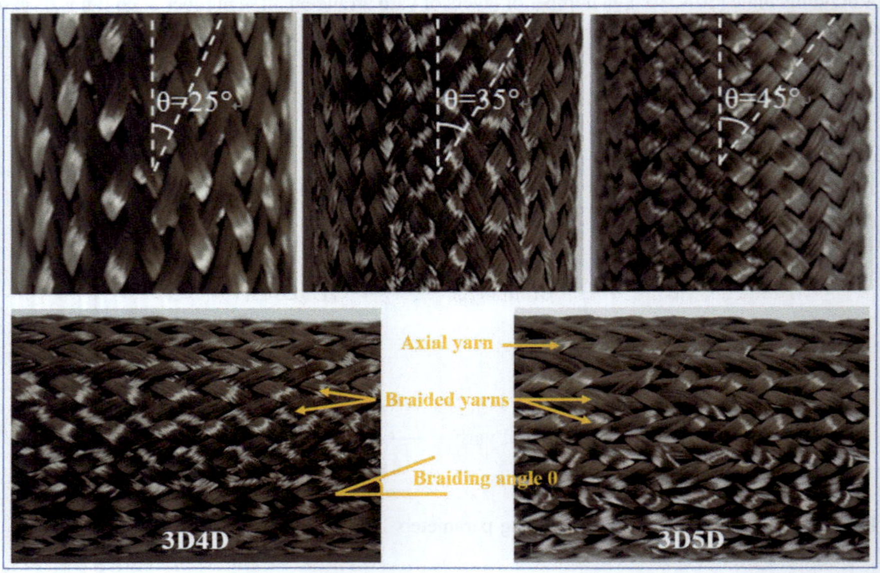

Fig. 5.3 Photograph of 3-D four-step circular braided preform

braiding parameter, where the other two braiding parameters are kept constant while discussing the effect of one of the braiding parameters on the transverse impact properties of the 3-D braided composite. The variation of the braiding angle is based on the three-layer structure without axial yarn. The variation of the number of braiding layers is based on the structure with 30° braiding angle without axial yarn. The five-directional structure is based on the addition of axial yarn to the four-directional structure with three layers and 30° braiding angle.

In order to explore the effect of different braided structure parameters on the impact compression behavior of 3-D braided composite tubes, we designed tubes with different structure parameters. Table 5.2 lists the yarn arrangement and total number of 3-D tubular braids with six different braiding structures. The preforms with braiding angles of 25°, 35°, and 45° and 3-D four-directional and five-directional preforms are compared in the photographs shown in Fig. 5.3.

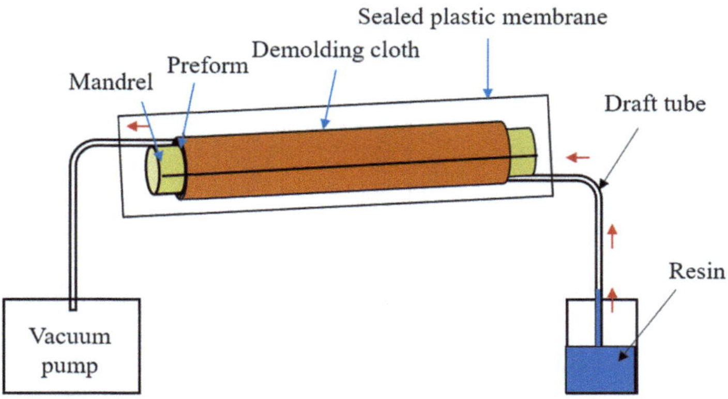

Fig. 5.4 Schematic diagram of VARTM molding process

The VARTM method was used in the molding process of the composite tube, and the silicone rod which was easy to demold after curing was used in the mandrel. Figure 5.4 shows the schematic diagram of the composite tube molding process. The mandrel is a solid silicone rod, coated with a layer of release agent on the surface. The composite and the mandrel can be separated smoothly after molding. The silicone mandrel was inserted into the tubular braided fabric, and the outer surface of the fabric was wrapped with demolding cloth. The function of the demolding cloth was to isolate the fabric and the vacuum bag and to ensure the surface quality of the composite. The matrix used in this study is epoxy resin (JC-02A) provided by Changshu Jiafa Chemical Co., Ltd. with low viscosity, good heat resistance, chemical corrosion resistance, and insulation, and the behavior parameters are shown in Table 4.2. The epoxy resin was mixed with curing agent JC-02B type and put into the vacuum drying oven for 2 h to de-bubble before injecting into the fabric to reduce the pores of the composite due to the existence of bubbles. The injection process was completed in the oven, and the temperature was maintained at 40 °C. The resin maintained good flow ability at this temperature. After the fabric was completely impregnated, the oven temperature was increased for curing, with curing conditions of 90 °C for 2 h, 110 °C for 1 h, and 130 °C for 4 h. After finishing the curing, it was allowed to cool naturally to room temperature and then the demolding process is carried out.

As shown in Fig. 5.5, the composite tubes were cut to the experimentally required size of 150 mm. The structural parameters of the various structural composite specimens are shown in Table 5.3. The mandrel used for curing was the same for all structural specimens. The wall thickness and fiber volume fraction of the composite tubes increased as the number of layers increased, the braiding angle increased, or the axial yarns were added.

High-pressure water jet was used to cut the composite tubes according to the aspect ratio of 1:1, as shown in Fig. 5.6. The photographs of the six specimens used

Fig. 5.5 Test specimens of 3-D braided composite tube

Table 5.3 Structural parameters of 3-D braided composite tubes

Type of structure	Inner diameter (mm)	Wall thickness (mm)	Fiber volume fraction (%)	Linear density (g/mm)
4D2L30A	20.52	1.75	41.4	0.192
4D3L15A	20.51	1.75	40.4	0.160
4D3L30A	20.54	2.20	43.3	0.213
4D3L45A	20.50	2.65	46.1	0.261
4D4L30A	20.51	2.75	45.0	0.268
5D3L30A	20.53	3.50	48.2	0.374

for compression tests are shown in Fig. 5.7, and the size and specifications of the specimens are shown in Table 5.4.

5.2 Testing Setup

The transverse impact test setup and principle are the same as Sect. 4.3. The holding part is shown in Fig. 5.8.

Fig. 5.6 Braiding composite tubes and cutting samples

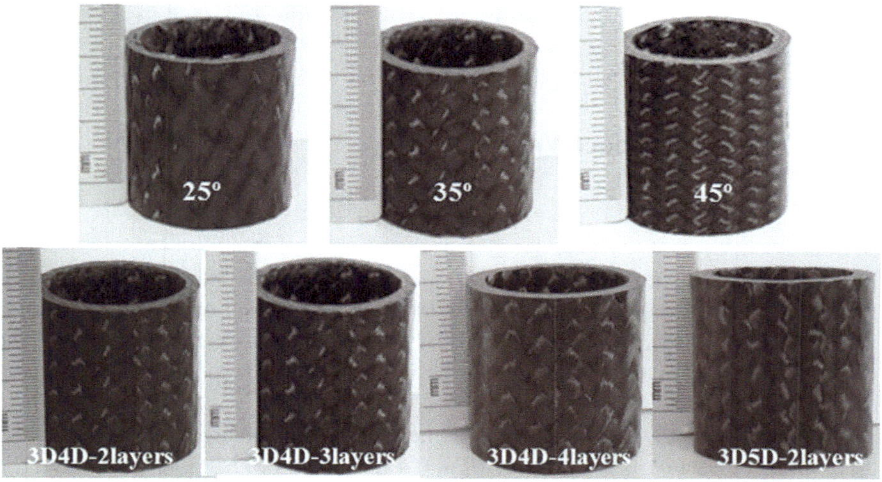

Fig. 5.7 Photograph of 3-D four-step braided composite tubes

Table 5.4 Size and specifications for braided composite tubes

Braided structure	Braiding angle (°)	Braided length (mm)	Outer diameter (mm)	Braided layer	Wall thickness (mm)	Length (mm)	Fiber volume fraction (%)
3D4D	25	8	25.21	3	2.21	25.0	42.65
3D4D	35	5	25.50	3	2.34	25.5	43.62
3D4D	45	3	25.82	3	2.50	25.5	44.82
3D4D	35	5	24.61	2	1.89	25.0	41.02
3D4D	35	5	26.48	4	2.83	26.5	46.31
3D5D	35	5	26.73	2	2.95	26.8	46.53

Fig. 5.8 Clamp part

5.3 Effects of Impact Velocities

5.3.1 Test Results

As the impact pressure increases, the stress amplitude of the one-dimensional stress wave generated by the incident bar increases, i.e., the impact energy of the impact stress wave increases. Specimen 4D3L30A was used as an example to explore the effect of impact pressure on the transverse impact loading response of 3-D braided composite tube.

Figure 5.9a–c shows the load-displacement curves of specimen 4D3L30A under three impact pressure loading at 0.2 MPa, 0.4 MPa, and 0.6 MPa, respectively. The fluctuation curves contain four impact stages, and each impact stress wave completes an incidence and reflection process. When the compressive stress wave reaches the contact surface between the incident bar and the specimen, one part enters the specimen, and another part is reflected by the end face as tensile stress wave. The tensile stress wave propagates in the opposite direction in the incident bar, and when it reaches the other end face of the incident bar, it is reflected again as compressive stress wave, and the specimen is impacted for the second time. This stage repeated many times until the impact energy dissipates or the material fails.

Figure 5.10 shows the peak load-impact cycle curves for specimen 4D3L30A at different impact pressures. As the impact pressure increases, the peak load increases, which is due to the larger impact pressure generating larger impact energy. As the impact cycle increases, the peak load at all three impact pressures decreases. On the one hand, the sample damaged to some extent in the previous impact; on another hand, the impact energy of the stress wave is continuously depleted. After four impact cycles, the peak load at 0.2, 0.4, and 0.6 MPa all decreased by about 30%.

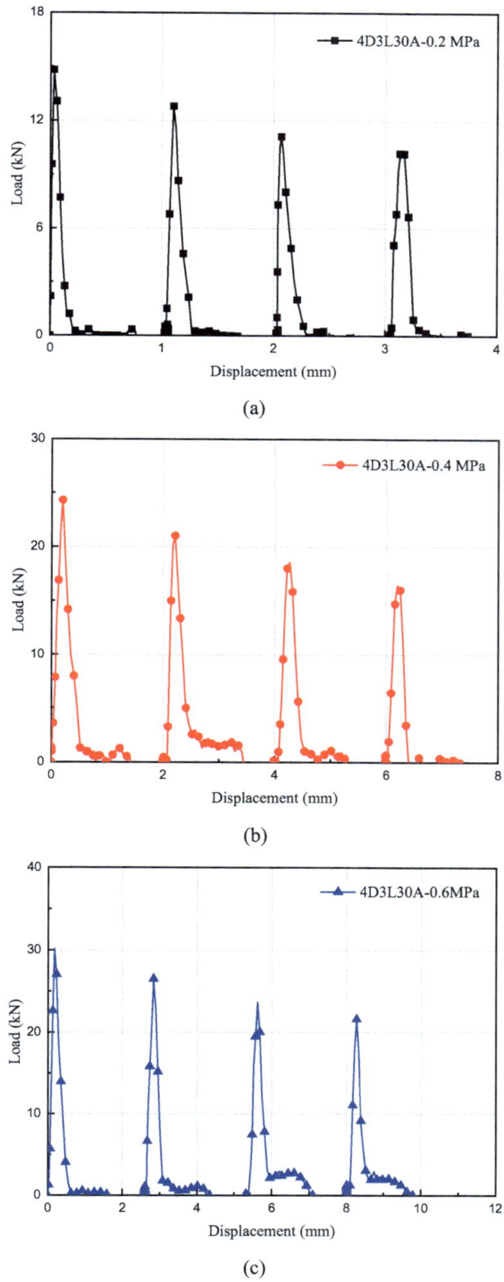

Fig. 5.9 Load-displacement curves for specimen 4D3L30A under different impact pressures. (**a**) 0.2 MPa, (**b**) 0.4 MPa, (**c**) 0.6 MPa

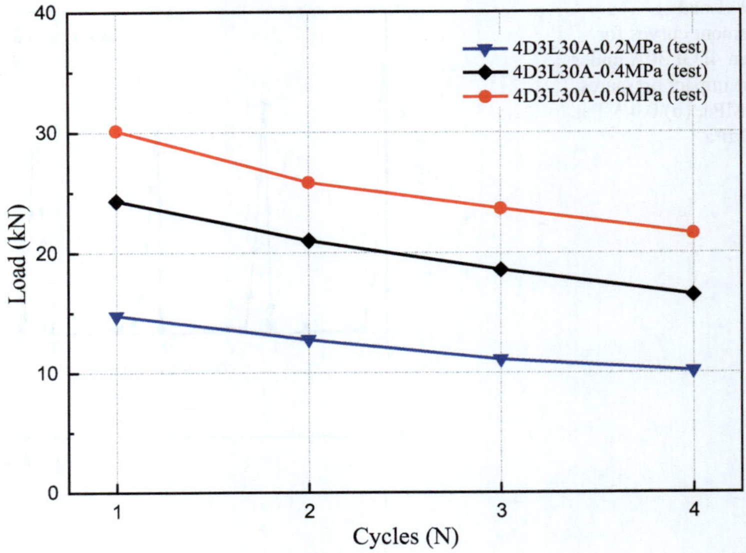

Fig. 5.10 Load-impact cycle curves under different impact pressures

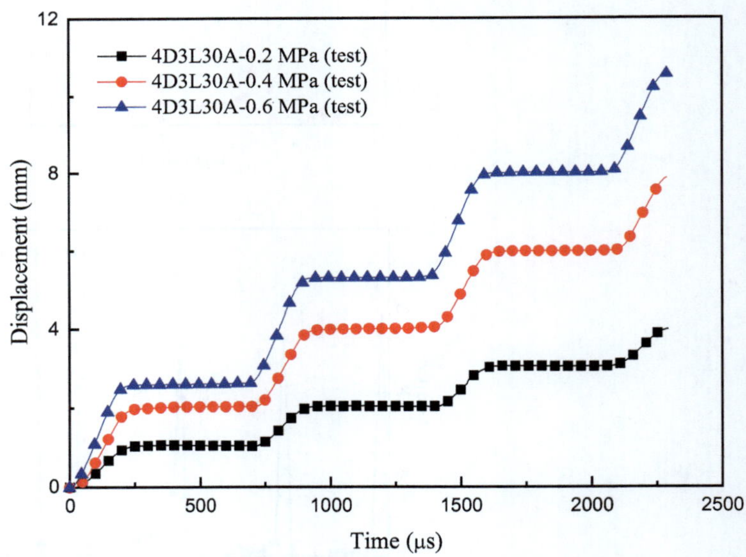

Fig. 5.11 Displacement-time curves for specimen 4D3L30A under different impact pressures

As shown in Fig. 5.11, the displacement-time curve also shows fluctuations under different impact pressures, with a plateau between two adjacent impacts. During this period, the stress wave is completely reflected and propagated in the incident bar

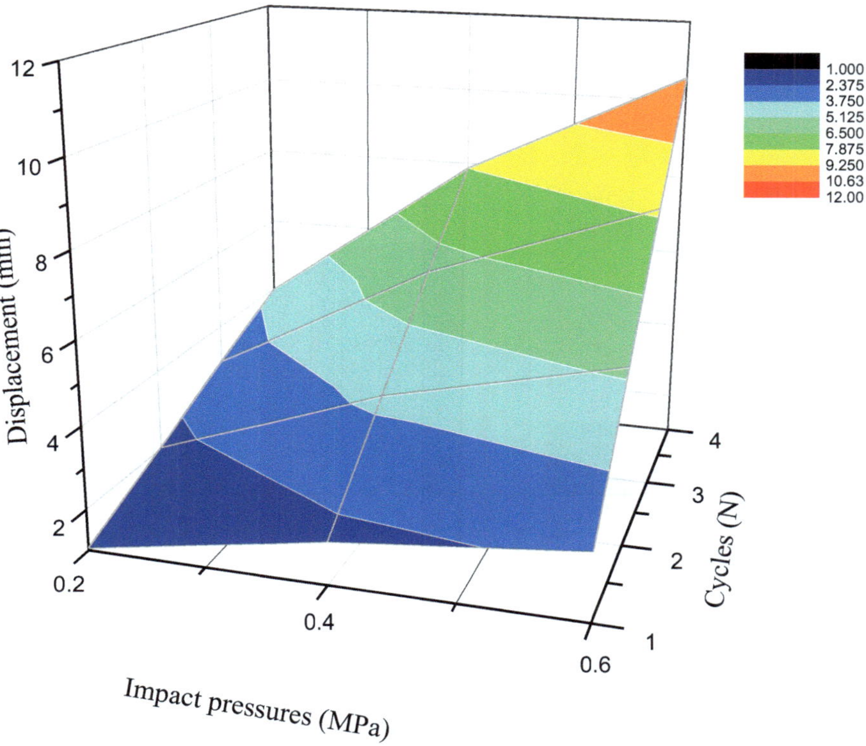

Fig. 5.12 Displacement of specimen 4D3L30A under different impact pressures and impact cycles

until the stress wave impacts the specimen again, and the displacement continues to rise. As shown in Fig. 5.12, the impact displacement increases as the impact pressure and impact cycles increase. When the impact pressure increases from 0.2 to 0.4 MPa, the displacement increment is larger; from 0.4 to 0.6 MPa, the displacement increment is relatively less. Meanwhile, the displacement difference under different impact pressures gradually increases as the impact cycle increases.

Figure 5.13 shows the total energy absorption of specimen 4D3L30A under different impact pressures. It can be seen that as the impact pressure increases, the total energy absorption of the material increases. When the impact pressure increases from 0.2 to 0.4 MPa and then to 0.6 MPa, the total energy absorption of the material increases by about 30 and 25 J. With the increase of impact pressure, the energy of impact wave increases, the damage degree of material increases, and the damage gradually expands from matrix crack to fiber tow fracture, which makes the energy absorption of 3-D braided composite tube increase.

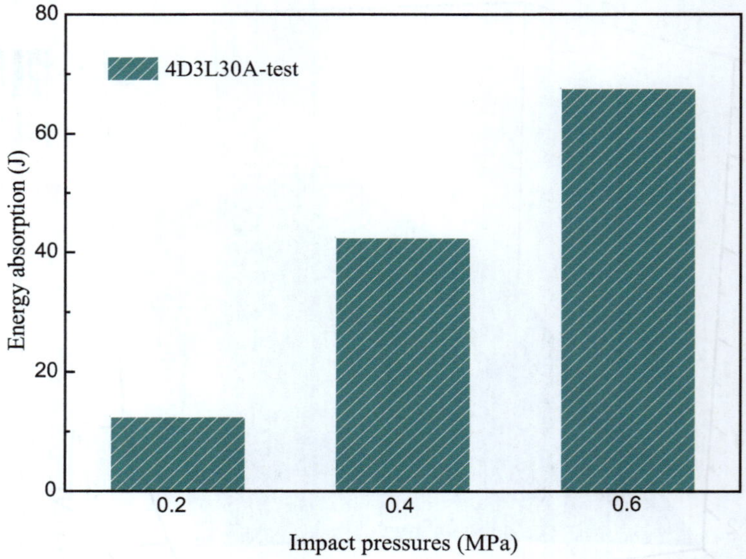

Fig. 5.13 Energy absorption of specimen 4D3L30A under different impact pressures

5.3.2 Damage Morphologies

Figure 5.14 shows the transverse impact damage morphologies of specimen 4D3L30A under different impact pressures. It can be observed that the damage of all specimens is concentrated in the impact area and the impacted surface is concave inward. At 0.2 MPa, the main failure mode is matrix cracking along the braiding angle. At 0.4 MPa, part of the matrix on the impact surface is broken into powder. At 0.6 MPa, fiber fracture can be seen obviously.

In order to observe the internal damage of the material, the μ-CT was used to show inner damages of impact area of the 3-D braided composite tube and reconstruct the 3D images. Two of the sections were selected for comparison, A-A for the radial section through the impact point and B-B for the axial section through the impact point. From the two sections, it can be seen that the damage under all impact pressures is mainly concentrated in the impacted area. The edge damage of the contact area with the incident bar is more serious, mainly in the form of shear failure. The shear failure becomes more obvious as the impact pressure increases. Under the impact pressure of 0.2 MPa, there was no obvious damage in the thickness direction of the impact surface. The cracking at the interface between the fiber tow and the matrix is mainly found around the impact surface, and gaps are formed between the fiber tows. At 0.4 MPa, the A-A section shows that there is an obvious gap between the yarns in the thickness direction of the impact surface, i.e., the interface cracking between fiber tows. There is obvious fracture of fibers at the edge of the impact surface in the B-B section, forming a shear zone. At 0.6 MPa, more severe damage

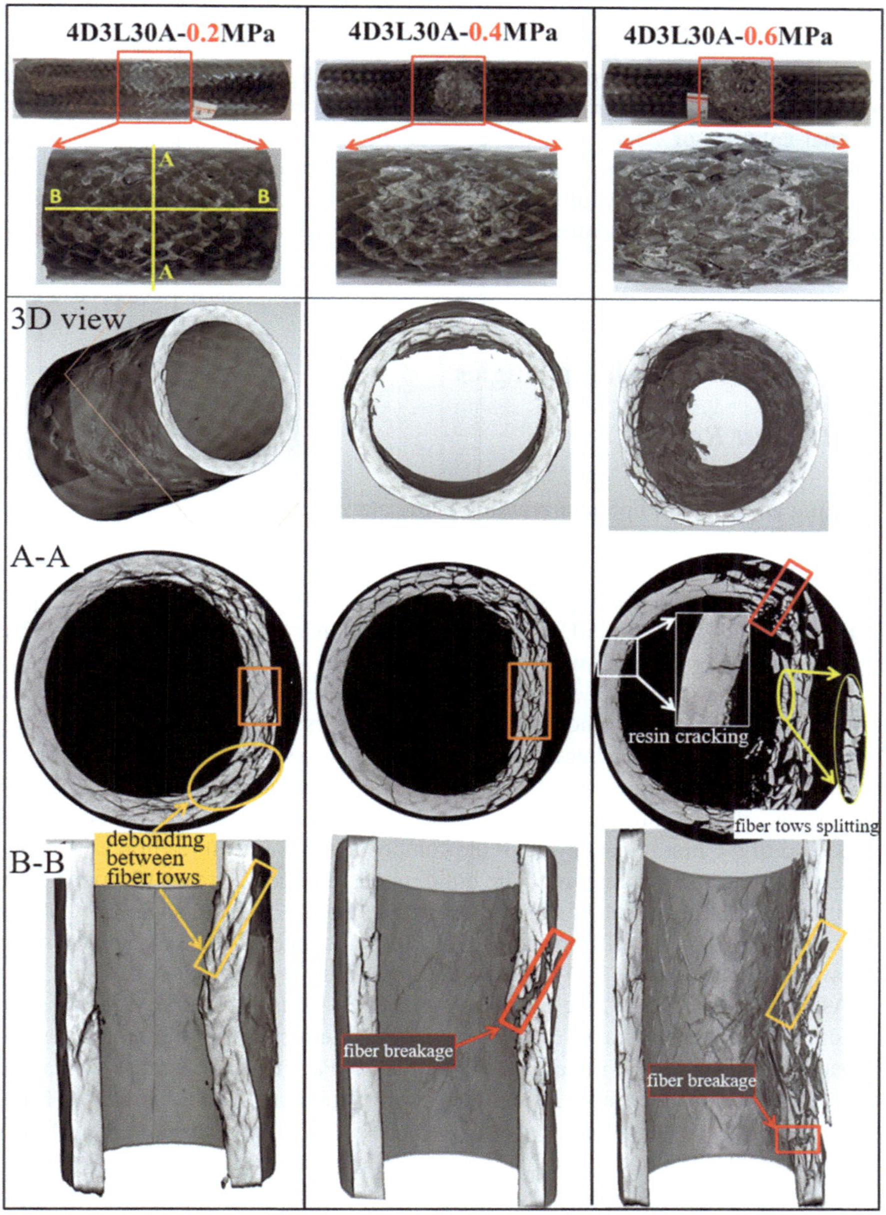

Fig. 5.14 μ-CT results of 4D3L30A after loading with different impact pressures

occurred throughout the impact area. Multiple failure modes could be observed, including matrix cracking, interfacial cracking, internal splitting of fiber tows, and fiber breakage, and the phenomenon of fiber tow hindrance to matrix crack expansion was observed.

Combining the above two methods, the external morphology and internal damage of the specimen after impact testing can be observed. The damage area is mainly concentrated around the impact area, and an elliptical shear zone appears. The failure modes mainly include matrix fragmentation, interface cracking, fiber breakage, and internal splitting of fiber tows. Under small impact pressure, the damage of the specimen was mainly matrix fragmentation and interface cracking. With the increase of impact pressure, a large number of fiber tow fracture occur, especially at 0.6 MPa. In addition to the shear fracture around the impact area, there is also obvious damage to the fibers in the thickness direction of the impact surface.

5.4 Effects of Braiding Angles

5.4.1 Test Results

Braiding angle is an important braiding parameter affecting the mechanical properties of 3-D braided composites. In this study, three braiding angles (15°, 30°, and 45°) were selected to investigate the effects on the transverse impact mechanical response of 3-D braided composite tubes. The specimens with three braiding angles 4D3L15A, 4D3L30A, and 4D3L45A were kept with the same number of braiding layers (three layers) and all without axial yarn.

Figures 5.15a–c show the load-displacement curves of the composite tubes with different braiding angles under the three impact pressures and the corresponding local enlargements at the beginning of the impacts, respectively. The comparison shows that the transverse impact load of 3-D braided composite tube increases with increasing braiding angle at all three impact pressures. Figure 5.16 compares the peak load of the first impact cycle of the composite tubes with three braiding angles under different impact pressures. The braiding angle increases from 15° to 30°, and the increment of peak load is slightly larger than that of braiding angle from 30° to 45°. With the increase of impact pressure, the influence of braiding angle on the transverse impact load of 3-D braided composite tube becomes more obvious.

As shown in Fig. 5.17, the deformation of three kinds of tubes decreases as the braiding angle increases, and the decrease amplitude increases with the increase of impact cycles. Under the impact pressure of 0.2 MPa, the deformation of the three structures is almost the same. As the impact pressure increases to 0.4 and 0.6 MPa, the more obvious the difference in deformation of the specimens with different braiding angles. This indicates that the resistance to transverse impact deformation of the 3-D braided composite tube increases with the increase of braiding angle.

As the braiding angle increases, the tightness of the yarn arrangement in the 3-D braided composite tube increases, which affects the fiber volume fraction of the composite tube. With the increase of braiding angle, the fiber volume fraction increases slightly. Here we used specific energy absorption to characterize the energy absorption properties of composite tubes with different braiding angles as follows:

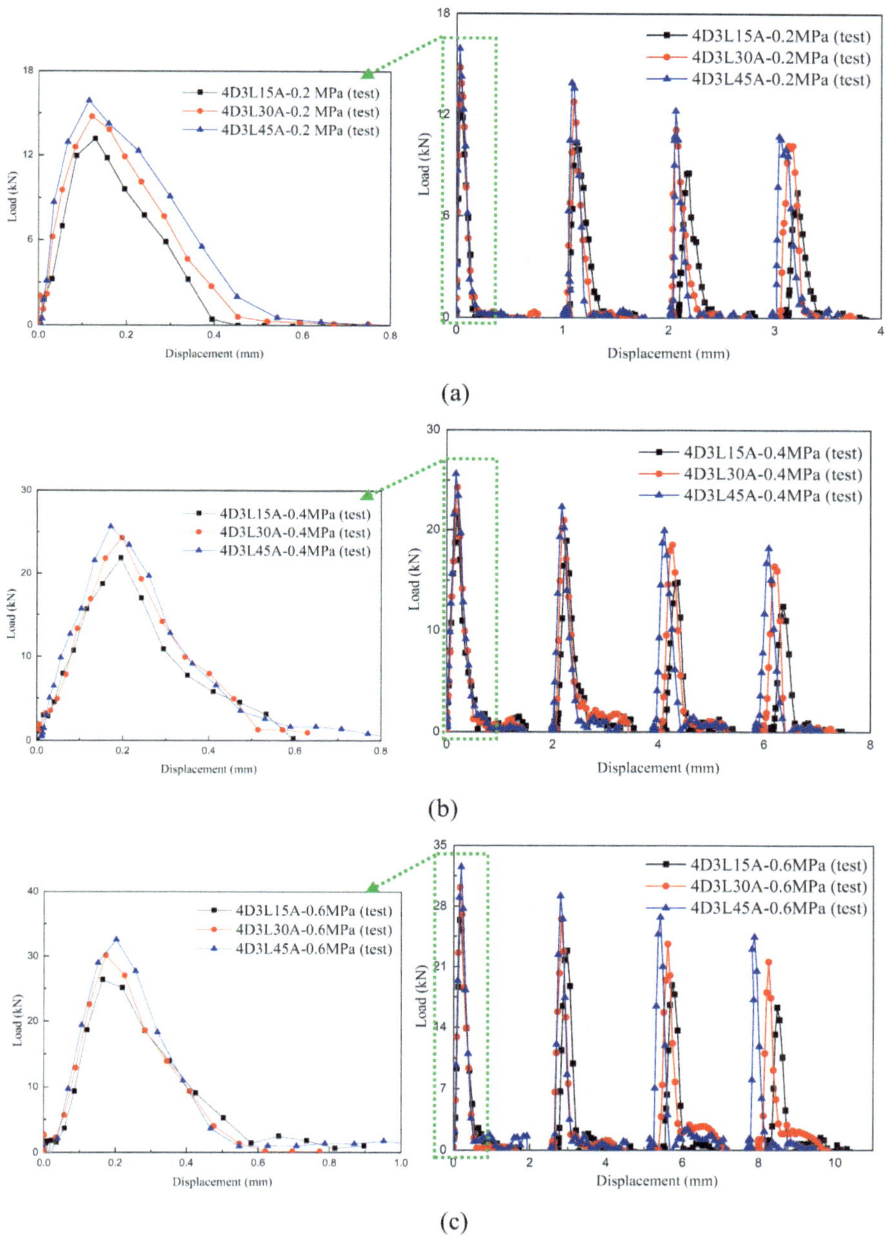

Fig. 5.15 Load-displacement curves with different braiding angles. (**a**) 0.2 MPa, (**b**) 0.4 MPa, (**c**) 0.6 MPa

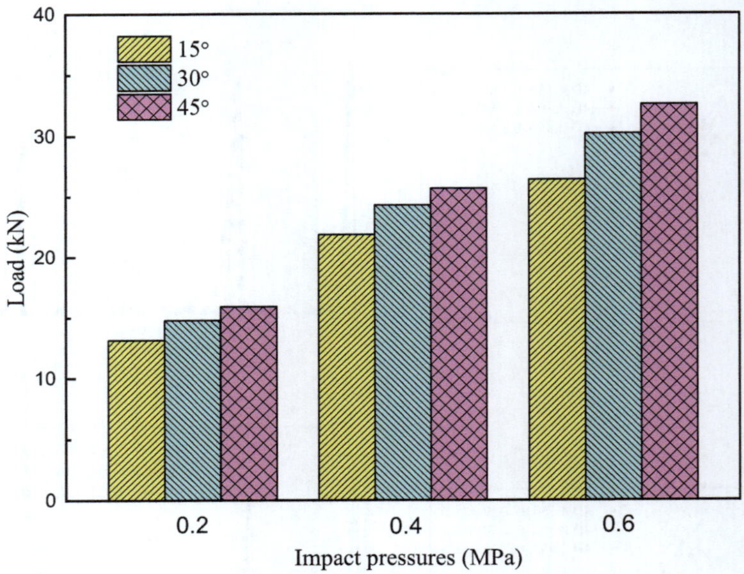

Fig. 5.16 Peak load of the first impact cycle with different braiding angles

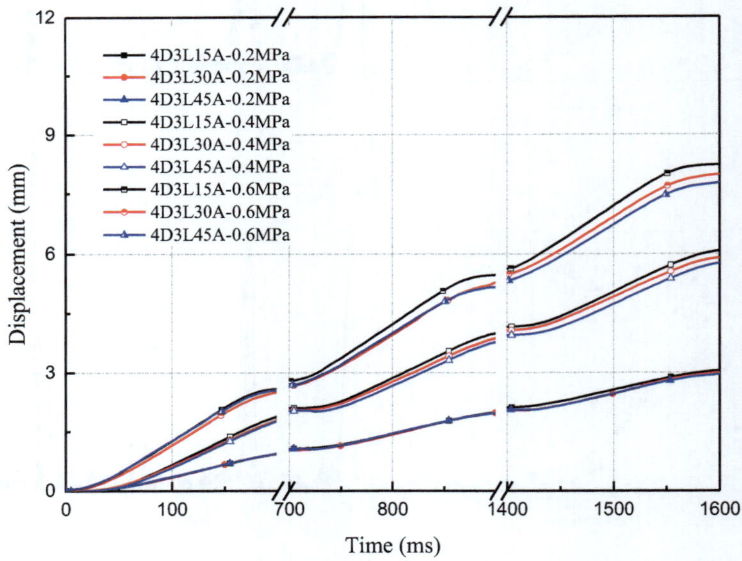

Fig. 5.17 Displacement-time curves with different braiding angles

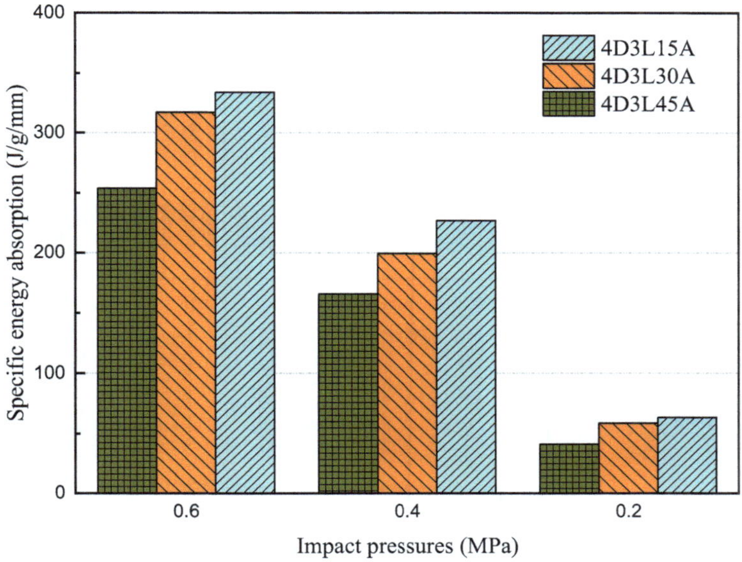

Fig. 5.18 Specific energy absorption with different braiding angles under three impact pressures

$$E_s = \frac{E_T}{\rho_t} \tag{5.3}$$

where E_s and E_T are the specific energy absorption and total energy absorption, respectively, and ρ_t is the linear density of the 3-D braided composite tubes. The linear densities of various structural composite tubes are shown in Table 4.6.

Figure 5.18 shows the specific energy absorption of the composite tubes with different braiding angles under three impact pressures [2]. With the increase of impact pressure, the specific energy absorbed by the tubes with the same braiding angle increases. With the increase of braiding angle, the specific energy absorbed by the tubes under different impact pressures reduces. The fiber tows are more inclined to axis direction for the tubes with lower braiding angle, the stress waves are faster and energy absorptions are higher than the larger braiding angle tube samples.

5.4.2 Damage Morphologies

Figure 5.19 shows the impact damage morphology of the composite tubes with different braiding angles under three impact pressures. As can be seen from the figure, the damage area is concentrated on the impact surface. The damage morphology of the impact area appearance of the specimen can be observed by the local magnification of the impact area. The damage degree of all specimens increases with the increase of impact pressure. The damage morphologies of the specimens with

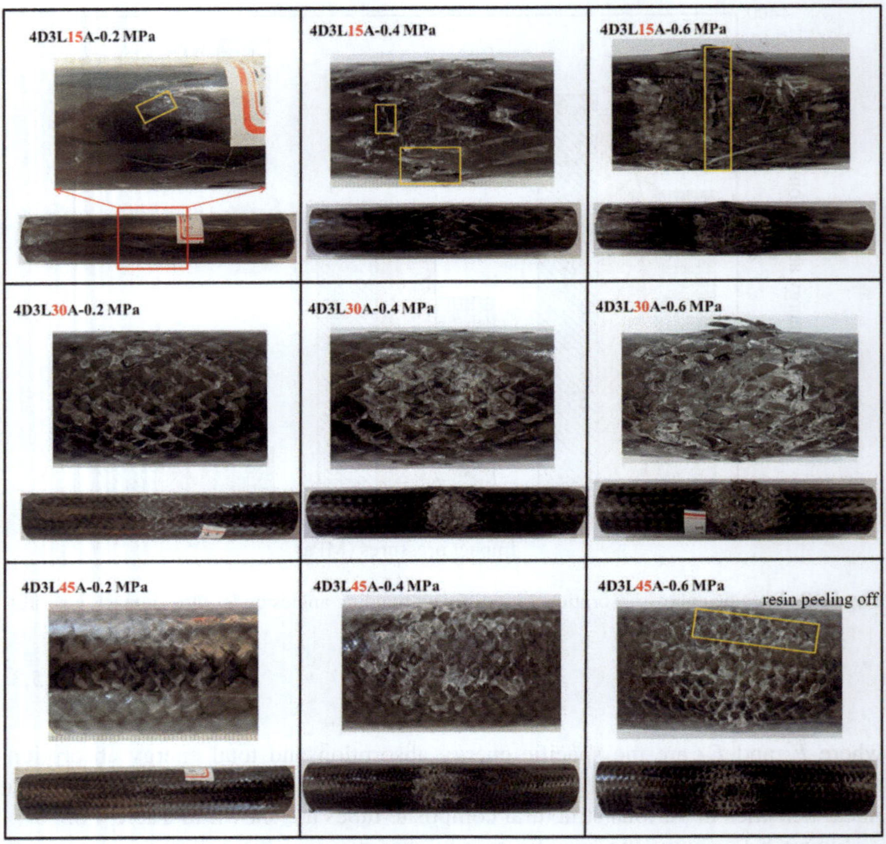

Fig. 5.19 Impact damage photographs of 3-D braided composite tubes with different braiding angles

three braiding angles are different. The specimens with the braiding angles of 15° and 30° have fiber tow fracture as the main failure mode, and the damage location is different. The specimens with the braiding angle of 45° have matrix cracking and peeling off as the main failure mode.

For the 15° braiding angle sample, because the fiber tows were oriented almost along the axis, the stress was rapidly propagated from the impact point to the holding area on both sides. The fiber tows on both sides of the impact point were rapid in a straightened state and had a weak resistance to deformation. The fiber tows broke transversely at the impact point as the impact pressure increased. For this lower braiding angle sample, the interaction regions decrease as the lower interweaving density. The deformation resistance and resin/fiber tows' bonding also decrease. Compared with the other two kinds of specimens with the braiding angles of 30° and 45°, the matrix damage in the 15° specimen is lower. The main failure mode is shear damage at the edge of the impact zone, and the fracture position of the fiber tows is mainly concentrated at the edge of the impact zone.

For the specimen with maximum braiding angle of 45°, the failure mode is different from the first two braiding angle specimens. From the local magnification, it can be seen that there is no obvious dent in the impact area and no obvious yarn fracture on the surface. The main failure mode is matrix fragmentation between fiber tows, and some matrix is peeling off at 0.6 MPa. This indicates that the 45° specimen has high impact damage resistance. For the higher braiding angle, the interweaving density increases, and the stress wave propagation is impeded along axial direction. The fiber tows from impact point to clamp point are difficult to be straightened. Furthermore, the fiber proportion along thickness increases and resists the shear damage.

The μ-CT were used to display inner damages of the specimens with three braiding angles at 0.6 MPa. The effect of different braiding angles on the internal damage of 3-D braided composite tubes was investigated. As shown in Fig. 5.20, two sections were also selected in the middle of the impact area for observation: section A-A along the radial direction and section B-B along the axial direction. For the 15° specimen, from the A-A section, it can be seen that the fiber tows in the impact area are severely delaminated to form large gaps. The fiber tows are split internally, and the fiber splits within the yarn on both sides of the impact area are especially obvious. Many matrix cracks and interfacial cracks also appear on the back of the tube. From the B-B section, it can be observed that the fiber tows in the impact area split along the axial direction and the fiber tows are delaminated. Along the impact direction, the fiber tows break in the center. The 30° specimen is dominated by shear fracture of fibers at the edge of the impact surface and interfacial cracking in the thickness direction, and there are also obvious splits within the fiber tow. In contrast, the damage of the 45° specimen differs greatly from the first two structures. As shown by the B-B section, there is no obvious fiber tow fracture in the impact region, mainly showing fiber tow deformation by extrusion and interfacial cracking. The splitting direction inside the yarn in the impacted area of the three structures was observed by the A-A section, and it was found that the splitting direction inside the yarn in the 15° and 30° specimens was along the short axis of the elliptical section, while the splitting direction inside the yarn in the 45° specimen was along the long-axis direction. Observe the back of the tube from section B-B; we can see that the crack propagation directions of 15° and 30° specimens tend to the axial direction, while many radial matrix cracks appear on the back of 45° specimen. This is due to the fact that the fiber tows with small braiding angle tend to be oriented axially, which hinders the radial expansion of the matrix crack. There are fewer interlacing points along the axial direction, which facilitate crack propagation along the axial direction. However, the fiber tows with large braiding angle have a larger rotation angle relative to the axial direction and more interlacing points along the axial direction, which is not conducive to crack propagation. The transverse matrix cracking between interlacing points is the predominant failure mode.

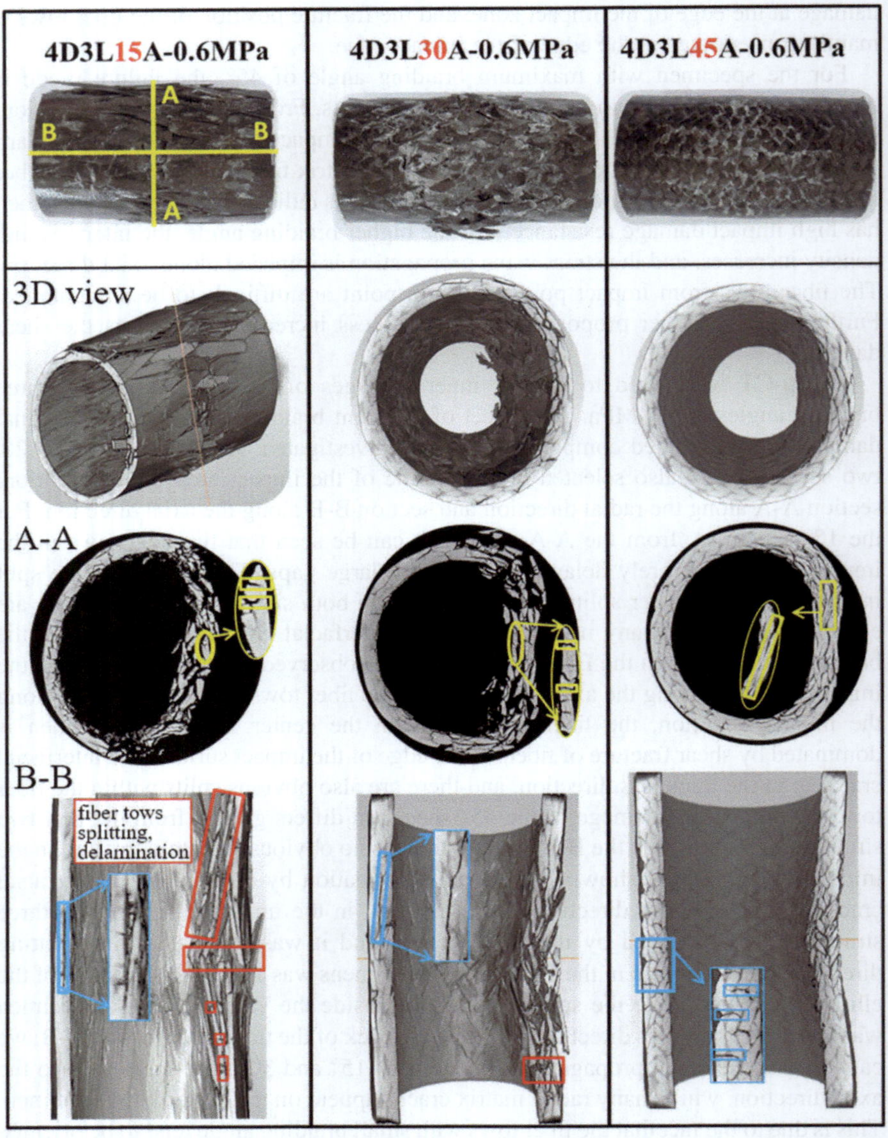

Fig. 5.20 μ-CT results with different braiding angles at 0.6 MPa

5.5 Effects of Braided Layers

5.5.1 Test Results

Three braided structures of two, three, and four layers were selected to investigate the effect of the number of braided layers on the transverse impact behavior of 3-D braided composite tubes. All three structures (4D2L30A, 4D3L30A, and 4D4L30A) maintained the same braiding angle of 30°, and none of them incorporated axial yarns.

Figures 5.21a–c show the load-displacement curves of the composite tubes with different numbers of braided layers under three impact pressures, respectively. The comparison shows that the number of braided layers has a greater effect on the transverse impact mechanical response of 3-D braided composite tube. Under the same impact pressure, the peak load increases, and the displacement decreases as the number of braided layers increases. With the increase of the number of braided layers, both the wall thickness of the tube and the number of yarn interlacing points along the direction of the wall thickness increases; therefore, the transverse impact deformation resistance of the composite tube is enhanced.

Figure 5.22 shows the first peak loads for the composite tubes with different braided layers under three impact pressures. The load increment for the three-layer to four-layer structure is greater than that for the two-layer to three-layer. The comparison shows that the higher the impact pressure, the more significant the difference in peak load for the tubes with different braided layers. As shown in Fig. 5.23, the lower the number of braided layers, the greater the deformation of the 3-D braided composite tube by impact, which is especially obvious at 0.4 and 0.6 MPa.

As shown in Fig. 5.24, the specific energy absorption of the transverse impact of 3-D braided composite tube increases with the increase of the number of braided layers under the same impact pressure, where the difference of specific energy absorption between the three structures at 0.2 MPa is not higher. However, the difference of energy absorption between different structures is obvious as the impact pressure increases.

5.5.2 Damage Morphologies

Figure 5.25 shows the damage morphologies of the composite tubes with different braided layers under three impact pressures. The damage of all three structural specimens increased with the increase of impact pressure. The more the number of braided layers, the smaller the overall damage of the composite tube under the same impact pressure. The two-layer specimens showed no significant impact concave and mainly matrix cracking at 0.2 and 0.4 MPa, while the impact zone fractured inward along the edges and showed significant brittleness at 0.6 MPa. The three-layer and four-layer specimens were damaged mainly with matrix fragmentation between fiber

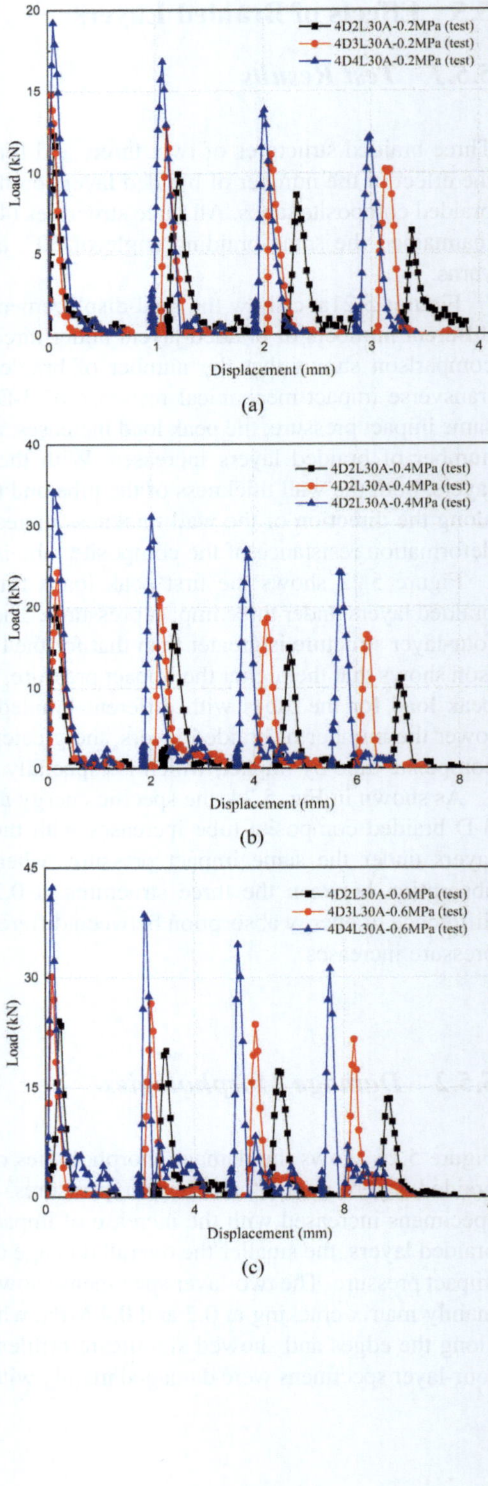

Fig. 5.21 Load-displacement curves with different braided layers under three impact pressures. (**a**) 0.2 MPa, (**b**) 0.4 MPa, (**c**) 0.6 MPa

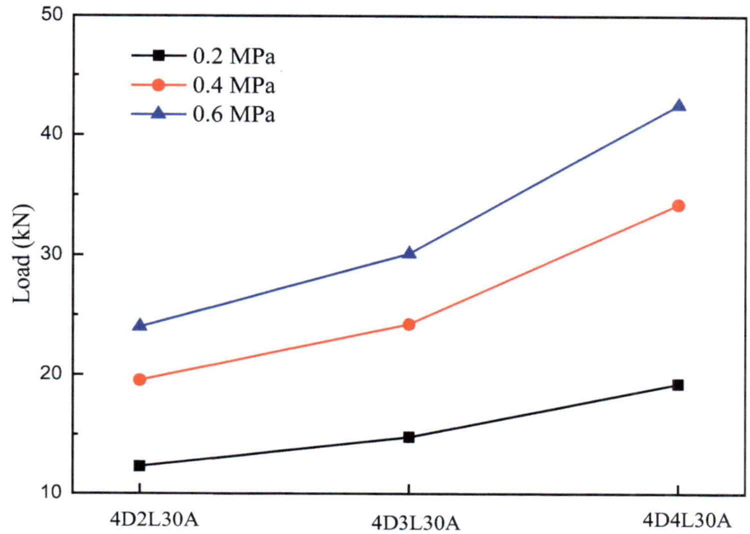

Fig. 5.22 Peak loads of the first impact cycle for circular tubes with different braided layers

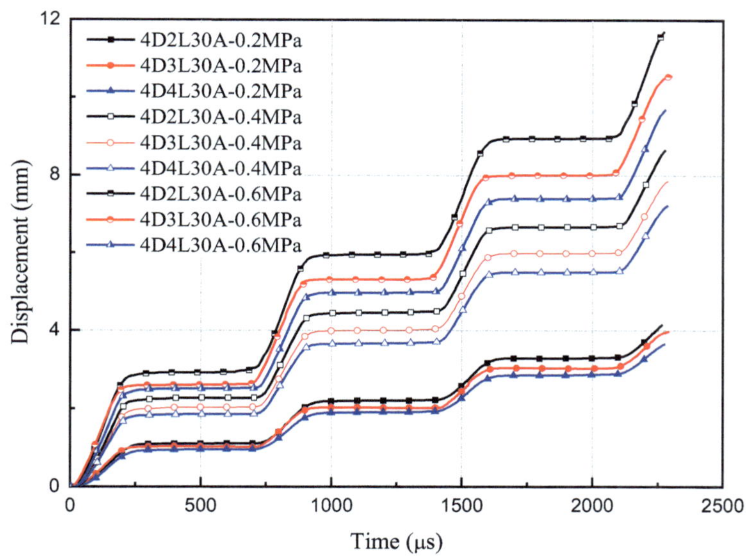

Fig. 5.23 Displacement-time curves of circular tubes with different braided layers under different impact pressures

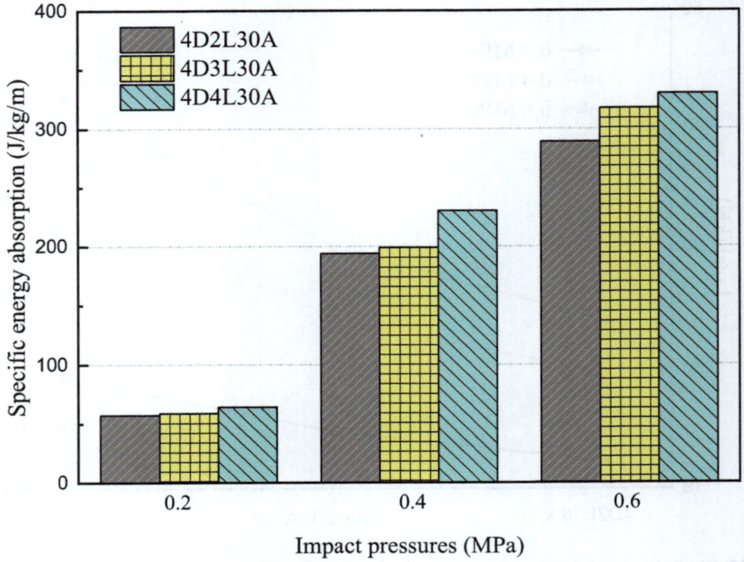

Fig. 5.24 Effect of different number of braided layers on the specific energy absorption of the circular tubes

tows at 0.2 and 0.4 MPa. At 0.6 MPa, the three-layer specimens showed obvious withdrawal fracture and shear damage along the edge of the impact zone, while the four-layer specimens still maintained good integrity with no obvious fiber breakage on the surface.

Figure 5.26 shows the μ-CT results of the composite tubes with different braided layers at 0.6 MPa. As seen from the B-B section along the axial direction, the impact surface of the two-layer specimen is concave to the inner wall of the back side, with yarn fracture mainly at the center. The axial deformation of both sides is mainly inward, and the internal damage is not obvious. Compared with the two-layer structure, the impact area of both the three-layer and four-layer specimens is less concave, but the internal damage of the whole impact area is more serious. The three-layer specimens have looser fragments in the damage area, with fiber breakage and interfacial cracking between fiber tows. Radial damage can be observed in A-A section, and shear failure occurred on both sides of the radial impact surface in all three structures of tubes. The center of the impact surface of the two-layer specimen is completely destroyed, with more serious damage on the back side, while the damage of the impact surface of the three-layer and four-layer specimens is less than that of the two sides. In the four-layer specimen, the cracks between the yarns are predominant. While in the three-layer specimen, in addition to the cracks between the yarns, there are obvious splits within the fiber tows.

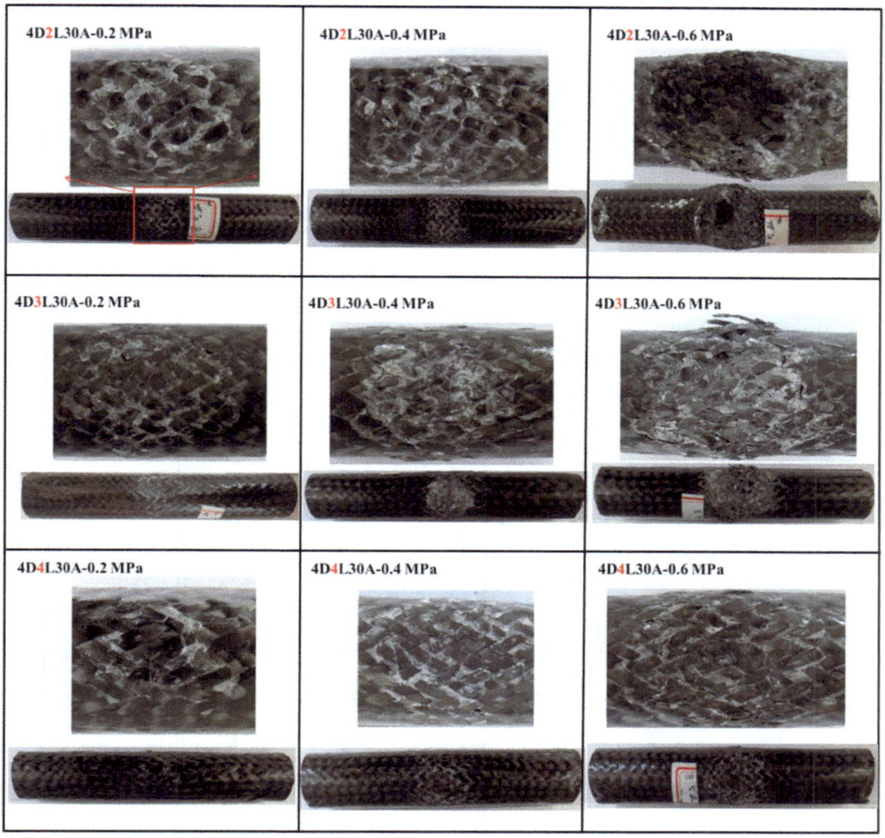

Fig. 5.25 Impact damage morphologies of 3-D braided composite tubes with different braided layers

5.6 Effects of Axial Yarns

5.6.1 Test Results

The 3-D five-directional braided structure is based on the 3-D four-directional braided structure by adding axial yarns along the braided direction. The specimen 4D3L30A and specimen 5D3L30A are 3-D four-directional braided composite tube and 3-D five-directional braided composite tube, respectively, maintaining the same number of braided layers (three-layer) and braiding angle (30°) for both structures.

Figure 5.27a, b show the load-displacement curves of the two structures of composite tubes under different impact pressures, respectively. As the impact pressure increases, the peak load and displacement of the specimens of both structures increase. The peak load of 3-D five-directional braided composite tube is much

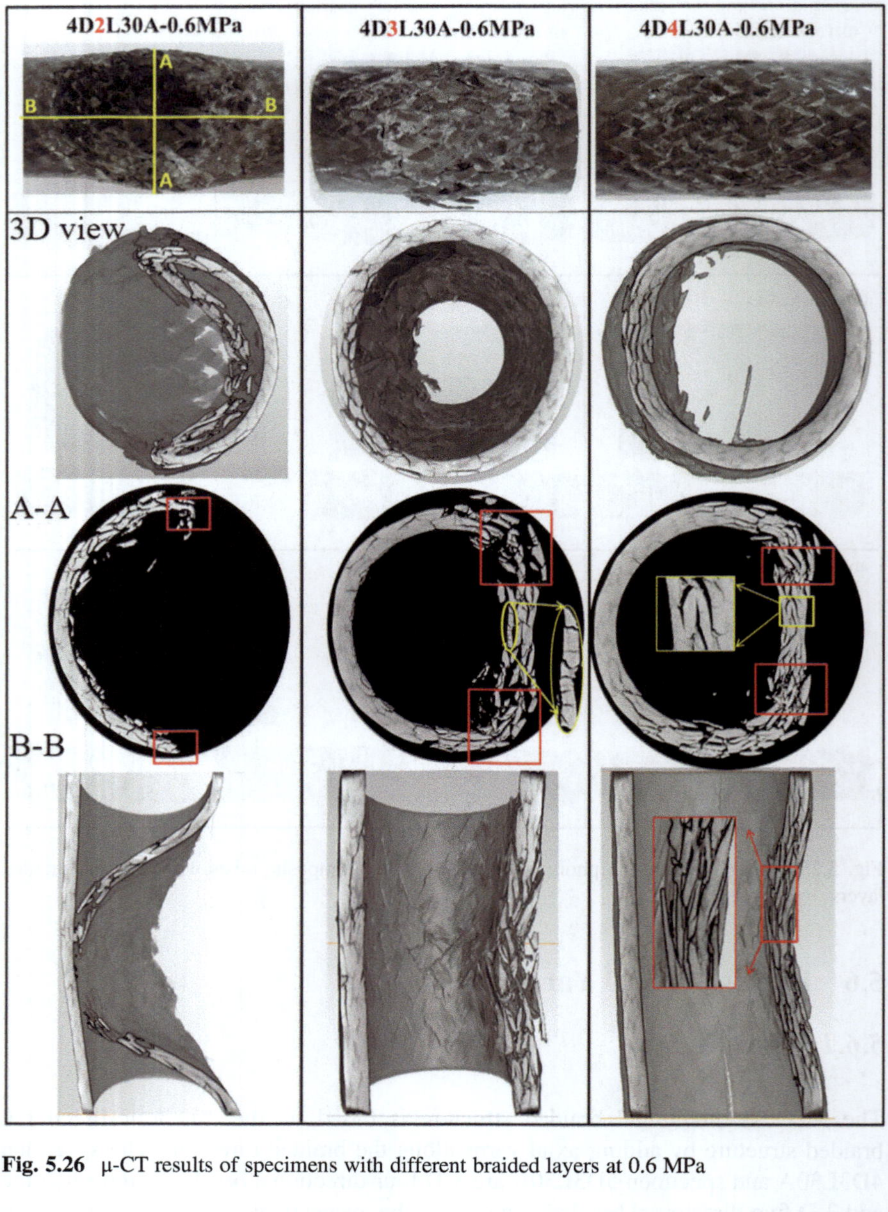

Fig. 5.26 μ-CT results of specimens with different braided layers at 0.6 MPa

larger than that of 3-D four-directional braided composite tube. The impact load of the 3-D five-directional braided composite tube remains at a higher level after four impact cycles.

Figure 5.28 compares of the first peak loads for the two structures. At each impact pressure, the peak load on the 3-D five-directional braided specimen approximates

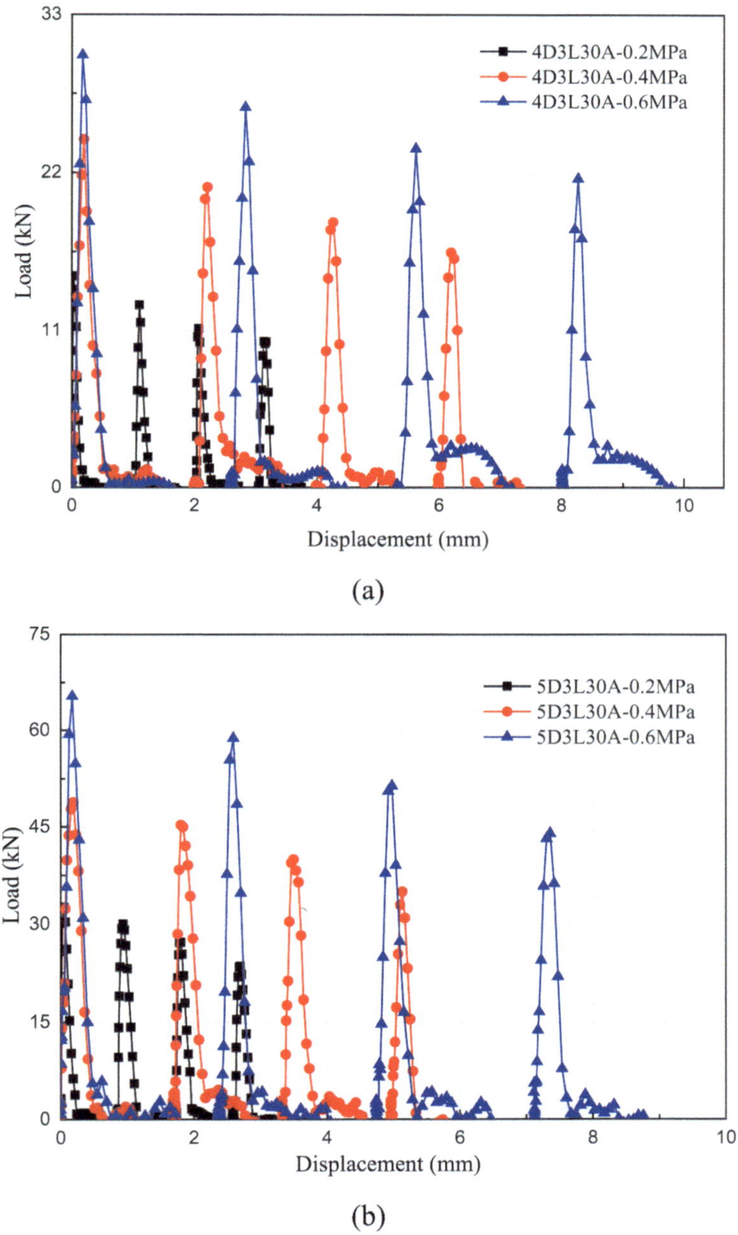

Fig. 5.27 Load-displacement curves for 3-D four-directional and 3-D five-directional braided composite tubes under different impact pressures. (**a**) 3-D four-directional, (**b**) 3-D five-directional

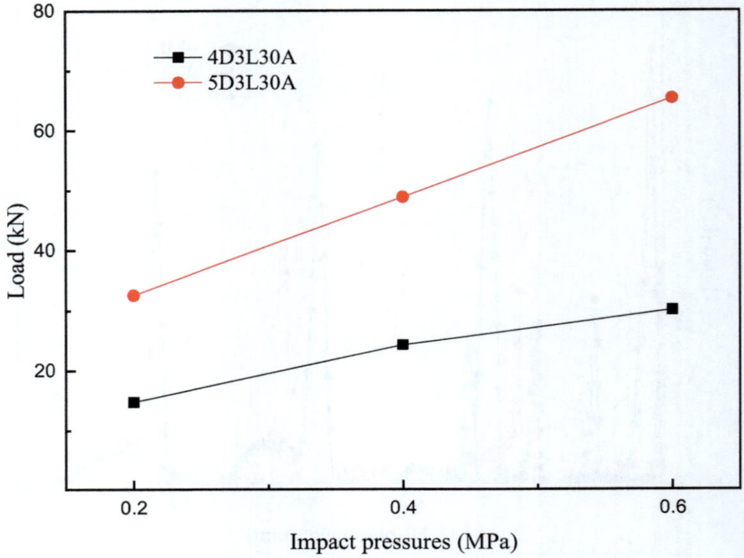

Fig. 5.28 First peak load for specimens 4D3L30A and 5D3L30A under different impact pressures

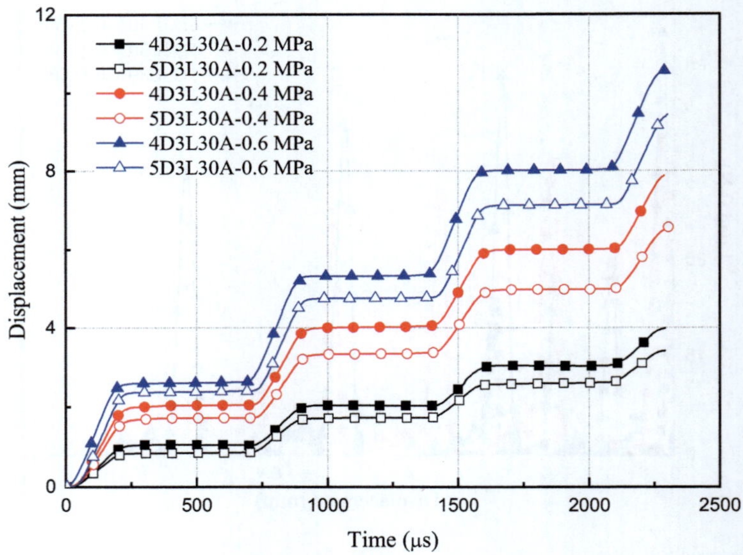

Fig. 5.29 Displacement-time curves for specimens 4D3L30A and 5D3L30A under different impact pressures

twice of the 3-D four-directional braided specimen. As the impact pressure increases, the increment of load on the 3-D five-directional braided specimen is greater than that on the 3-D four-directional braided specimen. Figure 5.29 shows the

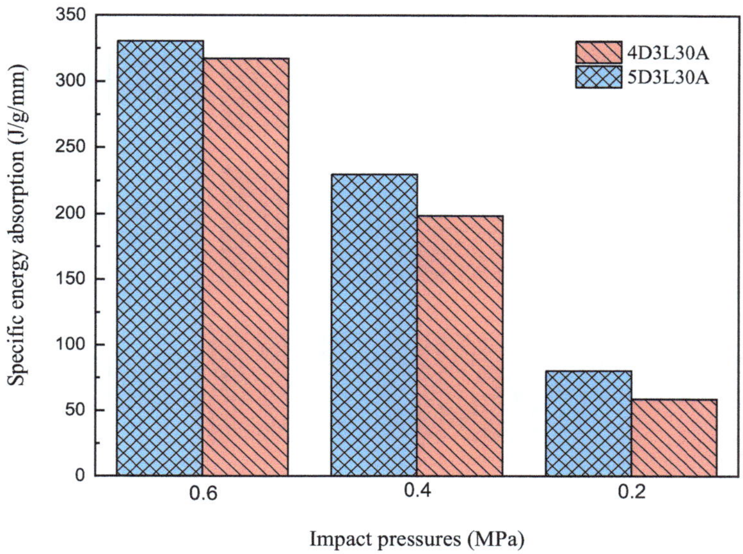

Fig. 5.30 Specific energy absorption of specimens 4D3L30A and 5D3L30A under different impact pressures

displacement-time curves for the two structural composite tubes under different impact pressures. Under the action of four continuous impact cycles, the displacement fluctuates upward with time. The displacements of 3-D five-directional braided specimens were smaller than 3-D four-directional braided specimens at each impact pressure.

The addition of the axial yarn made the fiber volume fraction of the 3-D five-directional braided specimen greater than that of the 3-D four-directional braided specimen. The specific energy absorption was used to compare the energy absorption characteristics of the two structures [3]. As shown in Fig. 5.30, the specific energy absorption of 3-D five-directional braided specimen was greater than that of 3-D four-directional braided specimen under three impact pressures. In summary, 3-D five-directional braided composite tube has stronger impact resistance and energy absorption capacity than 3-D four-directional braided composite tube.

5.6.2 Damage Morphologies

Figure 5.31 shows the damage morphologies of 3-D four-directional and 3-D five-directional braided composite tubes under the impact pressures of 0.2, 0.4, and 0.6 MPa. We can see from the figure that the damage of both structures of the composite tubes occurs mainly in the impact zone and the damage increases with the increase of impact pressure. The impact zone of 3-D four-directional braided

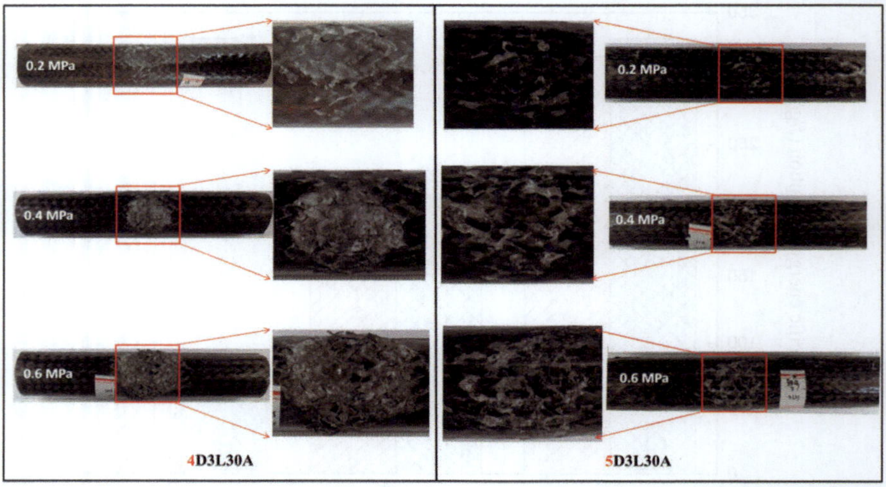

Fig. 5.31 Transverse impact damage photographs of 3-D four-directional and five-directional braided composite tubes

composite tube is obviously concave, and the failure mode of the composite tube changes from matrix cracking to fiber tow fracture with the increase of impact pressure. The damage of 3-D five-directional braided specimen is significantly smaller than that of 3-D four-directional braided specimen. There is no obvious concave in the impact area, and the material damage is mainly manifested by matrix cracking between yarns under three impact pressures. In summary, the addition of the axial yarns significantly improves the impact damage resistance of 3-D braided composite tube.

References

1. Sun, X.: Micro-geometry of 3-D braided tubular preform. J. Compos. Mater. **38**(9), 791–798 (2004)
2. Haili Zhou, Wei Zhang, Tao Liu, Bohong Gu, Baozhong Sun: Finite element analyses on transverse impact behaviors of 3-D circular braided composite tubes with different braiding angles. Compos. A: Appl. Sci. Manuf. **79**, 52–62 (2015)
3. Haili Zhou, Baozhong Sun, Bohong Gu: Responses of 3D four-directional and five-directional circular braided composite. Int. J. Crashworthiness. **21**, 353–366 (2016)

Chapter 6
Axial Impact Damages of Braided Tubes at Room Temperature

This chapter focuses on the axial compression response of 3-D braided composite tubes at room temperatures and the structural and strain rate effects on the axial compressions.

6.1 Testing Setup

6.1.1 Quasi-static Compression Tests

The quasi-static axial compression tests on the 3-D braided composite tubes were carried out on the MTS (materials testing system) 810.23, a universal materials testing system; the MTS and specimen clamping are shown in Fig. 6.1. The compression test velocity was 2 mm/min, and the compression strain rate was calculated as:

$$\dot{\varepsilon} = \frac{\varepsilon}{t} = \frac{\nu}{L} \tag{6.1}$$

where $\dot{\varepsilon}$ is the strain rate (s^{-1}), ε is the strain, t is the time (s), ν is the compression velocity (mm/s), and L is the original length of the specimen (mm). The quasi-static compression strain rate in this test can be estimated to be approximately 0.001 s^{-1} based on the original length of the specimen.

6.1.2 Impact Compression Test

The SHPB apparatus designed by authors was used as the impact compression test device. The SHPB test device consists of the following parts: high-pressure gas

M. Hu, B. Gu, *Impact Damages of Braided Composites*, Engineering Materials,
https://doi.org/10.1007/978-981-16-5703-0_6

Fig. 6.1 Materials testing system

Fig. 6.2 Photograph of the SHPB apparatus

launcher (gas gun), impact bar (bullet), input bar (incident bar), output bar (transmission bar), absorbing bar, signal acquisition system, and dampers. The photograph of the SHPB apparatus used for the test is shown in Fig. 6.2. The lengths of the impact bar, incident bar, transmission bar, and absorbing bar are 400 mm, 2000 mm, 2000 mm, and 600 mm, respectively, all with the diameter of 30 mm and made of cold-rolled GCr15 bearing steel. The physical parameters are shown in Table 6.1.

The specimen was clamped between the incident and transmission bars, and the contact end surfaces were kept smooth and flat during the tests to enhance the stress wave transmission effect and the reliability of the collected signals. The incidence

Table 6.1 Physical parameters of SHPB compression bar

Materials	Cold-rolled GCr15 bearing steel
Density (g/cm^3)	7.8
Young's modulus (GPa)	210
Ultimate tensile strength (MPa)	1965
Poisson's ratio	0.3
Brinell hardness	580

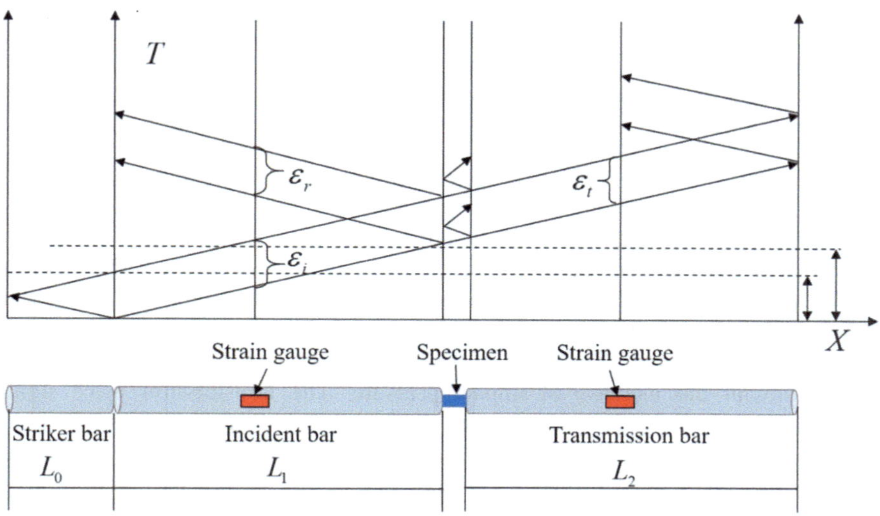

Fig. 6.3 Lagrange X-T diagram of stress wave propagation in the SHPB bars

velocity of the impact bar was controlled by adjusting the pressure of high-pressure nitrogen to obtain the corresponding range of strain rate. Three impact pressures of 0.3, 0.5, and 0.7 MPa were used in this study, with strain rates ranging from 416 to 936 s^{-1}.

Figure 6.3 is the principle of the SHPB test system. The detailed test principle can be found in the literature [1]. Based on the theory of one-dimensional stress wave propagation in a uniform slender rod and the assumption of stress and strain uniformity in the specimen, the equations for calculating the average stress, average strain, and average strain rate of the composite specimen during impact loading can be obtained as follows:

$$\sigma_s = \frac{E_b A_b}{A_s} \varepsilon_T \tag{6.2}$$

$$\dot{\varepsilon} = -\frac{2C_0}{l_s} (\varepsilon_I - \varepsilon_T) \tag{6.3}$$

$$\varepsilon = -\frac{2C_0}{l_s}\int(\varepsilon_I - \varepsilon_T)dt \tag{6.4}$$

where σ_s, $\dot{\varepsilon}$, and ε are the stress, strain rate, and strain of the specimen, respectively; ε_I and ε_T are the strains of the incident and transmitted waves; l_s, A_s, and A_b are the length of specimen, the cross-sectional area specimen, and the cross-sectional area of the bar, respectively; and $C_0 = \sqrt{E_b/\rho_b}$, E_b, and ρ_b are the elastic wave velocity in the bar, the elastic modulus of the bar, and the density of the bar, respectively.

6.1.3 Test Signal for Impact Compression

Figure 6.4 shows the test signals of 3-D four-directional braided composite tube with the braiding angle of 25° under three impact pressures of 0.3, 0.5, and 0.7 MPa and the SHPB test signals of 3-D four-directional braided composite tube with three braiding angles of 25°, 35°, and 45° at the same impact pressure (0.5 MPa). As seen in Fig. 6.4, the higher the impact pressure, the stronger the incident wave signal. When the braiding angles are the same, the transmission waveforms at 0.3, 0.5 and 0.7 MPa are compared. The slope of the rising section of the transmission wave increases with the increase of impact pressure. The transmission wave signal responds to the mechanical response of the material, indicating that the Young's modulus of the material increases with the increase of impact pressure. Under the same impact pressure, the impact strength is the same, that is, the incident wave signal is the same. Comparing the transmitted waves of 25°, 35°, and 45° specimens at 0.5 MPa, the slope of the rising section of the transmitted wave is the largest for the 25° specimen and the smallest for the 45° specimen, indicating that the Young's modulus decreases as the braiding angle increases.

From Eq. (6.3), the strain rate of material is determined by the difference between the strain signal ε_I of the incident bar and the strain signal ε_T of the transmission bar. For the same specimen, there is a tendency for the strain rate to increase as the impact pressure increases. There is a small difference in strain rate for the specimens with different braiding angles under the same impact pressure.

6.2 3-D Four-Directional Braided Composite Tubes

6.2.1 Braiding Angle Effect

The axial compression stress-strain curves of the three-layer 3-D four-directional braided composite tube with the braiding angles of 25°, 35°, and 45° under quasi-static conditions and three impact pressure of 0.3, 0.5, and 0.7 MPa are shown in Fig. 6.5. The sample exhibits elastic-plastic characteristics. Under quasi-static

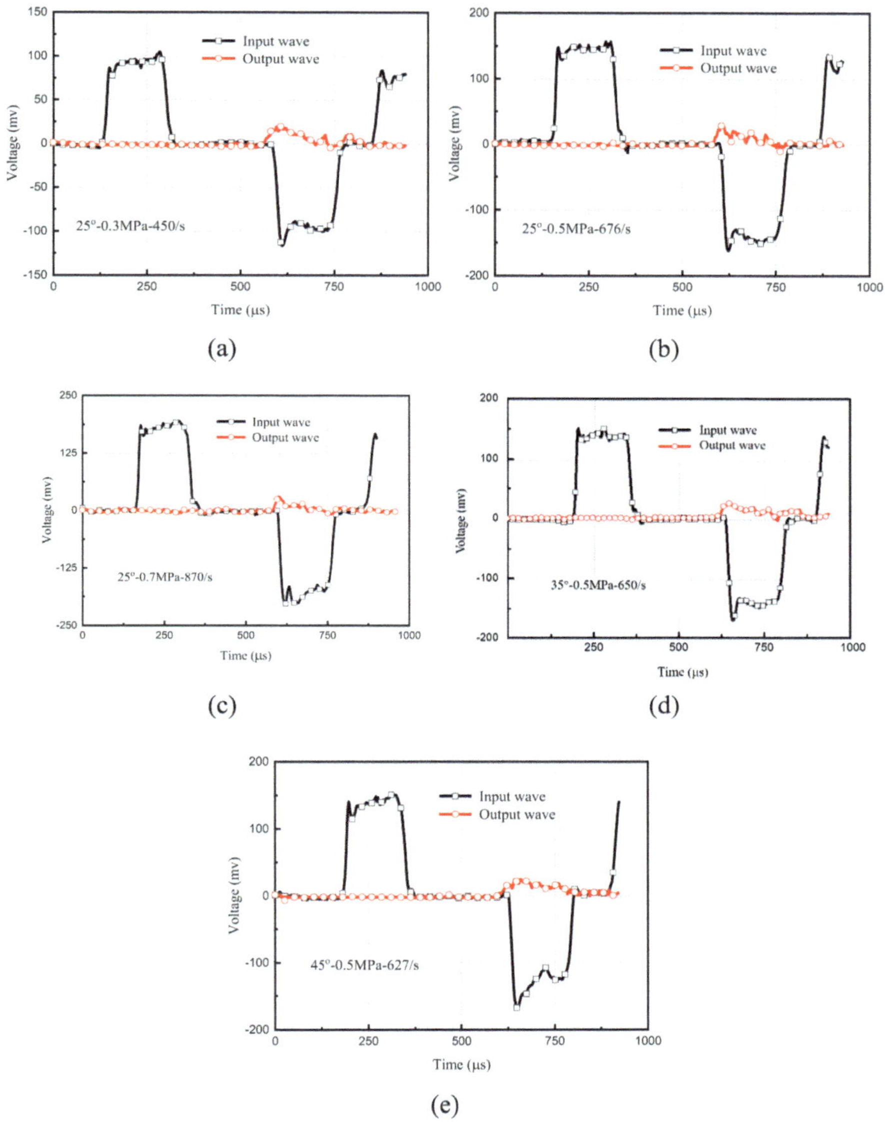

Fig. 6.4 Test signals of different braiding angles under various impact pressures

compression, for the 25° specimens, the plastic stage is short. The stress-strain curve increases almost linearly until it reaches the maximum stress and then decreases rapidly. The specimens with braiding angles of 35° and 45° have more obvious plastic stage than those with braiding angles of 25° and decrease slowly after reaching the peak stress. The failure strain and strain range of materials increase with the increase of braiding angle. At 0.3 MPa, the strain rates of the specimens

Fig. 6.5 Effect of braiding angle on the compressive stress-strain curves of 3-D braided composite tubes

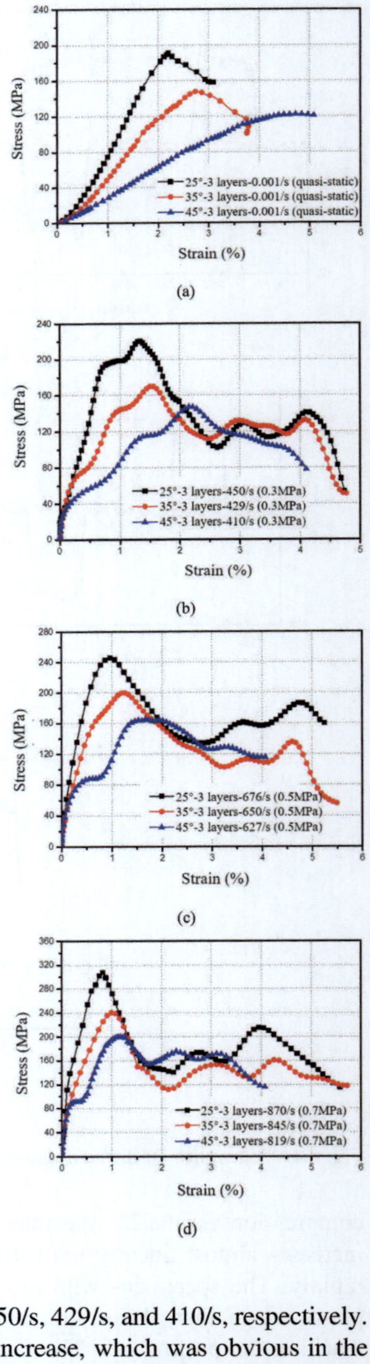

with three braiding angles were measured to be 450/s, 429/s, and 410/s, respectively. The stress-strain curves all showed a nonlinear increase, which was obvious in the yielding stage. The stress-strain curves of the 35° and 45° specimens showed bilinear characteristics. At 0.5 and 0.7 MPa, the larger the braiding angle, the more significant the plastic characteristics of the materials. The stress-strain curves of the 45° specimen showed an obvious inflection point in the rising stage with the smallest strain range, while the stress-strain curve of the 25° specimen increased almost linearly in the rising stage.

6.2.2 Braided Layer Effect

Figure 6.6 shows the axial compressive stress-strain curves of the two-layer, three-layer, and four-layer 3-D four-directional braided composite tubes with the same braiding angle (35°) under quasi-static conditions and three impact pressures of 0.3, 0.5, and 0.7 MPa. Under quasi-static conditions, the specimens with different braided layers exhibit similar elastic-plastic characteristics, with the stresses going through the initial elastic stage and the yielding stage and then slowly increasing after entering the plastic section. The larger the number of braided layers, the more obvious the plastic stage is. At 0.3 MPa, the yield point and plastic stage of three-layer and four-layer specimens are more obvious than that of two-layer specimen. At 0.5 MPa, the plastic characteristics of two-layer specimen are not obvious, while the plastic characteristics of three-layer and four-layer specimens are still very obvious. At 0.7 MPa, all three kinds of specimens exhibited elastic-plastic characteristics. The slope of the stress-strain curve is large at the initial stage, and the stress increases rapidly to the maximum and then drops rapidly to a stable value, where it vibrates until the end of loading. At high strain rate, the strain range of two-layer specimen is the largest, and the four-layer specimen is the smallest. This is because with the increase of the number of braided layers, the thickness of the composite increases and the impedance increases, so the compression deformation of the specimen decreases.

6.2.3 Strain Rate Effect

The compressive stress-strain curves of 3-D four-directional braided composite tubes with the same braiding angles under different strain rates are shown in Fig. 6.7. It can be seen that the 3-D four-directional composite tubes with different structural parameters exhibit significant strain rate effects. The mechanical properties of the composite under quasi-static compression are significantly less than those under high strain rate compression for the same specimen. The failure strain increases with increasing strain rate. Since carbon fiber is strain rate-insensitive, the strain rate sensitivity of 3-D braided composite tube is caused by the strain rate sensitivity of the epoxy resin and the structural effect of 3-D braided composite tube.

When the braiding angle is small, namely, 25° and 35°, the stress-strain curves of the braided composites have similar characteristics. With the increase of strain rate,

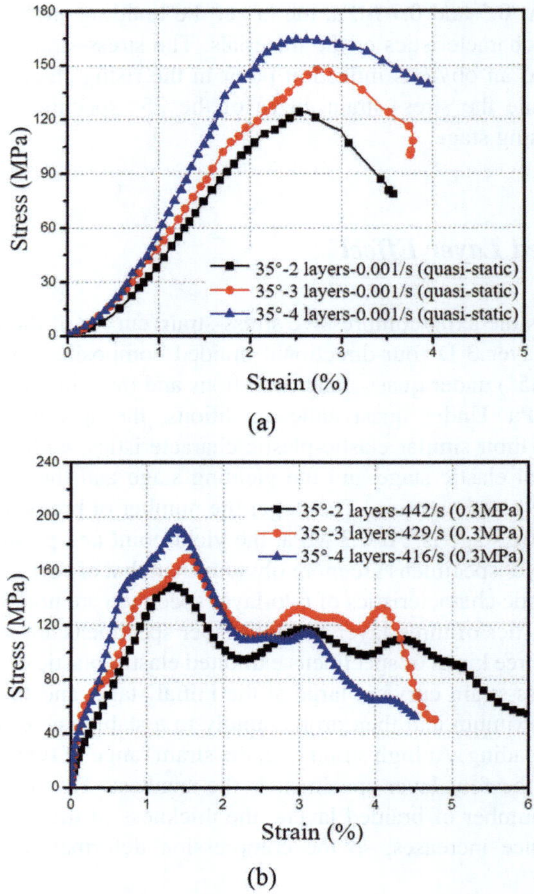

Fig. 6.6 Effect of number of braid layers on compressive stress-strain curves of braided composite tubes

the plastic properties of the composite decrease. At large strain rate, the stress-strain curve increases essentially linearly until the maximum stress is reached, then decreases rapidly after reaching the maximum stress, and finally vibrates around an intermediate value as deformation increases. As the braiding angle increased to 45°, the stress-strain curve differs markedly from the first two specimens, with a yield point and a distinctly nonlinear stage at all three strain rates. The failure strain of the 45° specimen under quasi-static compression is also significantly greater than the other two kinds of composites with small braiding angles, showing that the composite with large braiding angle has a more pronounced viscoelastic effect. As seen in Fig. 6.8, in addition to the braiding angle, the number of braided layers also has an effect on the stress-strain curve of the composite. When the number of braided layers is two, the stress-strain curve of the composite increases linearly, with no

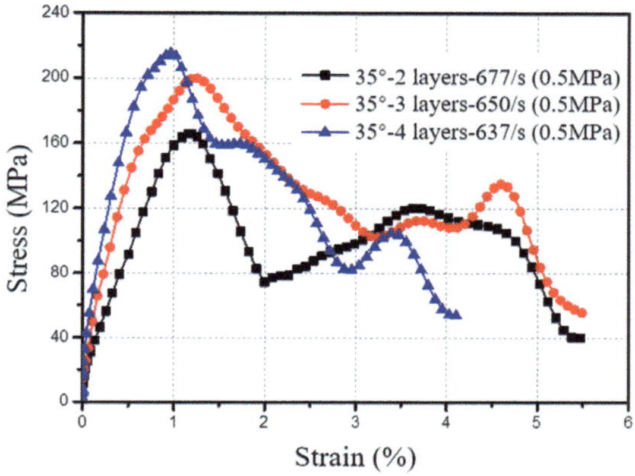

(c)

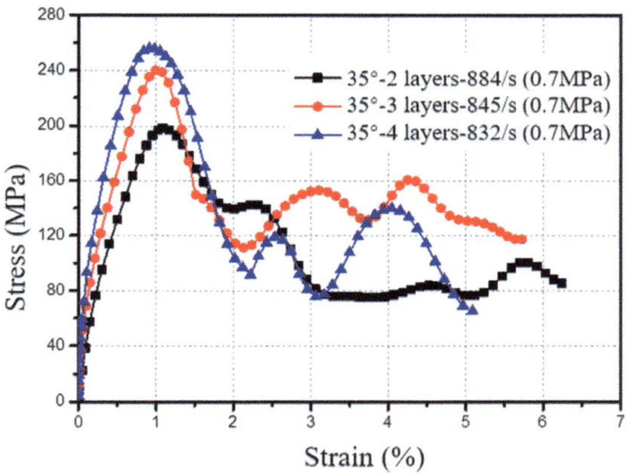

(d)

Fig. 6.6 (continued)

obvious yielding phenomenon and plastic stage. When the number of braided layers is four, the stress-strain curve are similar to those of the three-layer specimens, both showing significant elastic-plastic behavior at small strain rates.

6.2.4 Structural and Strain Rate Effects

Figures 6.8 and 6.9 show the trends of compressive stiffness, strength, and specific energy absorption with strain rate for 3-D four-directional braided composite tubes with different braiding angles and different layers. The compressive stiffness and strength can be directly calculated from the compressive stress-strain curve, and the

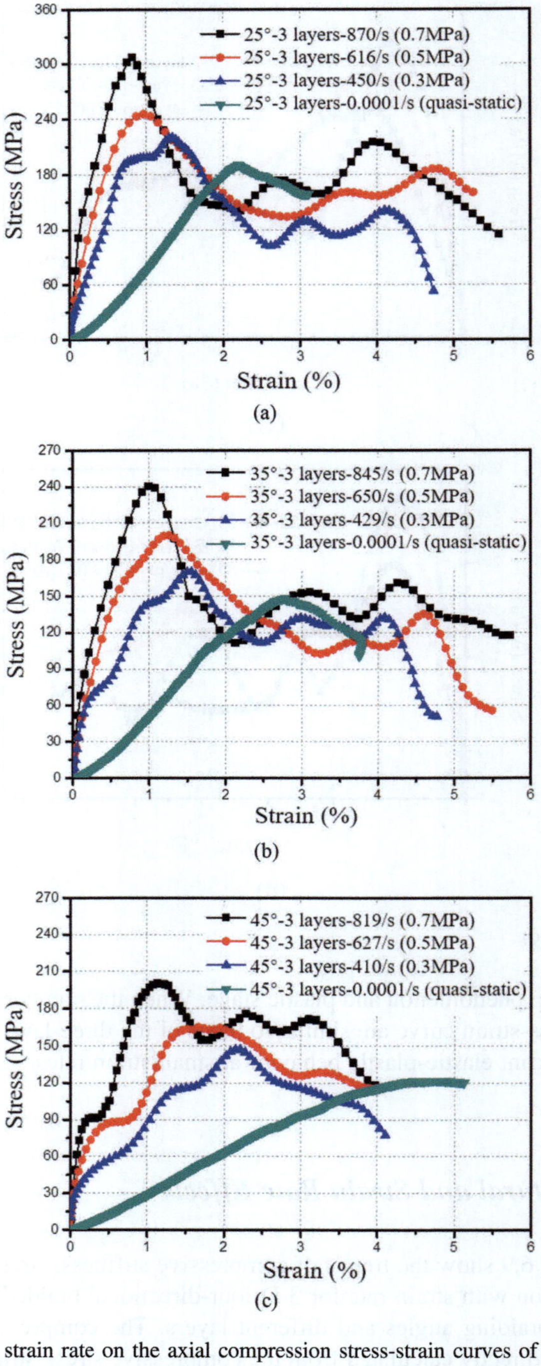

Fig. 6.7 Effect of strain rate on the axial compression stress-strain curves of braided composite tubes

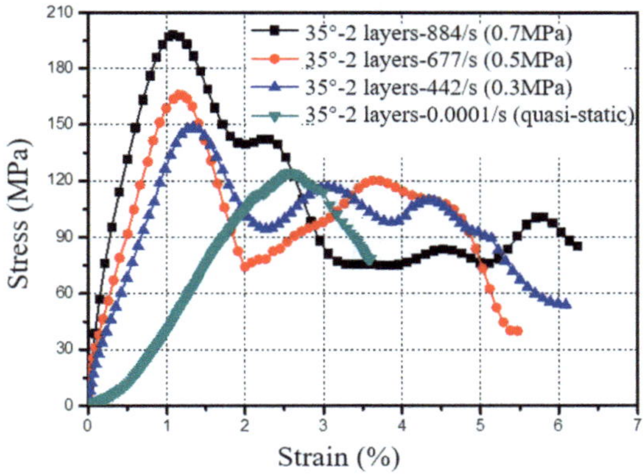

(d)

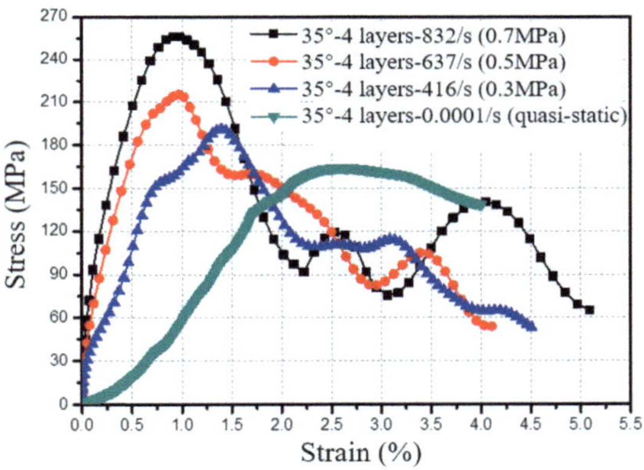

(e)

Fig. 6.7 (continued)

specific energy absorption (SEA) is defined as the energy absorbed per unit mass of material during compression damage, which is calculated by Eq. (6.5):

$$\text{SEA} = \text{total energy absorptionmass of material} = \frac{\int_0^{d_c} P dx}{\rho_t A_t d_c} \tag{6.5}$$

where ρ_t is the density of material, d_c is the failure displacement of material, A_t is the cross-sectional area of material, and P is the compressive load.

From Figs. 6.8 and 6.9, it can be seen that the compressive stiffness, strength, and specific energy absorption all increase with increasing strain rate for three different braiding angles and three different braided layer tubes, showing a significant strain rate dependence. The compressive strength is more sensitive to strain rate than compressive stiffness. The compressive stiffness and strength decrease with increase in the braiding angle and braiding layers under quasi-static and high strain rate compression. For quasi-static compression, the specific energy absorption increases

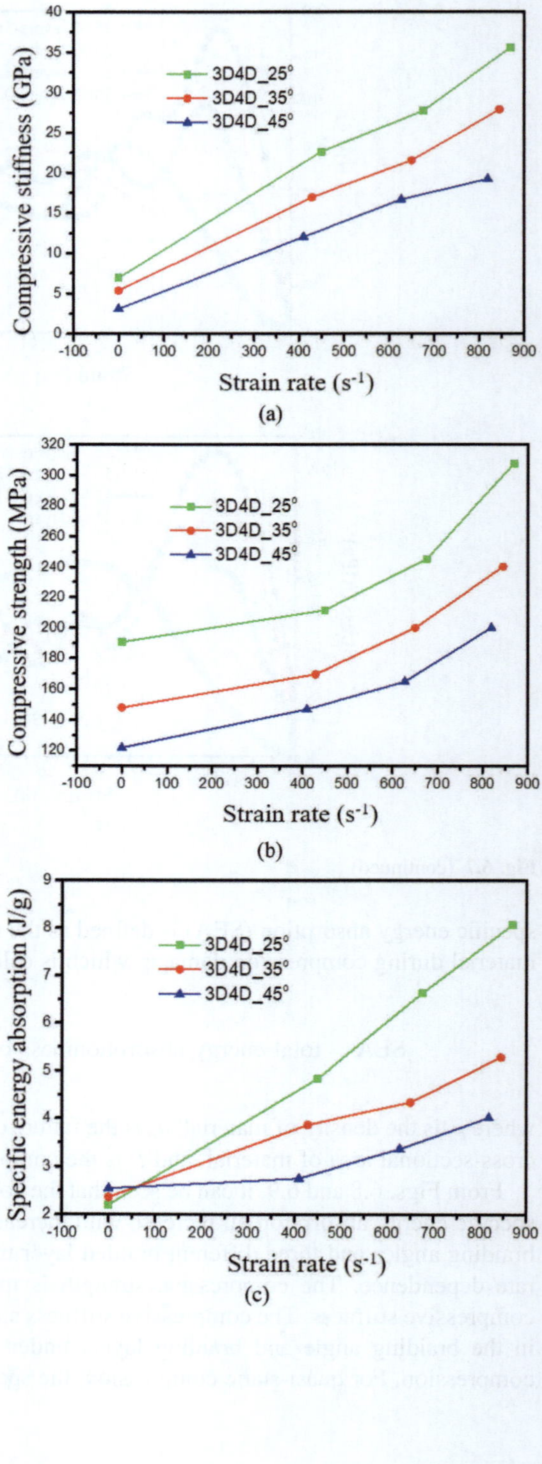

Fig. 6.8 Curves of compressive stiffness, strength, and specific energy absorption with strain rate for 3-D braided composite tubes with different braiding angles

Fig. 6.9 Curves of compressive stiffness, strength, and specific energy absorption with strain rate for 3-D braided composite tubes with different braided layers

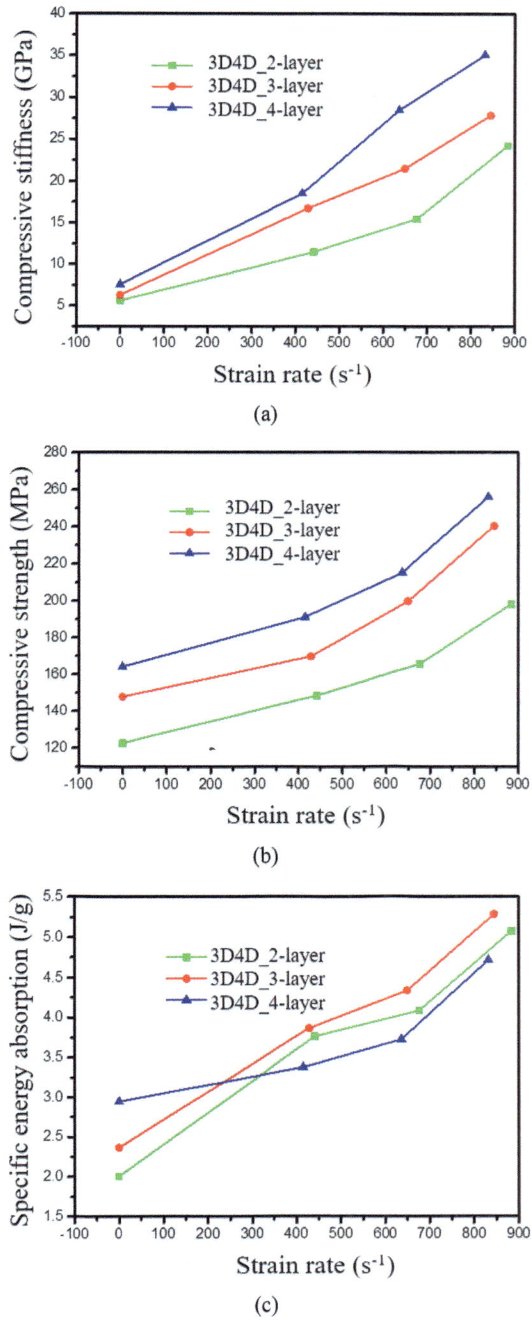

with increase in the braiding angle and the number of braided layers. For high strain rate, the specific energy absorption decreases with increase in the braiding angle and increases and then decreases with increase in the number of braided layers.

6.2.5 Failure Modes

Figure 6.10 shows the comparison of macro-failure modes of 3-D four-directional braided composite tubes with different braiding angles and braided layers under quasi-static axial compression. The main failure mode is shear damage and fine cracks in resin. When the number of braided layer is the same (three-layer), a clear triangular shear zone is formed along the braided yarn direction for the 25° specimen due to yarn fracture. For the 35° specimen, there is a clear unidirectional shear zone along the braided yarn direction from the loading end to the bottom of the specimen, where the braided yarn is obviously fractured. For the 45° specimen, the braided yarn is subjected to less load along the axial direction during the loading process, and the braided yarn is no longer the main bearing part, so there is no obvious fracture. At this time, the failure mode of the specimen is mainly resin compression deformation

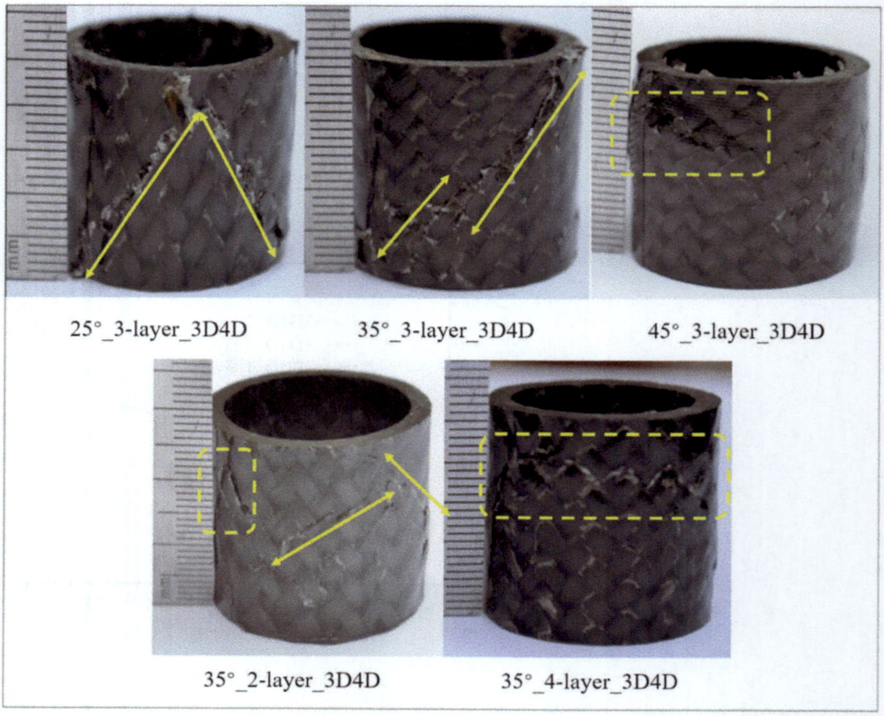

Fig. 6.10 Damage morphologies under quasi-static compression at room temperature

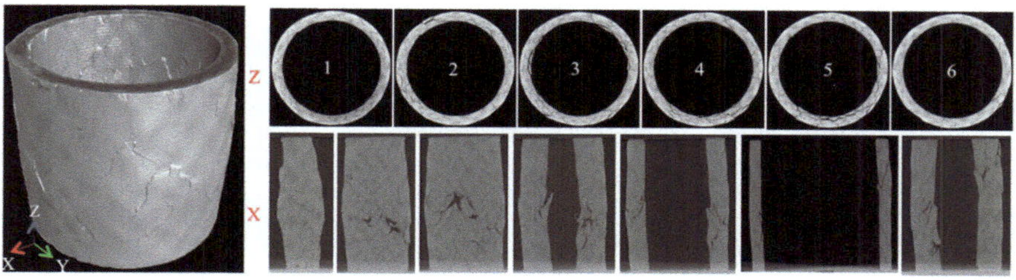

Fig. 6.11 CT results of damage morphology of 3-D four-directional braided composite tube under quasi-static compression

and the specimen buckles. After the experiment, the compression deformation has a slight recovery.

When the braiding angle is 35°, the macroscopic damage morphologies of the 3-D four-directional braided composite tubes with different braided layers are compared. For the two-layer and three-layer specimens, the failure mode of the specimens is mainly yarn fracture. With the increase of the braided layers, the damage of the specimens is mainly along the interface between the yarns and resin. For the four-layer specimen, there is no obvious yarn fracture, and only many cracks between resin and yarns are seen on the surface of the specimens.

In order to observe the internal damage of the composite, the CT pictures were photographed on the two-layer 3-D four-directional braided composite tube with the braiding angle of 35°. The results of the extracted sections along the length direction (Z-axis) and thickness direction (X-axis) μ-CT are shown in Fig. 6.11. 1 and 6 are sections near the ends of the specimen with fewer cracks, and the remaining four sections located in the middle of the specimen are visible with more small cracks. From the X-axis and Z-axis scans, it can be seen that some of the braided yarns have small cracks inside, some of the fibers are broken, and there are obvious cracks at the interface between the braided yarns and resin.

The macroscopic damage morphologies of 3-D four-directional braided composite tubes with different braiding angles under high strain rate impact loading under axial compression are shown in Fig. 6.12. As can be seen in Fig. 6.12a, with the increase of the strain rate, the failure mode of the 25° specimen changes from local shear damage to almost complete fragmentation. From Fig. 6.12b, with the increase of the strain rate, the crushing degree of fibers and resin on the surface of 35° specimen is more and more obvious, but the specimen still keep a complete form. A large amount of resin cracking results in large compression deformation of the specimens at a strain rate of 845/s. From Fig. 6.12c, as the strain rate increases, more and more resin cracking occur on the surface of the 45° specimen. The specimen maintains good integrity at all three strain rates without significant compressive deformation. Comparing the damage morphologies of the specimens with different braiding angles at the same strain rate, it can be seen that the braiding angle has a significant effect on the failure mode of the composite tube. The composite tube shows obvious brittleness when the braiding angle is small, and the damage is

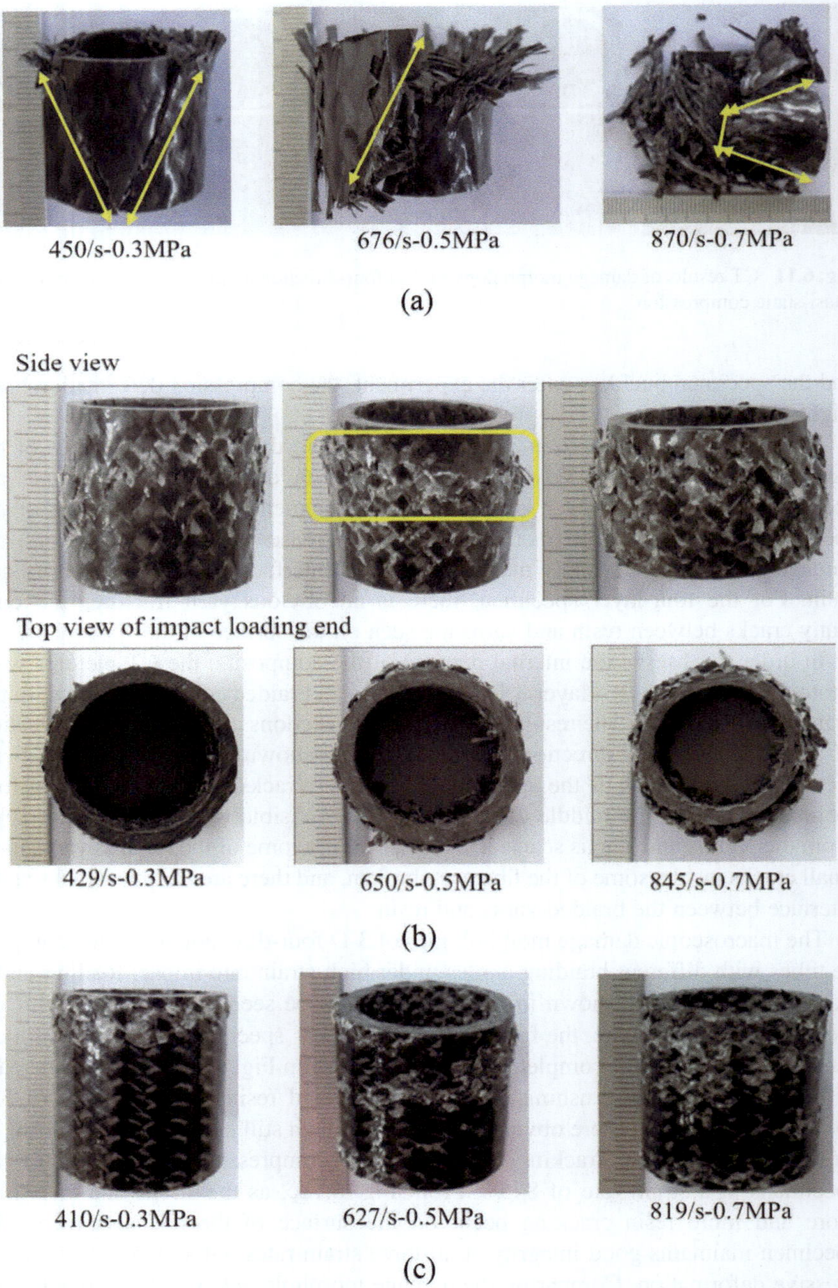

Fig. 6.12 Damage morphologies of 3-D four-directional composite tubes with different braiding angles under high strain rate compression at room temperature. (**a**) 25°_3-layer_3D4D, (**b**) 35°_3-layer_3D4D, (**c**) 45°_3-layer_3D4D

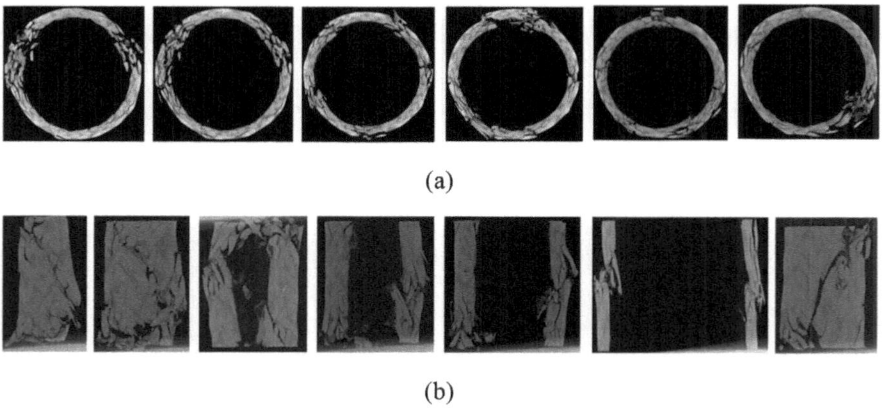

Fig. 6.13 μ-CT results of different orientations of damage morphology at 450/s strain rate for 25° specimen. (a) Length direction_Z-axis, (b) thickness directional_X-axis

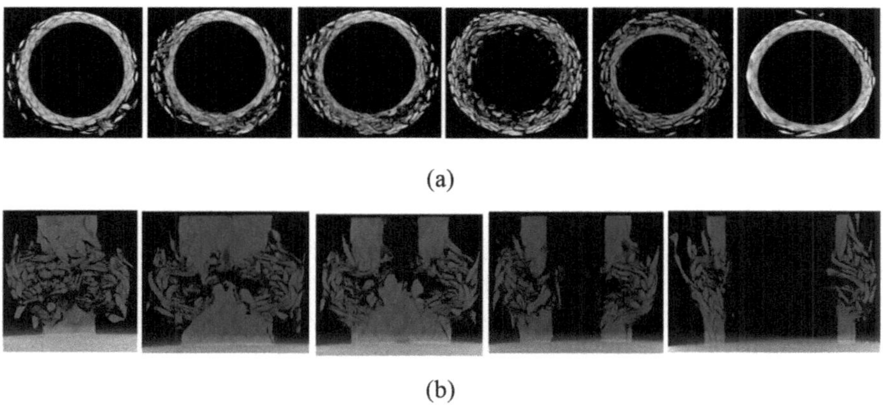

Fig. 6.14 μ-CT results of different orientations of damage morphology at 845/s strain rate for 35° specimen. (a) Length direction_Z-axis, (b) thickness directional_X-axis

dominated by shear damage, buckling failure, fiber pullout, and fracture. The failure mode is mainly resin cracking and some fiber breakage when the braiding angle is large, and there is no obvious shear damage.

Figures 6.13, 6.14, and 6.15 show the results of the composite tubes with the braiding angles of 25°, 35°, and 45° extracted from μ-CT 450/s, 845/s, and 819/s along the Z-axis in the length direction and X-axis in the thickness direction, respectively. From Fig. 6.13a, it can be clearly seen that 25° specimen shows significant fiber cracking and breakage at several locations along the length direction in the cross section, which is caused by shear failure in the thickness direction inside the specimen (Fig. 6.13b). Some fibers on the cross section inside the 35° specimens are debonded with the resin, and a large amount of resin is peeling off. At both ends

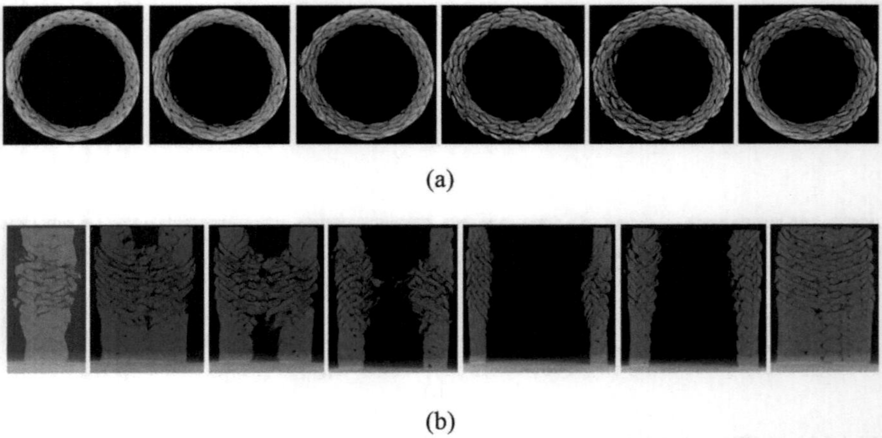

Fig. 6.15 μ-CT results of different directions of damage morphology at 819/s strain rate for 45° specimen. (a) Length direction_Z-axis, (b) thickness directional_X-axis

of the specimen, only part of the outer layers of fibers are debonded. In the middle of the specimen, each layer of fibers and resin is debonded. The images in the thickness direction show that a large amount of resin is cracked or even peeling off due to the debonding of the fiber-resin interface. The fibers in the middle of the specimen are obviously bent and deformed. The scanned images of 45° specimen show that the load-bearing capacity of the fiber along the axis weakened when the braiding angle is large. The resin bears the main load, resulting in a large amount of resin cracking inside the specimen. The cracks at the fiber-resin interface are clearly visible, and uniform cracking occurs at the interface along the circumference. Compared with 35° specimen, the bending deformation of fibers is not obvious because there is no obvious resin block peeling off.

Figure 6.16 shows the macroscopic damage morphologies of 3-D four-directional braided composite tubes with different braided layers under high strain rate axial impact compression loading. Both the two-layer and four-layer specimens with the same braiding angle exhibit similar strain rate effects as the three-layer specimen (Fig. 6.12b), i.e., the damage of composite becomes more pronounced as the strain rate increases. The two-layer specimen has more different failure modes at various strain rates, with resin cracking and a small amount of fiber fracture on the surface of the composite tube when the strain rate is small. With the increase of strain rate, more resin and fiber are fractured, and a large number of yarns are broken and pulled out at the head and middle of the tube. The composite tube exhibits buckling failure. When the strain rate is large, complete crushing damage of the tube occurs. Comparing the damage morphologies of the three-layer and four-layer specimens at various strain rates, it can be seen that there is no powdery failure mode in both of them and the tubes maintain good integrity. For the four-layer specimen, due to the increase in the number of braided layers, the ability of tube to resist compression deformation is

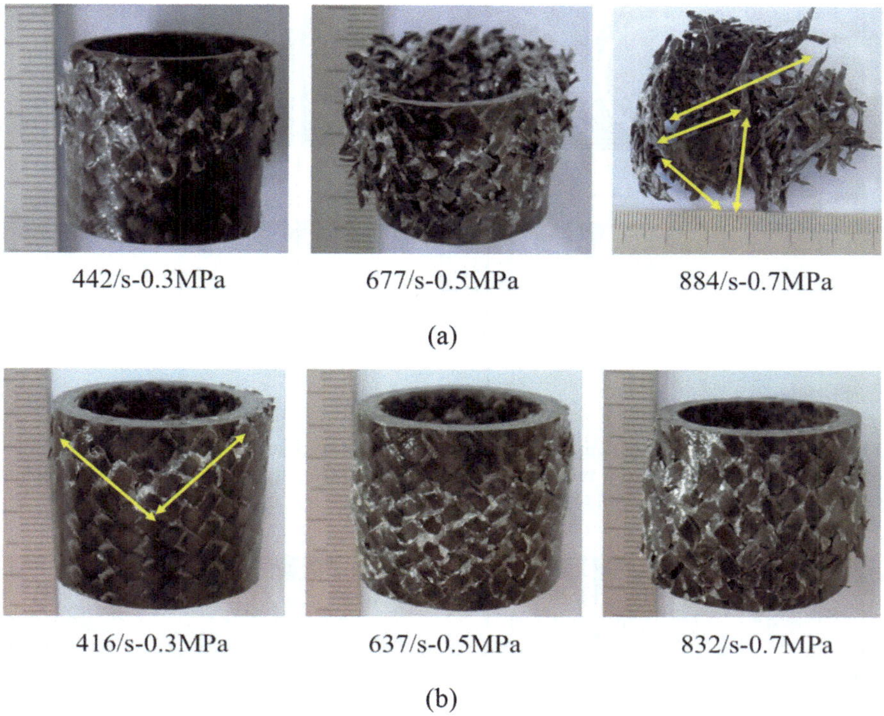

Fig. 6.16 Damage morphologies of 3-D four-directional composite tubes with different braided layers under high strain rate compression. (**a**) 35°_2-layer_3D4D, (**b**) 35°_4-layer_3D4D

enhanced. Compared to the three-layer specimen, yarn fracture is not obvious, and resin cracking on the surface is more obvious. There is no obvious buckling failure.

Figures 6.17 and 6.18 show the μ-CT results along the length and thickness directions for the two-layer and four-layer specimens with the braiding angle of 35° at 677/s and 637/s, respectively. From the damage morphologies of different cross-sections in the length direction, the area of debonding at the fiber-resin interface within the two-layer specimen gradually increases from the bottom to the loaded end, accompanied by fiber breakage. From the thickness direction, due to the resin peeling off in a large area, fiber breakage, and less braided layers, the specimen is inclined after multiple internal splits and fractures. For the four-layer specimen, due to the increase of braided layers, although the fiber-resin interface also cracks, the degree of debonding is not as serious as that of two-layer specimen. The cracks are relatively evenly distributed in the cross-section of the specimen. There is shear damage inside the specimen. Some fibers have bending deformation, the degree of deformation is less than three-layer and two-layer specimens, and there is no inclination after fracture.

Figure 6.19 is SEM images of typical failure modes of 3-D four-directional braided composite tubes under compression at various strain rates, where it can be

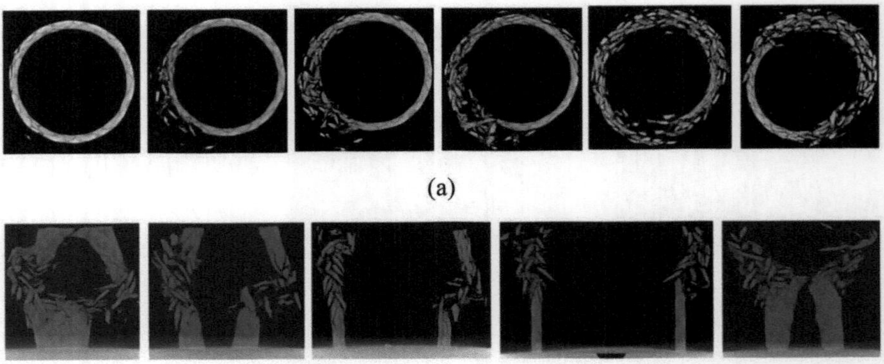

(a)

(b)

Fig. 6.17 CT scan results of damage morphologies in different orientations of two-layer 3-D four-directional specimen with the braiding angle of 35° at 677/s strain rate. (**a**) Length direction (Z-axis), (**b**) thickness direction (X-axis)

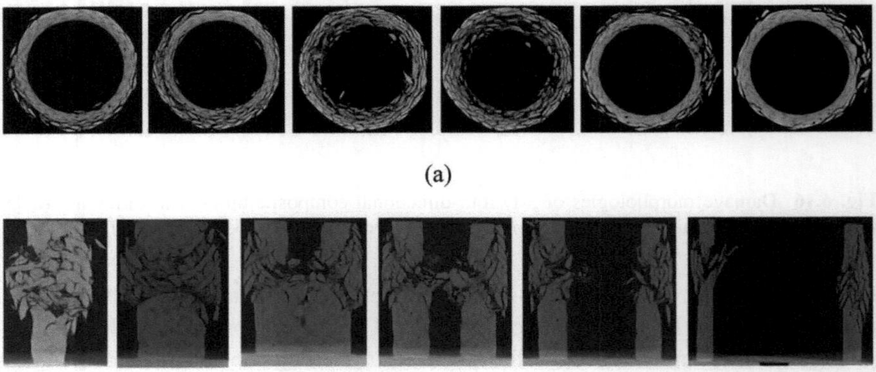

(a)

(b)

Fig. 6.18 CT scan results of damage morphologies in different orientations of four-layer 3-D four-directional specimen with the braiding angle of 35° at 637/s strain rate. (**a**) Length direction (Z-axis), (**b**) thickness direction (X-axis)

seen more clearly the fiber-resin interface debonding, fiber breakage in different forms, and resin cracking and peeling off.

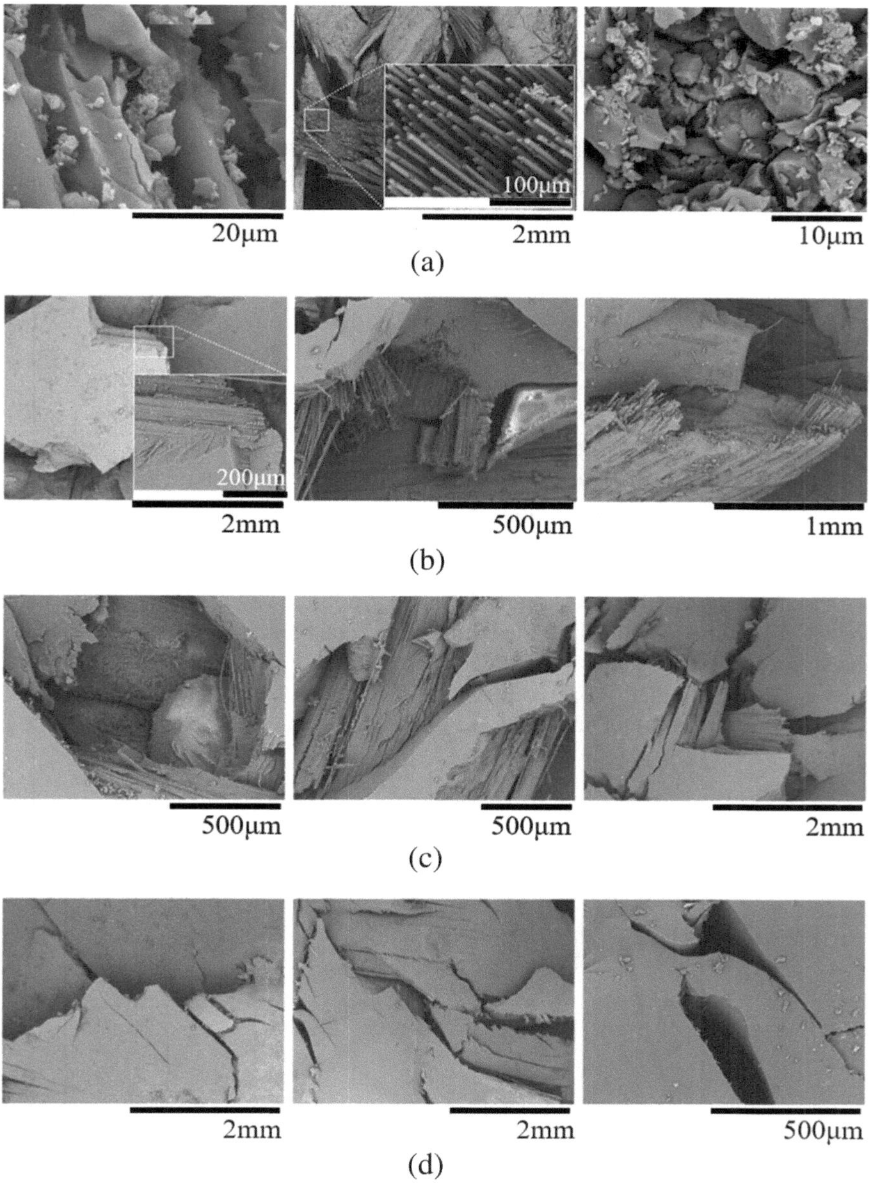

Fig. 6.19 SEM diagram of typical failure modes. (**a**) Fiber-resin interface debonding, (**b**) fiber breakage, (**c**) resin peeling off, (**d**) resin cracking

6.3 3-D Five-Directional Braided Composite Tubes

6.3.1 Strain Rate Effect

Figure 6.20 shows the stress-strain curves of 3-D five-directional braided composite tube at various strain rates. It is shown that the stress-strain curves also exhibit a significant strain rate effect and an obvious linear characteristic under high strain rate compression.

6.3.2 Axial Yarn Effect

Figure 6.21 shows the comparison of axial compression stress-strain curves for 3-D four-directional and five-directional braided composite tubes at quasi-static and the same impact pressure. The results show that when the braiding angle, number of braided layers, and loading conditions are the same, the peak stress and compressive stiffness of 3-D five-directional braided composite tube are much larger than that of 3-D four-directional braided composite tube, while the failure strain is lower than that of 3-D four-directional braided composite tube. Under quasi-static conditions, the strain range of the stress-strain curve of 3-D five-directional braided composite tube is larger than that of 3-D four-directional braided composite tube. At 0.3, 0.5, and 0.7 MPa, the strain range of the stress-strain curve of 3-D five-directional

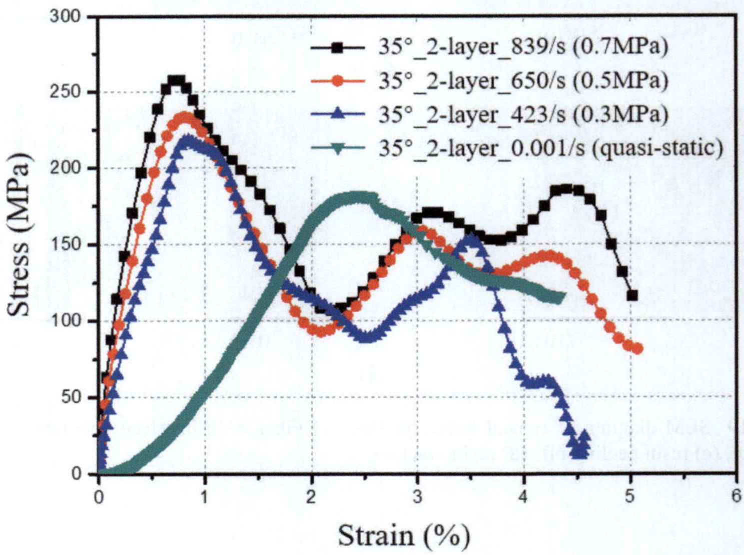

Fig. 6.20 Compressive stress-strain curves for 3-D five-directional braided composite tube

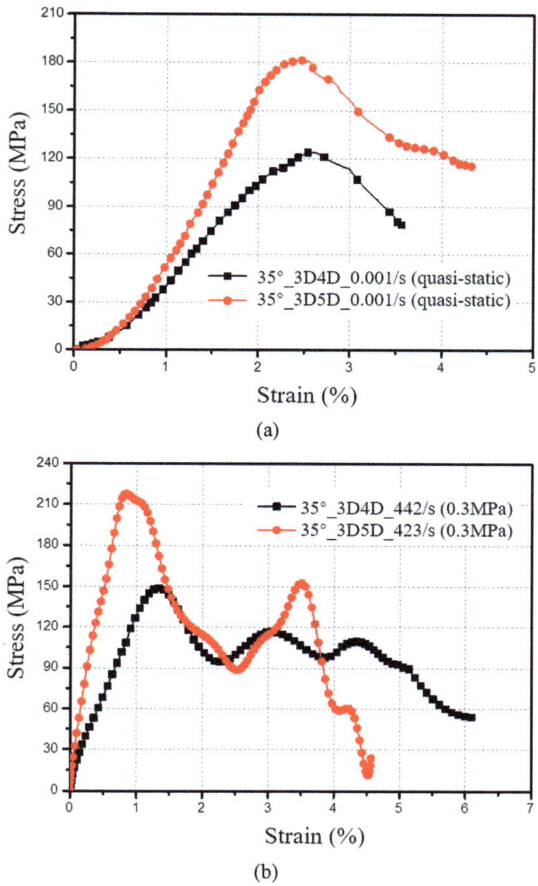

Fig. 6.21 Comparison of stress-strain curves for 3-D four-directional and 3-D five-directional braided composite circular tubes

braided composite tube is smaller than that of 3-D four-directional braided composite tube.

6.3.3 Structural and Strain Rate Effects

Figure 6.22 compares the trends in compressive stiffness and strength with strain rate for 3-D five-directional and four-directional braided composite tubes. Similar to 3-D four-directional braided composite tube, the compressive stiffness and strength of 3-D five-directional braided composite tube also exhibit significant strain rate dependence, with both increasing almost linearly as the strain rate increases. From

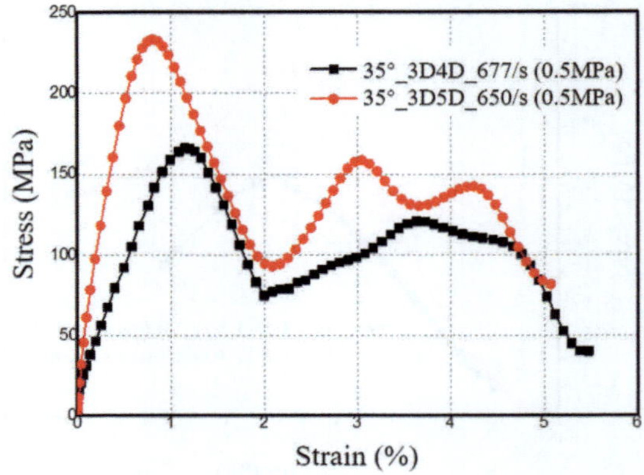

(c)

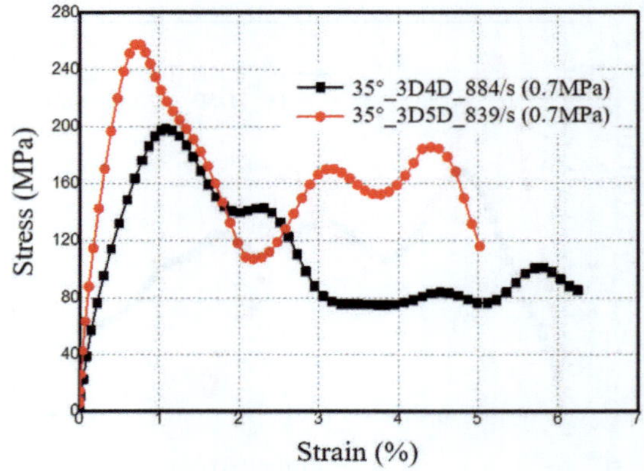

(d)

Fig. 6.21 (continued)

Fig. 6.22, it can be seen that the compressive strength and stiffness of 3-D five-directional braided composite tube are significantly greater than that of 3-D four-directional braided composite tube when the strain rates are close to each other. Under quasi-static compression, the compressive stiffness and compressive strength of 3-D five-directional braided composite tube are 164.9% and 147.9% of that of 3-D four-directional braided composite tube. Under high strain rate compression, the compressive stiffness of 3-D five-directional braided composite tube is 167.1–232.5% of that of 3-D four-directional braided composite tube, and the compressive strength is 129.8–146.5%.

The histogram of specific energy absorption with strain rate for 3-D five-directional and four-directional composite tubes is shown in Fig. 6.23. The 3-D five-directional braided composite tube has higher specific energy absorption than 3-D four-directional braided composite tube when the strain rates are close to each other.

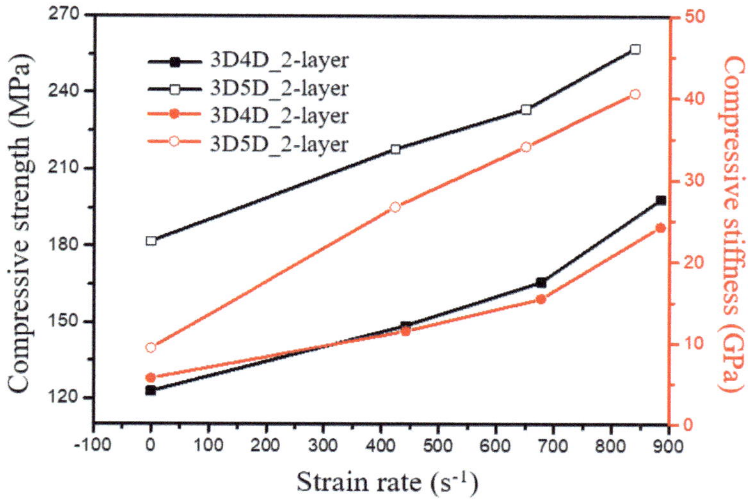

Fig. 6.22 Comparison of compressive stiffness and strength curves with strain rate for 3-D five-directional and four-directional composite tubes

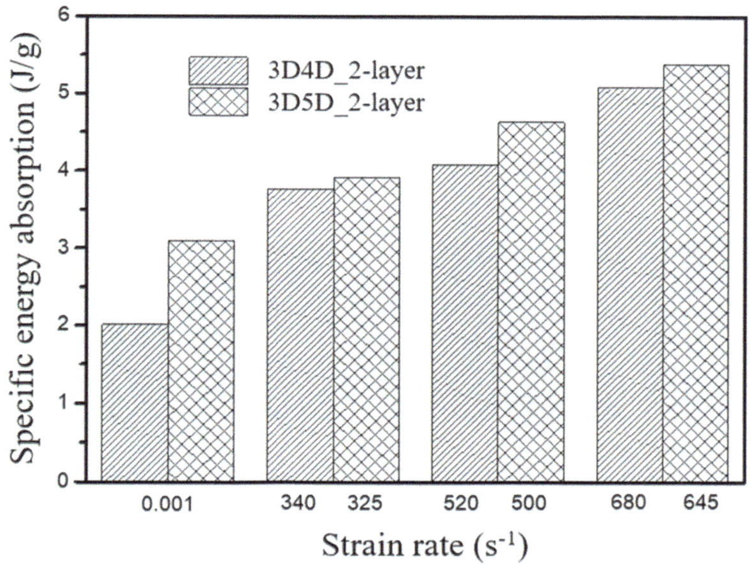

Fig. 6.23 Comparison of specific energy absorption with strain rate for 3-D five-directional versus four-directional braided composite tubes

6.3.4 Failure Modes

The macroscopic damage morphologies of 3-D five-directional braided composite tube under quasi-static and high strain rate loading are shown in Fig. 6.24. It can be seen that the damage morphologies of 3-D five-directional braided composite tube is also sensitive to the strain rate. When the strain rate is small, only a partial fiber breakage and a small amount of resin cracking occur at one end of the tube. As the strain rate increases, the resin cracking expands toward the middle and the other end. At the high strain rate, debonding between the yarns can be seen at one end of the tube due to the large amount of resin peeling off. When the braiding angle (35°) and the number of braided layers (two-layer) are the same, comparing the damage morphologies of 3-D four-directional (Fig. 6.16a) and 3-D five-directional specimens, it can be seen that the axial load-bearing capacity of the 3-D five-directional braided composite tube is higher. The tube does not show complete crushing damage, and maintain good integrity under different strain rates. The yarn fracture is not obvious, and the damage mainly occurs on the surface resin along the direction of braided yarn. This is due to the addition of axial yarns in the axial direction, which improves the compressive capacity of the tube.

The μ-CT results of the damage morphologies of the specimen at 650/s strain rate are extracted along the length direction and thickness direction in Fig. 6.25. Compared with the 3-D four-directional braided composite tube (Fig. 6.17), the internal μ-CT results in the length direction of the tube show that the fiber-resin interface debonding is more severe only at one end of the tube. The μ-CT results in the

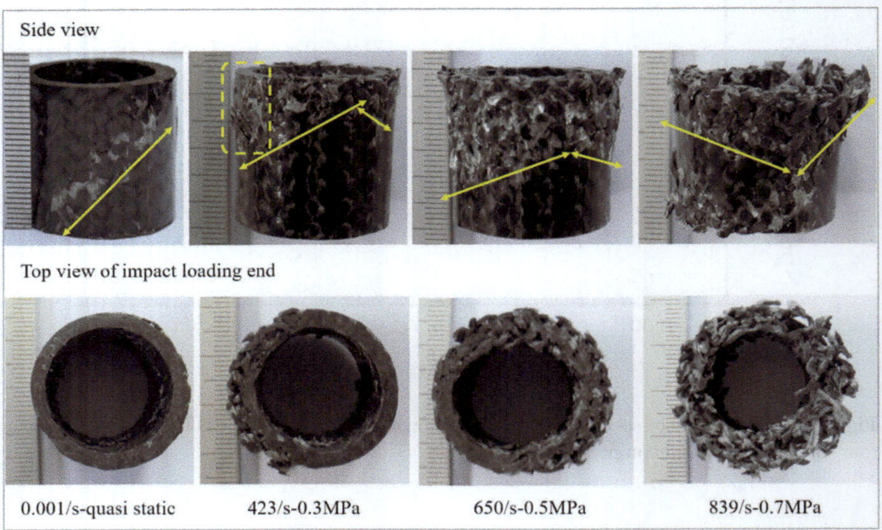

Fig. 6.24 Compression damage of 3-D five-directional braided composite tube under different strain rates

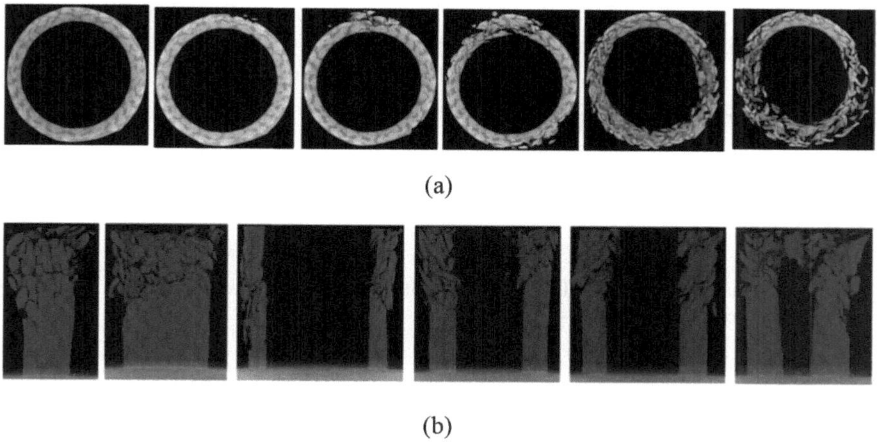

Fig. 6.25 μ-CT results of damage morphologies at strain rate of 650/s. (**a**) Length direction (Z-axis), (**b**) thickness direction (X-axis)

Fig. 6.26 SEM results of damage morphologies at strain rate of 650/s. (**a**) Fiber breakage, (**b**) fiber-resin interface debonding

thickness direction show that the tube has an enhanced resistance to deformation without fracture and significant fiber bending deformation.

The local SEM images of the damage morphologies of 3-D five-directional braided composite tube at 650/s are shown in Fig. 6.26. The damage morphologies of the inner wall of the tube is shown on the first left and the first right in Fig. 6.26a, where the fiber breakage form is different from that of the outer wall fibers (second from the left), and there is warpage after fiber breakage. Figure 6.26b shows clearly visible fiber-resin interface debonding and fragmented resin chips.

Reference

1. Sun, B.Z., Liu, F., Gu, B.H.: Influence of the strain rate on the uniaxial tensile behavior of 4-step 3D braided composites. Compos. Part A. **36**(11), 1477–1485 (2005)

Chapter 7
Axial Impact Damages of Braided Tubes at Low Temperature

This chapter focuses on the axial compression response of 3-D braided composite tubes at low temperatures and the strain rate and temperature effects on the axial compressions.

7.1 Testing Setup

7.1.1 Test Setup and Principles

In order to study the effect of low temperature on the quasi-static axial compression properties of composites, a liquid nitrogen cryogenic device was built on the above materials testing system. The schematic diagram of the cryogenic device is shown in Fig. 7.1a. The cryogenic device consists of a liquid nitrogen tank with liquid nitrogen, an insulation box, a temperature controller, an alternating current (AC) contactor, a transformer, a resistance temperature detector (RTD), and a toroidal heating coil. The modified materials testing system is shown in Fig. 7.1b. The liquid nitrogen tank is YDS-35 Dewar cryogenic storage-type liquid nitrogen tank. The insulation box is a polytetrafluoroethylene (PTFE) hollow rod which is easy to process and low temperature-resistant. The temperature controller is an electronic temperature controller with a temperature control range of 400 °C. The AC contactor is CJX2-09 AC contactor. The transformer is a single-phase 0–300 V contact adjustable transformer. The RTD is a PT100 platinum resistance with a temperature range of −199–500 °C. The toroidal heating coil is a hot runner spring heating coil with heating power of 200 W.

The low-temperature device is cooled down by liquid nitrogen. The liquid nitrogen in the liquid nitrogen tank (−196 °C) is heated by the heating coil and transformed from liquid to gas. As the pressure in the liquid nitrogen tank increases, the cryogenic nitrogen is ejected from the liquid nitrogen tank through the low temperature-resistant PTFE pipe into the insulation box to cool down the specimen.

M. Hu, B. Gu, *Impact Damages of Braided Composites*, Engineering Materials,
https://doi.org/10.1007/978-981-16-5703-0_7

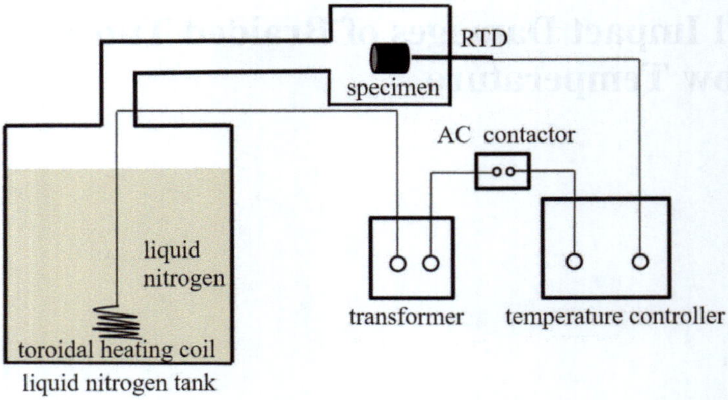

(a)

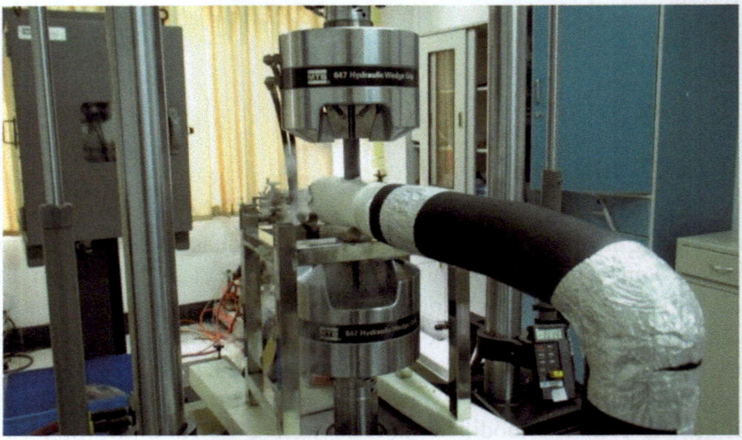

(b)

Fig. 7.1 Cryogenic device. (**a**) Schematic diagram, (**b**) mounted on materials testing system

If the ambient temperature of the insulation box does not reach the set temperature, the RTD senses the temperature and feeds back to the temperature controller, which controls the connection between the heating coil and the power supply through an AC contactor, thus ensuring that the heating coil is in working condition and controlling the continuous generation of cryogenic nitrogen gas. At the same time, the temperature controller uses proportional integral derivative (PID) control mode to control the opening and closing of the AC contactor according to the feedback of the RTD to achieve stable control of the test temperature and ensure that the specimen can be stable at the set temperature for a long time. The transformer is used to adjust the heating power of the heating coil and change the cooling time of specimen by controlling the gasification rate of liquid nitrogen [1].

In order to explore the temperature effect of the axial impact compression properties of 3-D braided composite tube at low-temperature field, a cryogenic device matched with the SHPB device was developed, which has the same composition and temperature control principle as the cryogenic device used under quasi-static compression. The physical diagram and test principle of the modified SHPB with cryogenic device are shown in Fig. 7.2.

7.1.2 Test Signals

Figure 7.3 presents the wave signals of 3-D braided composite tube obtained from SHPB test at 23, −50, and −100 °C under the same impact pressure (0.5 MPa). The peak of the transmitted wave signal at −100 and −50 °C is higher than that at 23 °C, indicating that the strength of material increases at the lower-temperature field. The slope of the initial segment also increases with decreasing temperature, indicating that the wave impedance of material increases and the material hardens at the lower-temperature field.

7.2 Quasi-static Stress-Strain Response

Figure 7.4 shows the quasi-static axial compressive stress-strain curves for the 3-D braided composite tubes at room temperature and low temperature. The stress-strain curve of the tube increases rapidly to the maximum stress after the yielding stage at −100 °C, while the stress-strain curve rises slowly at a strain of 2% and reaches the peak stress slowly at −50 and 23 °C. The peak stress is maximum at −100 °C and minimum at 23 °C.

7.3 Stress-Strain Curves at High Strain Rates

7.3.1 Strain Rate Effect

The stress-strain curves of 3-D four-directional braided composite tube under quasi-static and high strain rate compression at the same temperature are shown in Fig. 7.5. From the figure, it can be seen that the 3-D braided composite tube exhibits similar strain rate effects at −50 and −100 °C as that at room temperature. The compressive strength and stiffness increase with the increase of strain rate. At the same temperature, the larger the strain rate, the larger the strain range of the tube, indicating the greater the deformation of the tube.

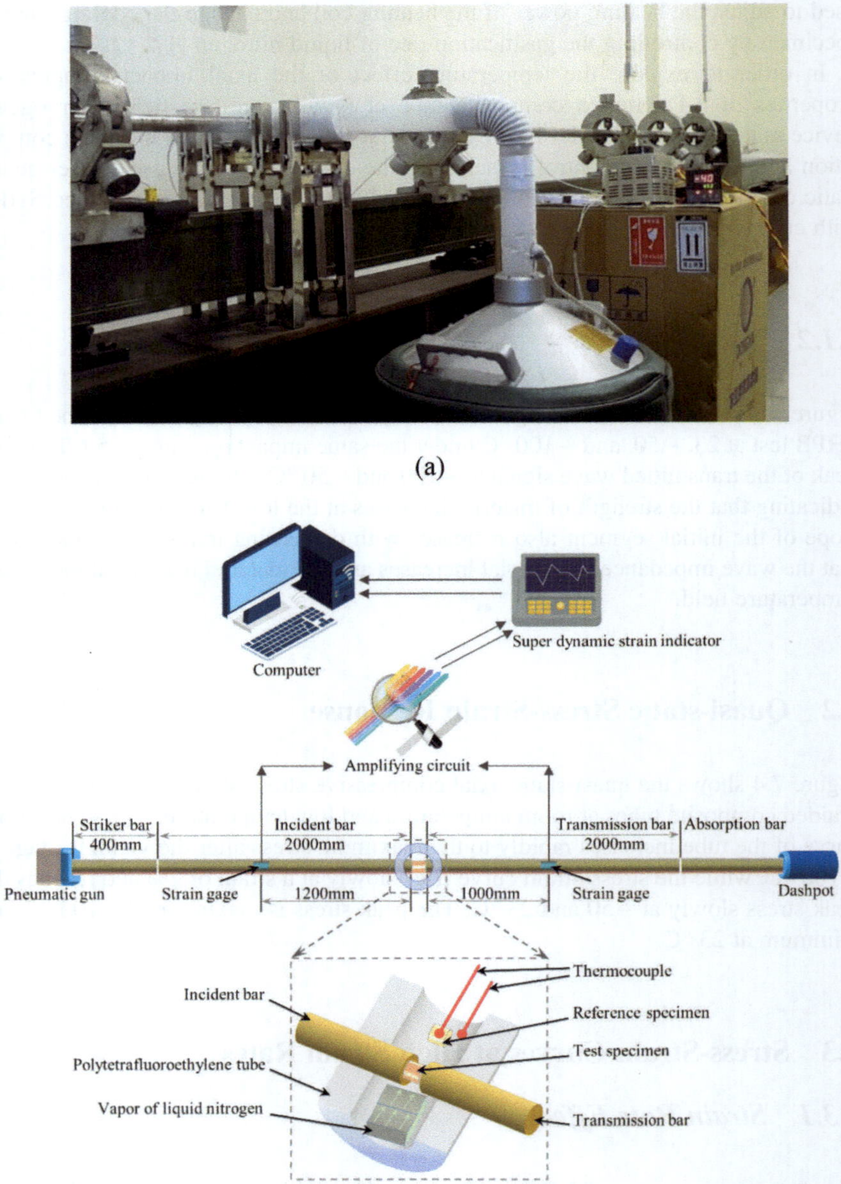

Fig. 7.2 Cryogenic device on SHPB mounted diagram and test principle. (**a**) Cryogenic device on SHPB test system mounted diagram, (**b**) schematic diagram of the test principle of SHPB at low temperature

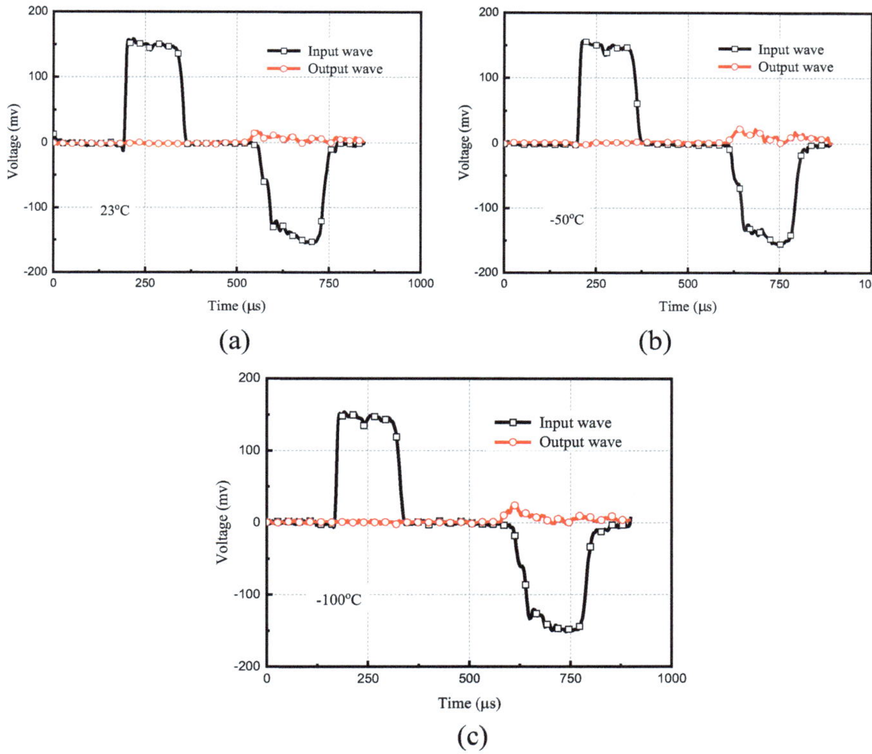

Fig. 7.3 Wave signals obtained from SHPB test at different temperatures and under the same impact pressure

7.3.2 Temperature Effect

The stress-strain curves of 3-D four-directional braided composite tubes at different temperatures under impact pressure loading of 0.3, 0.5, and 0.7 MPa are shown in Fig. 7.6. At 0.3 and 0.5 MPa, there is a clear yielding stage in the stress-strain curve at low-temperature fields of −50 and −100 °C compared to room temperature. At 0.7 MPa, the brittleness characteristic is obvious at low temperature, and there is no yielding stage. Under the same impact pressure, the shape and trend of the stress-strain curve are similar. With the decrease of temperature, the compressive strength and stiffness of the tube increase slightly.

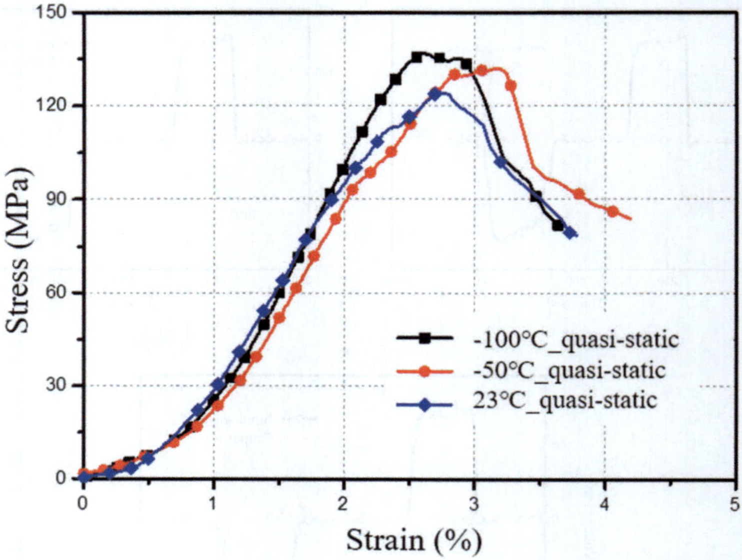

Fig. 7.4 Quasi-static compressive stress-strain curves for 3-D braided composite tubes at different temperatures

7.4 Strain Rate and Temperature Effects

Figures 7.7, 7.8, and 7.9 show 3-D diagrams of the influence of strain rate and temperature on the compressive properties of 3-D braided composite tubes. Compressive strength (Fig. 7.7), compressive stiffness (Fig. 7.8), and specific energy absorption (Fig. 7.9) increase with decreasing temperature and increase with increasing strain rate. It can be seen that the axial compressive mechanical properties of 3-D braided composite tube at low temperature depend on both temperature and strain rate. The compressive strength, compressive modulus, and specific energy absorption are more sensitive to strain rate than to temperature.

7.5 Compression Failure Modes

7.5.1 Macroscopic Failure Modes

The damage morphologies of 3-D braided composite tube at room temperature and low temperature under quasi-static compression are shown in Fig. 7.10, and those under high strain rate compression are shown in Fig. 7.11. The results show that the failure mode of the tube depends on the ambient temperature and strain rate. Under quasi-static compression at room temperature, the failure mode is fiber breakage

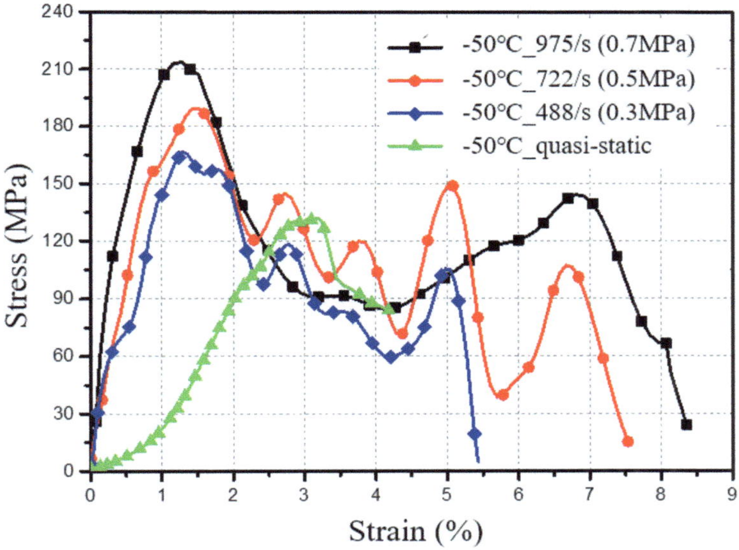

(a)

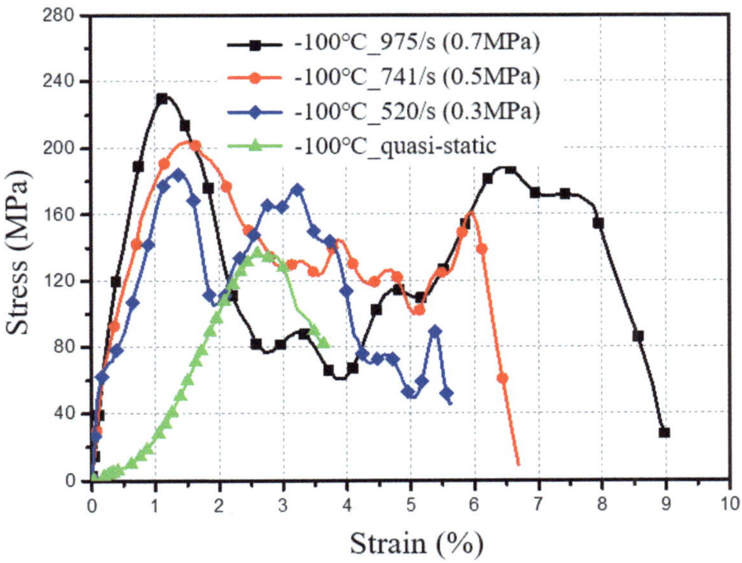

(b)

Fig. 7.5 Compressive stress-strain curves for 3-D braided composite tubes under different strain rates at low temperatures. (**a**) −50 °C, (**b**) −100 °C

Fig. 7.6 Test stress-strain curves for composite tubes at different temperatures under quasi-static and impact compression. (**a**) 0.3 MPa, (**b**) 0.5 MPa, (**c**) 0.7 MPa

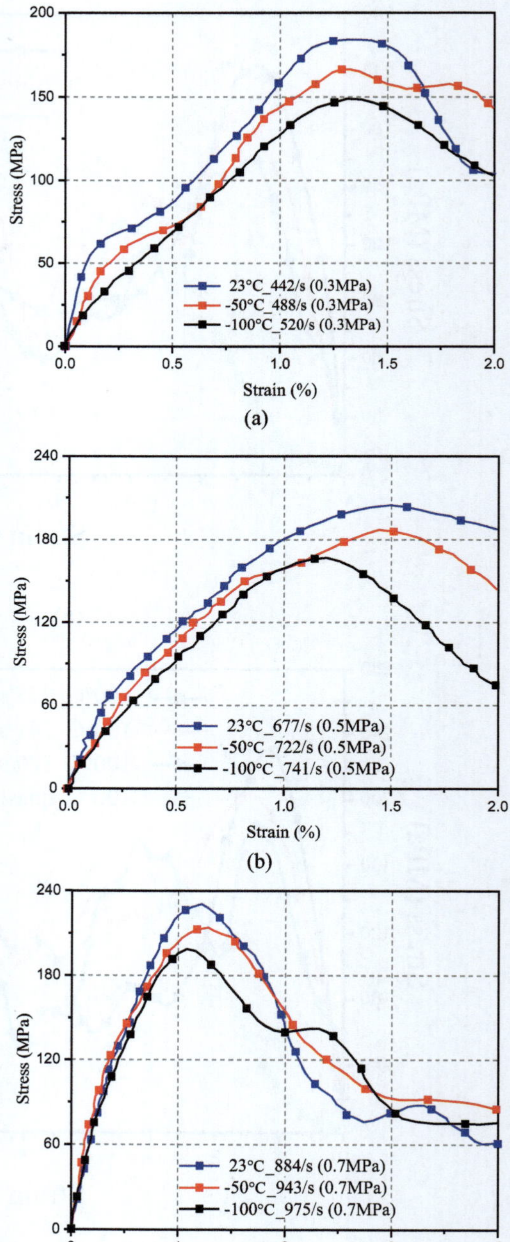

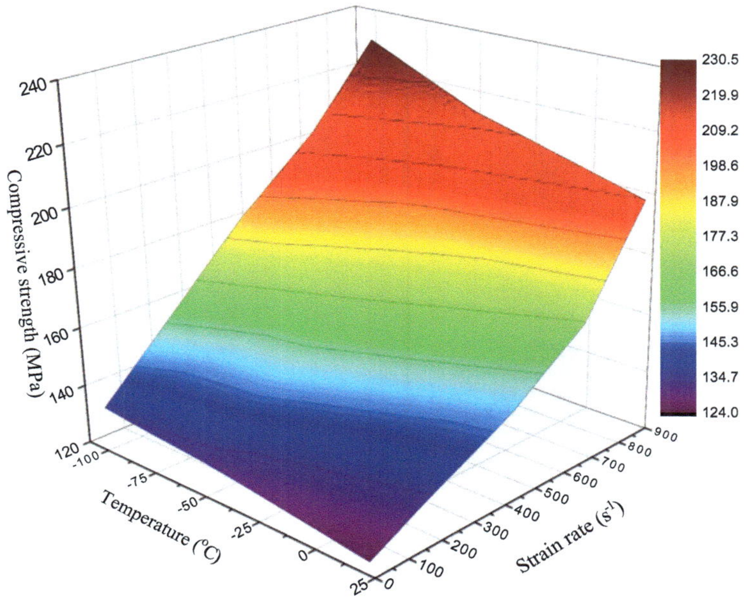

Fig. 7.7 Strain rate effect and temperature effect on the compressive strength of 3-D braided composite tube

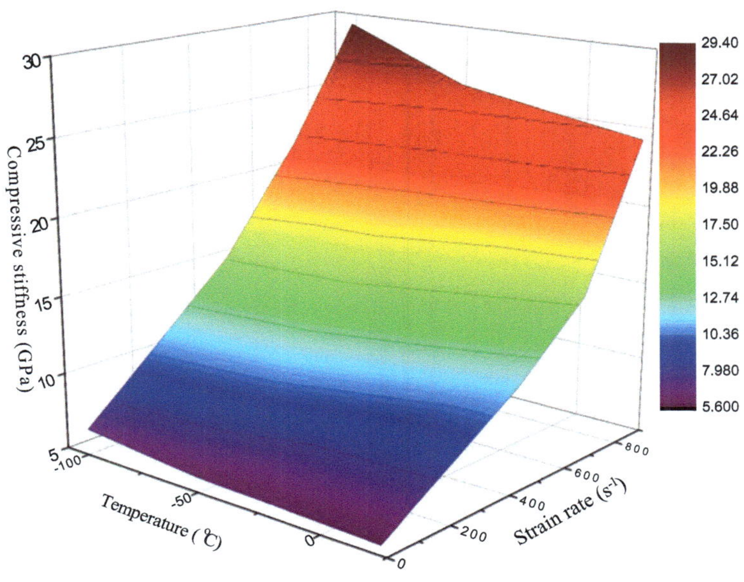

Fig. 7.8 Strain rate effect and temperature effect on the compressive stiffness of 3-D braided composite tube

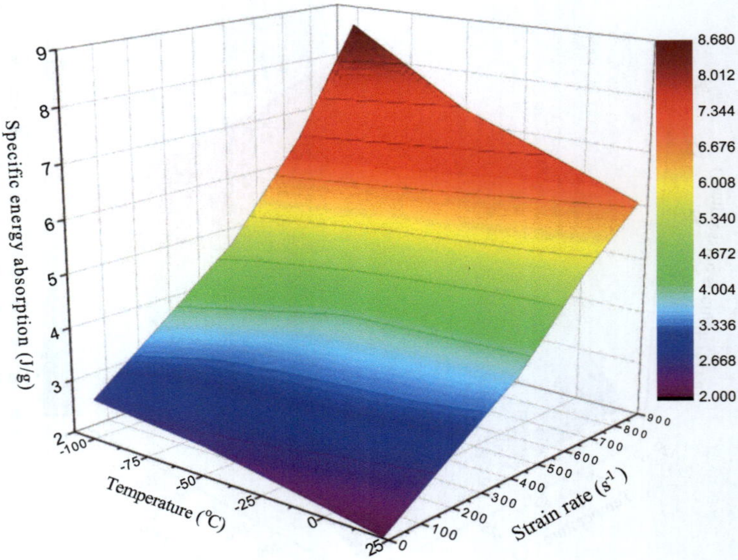

Fig. 7.9 Strain rate effect and temperature effect on the specific energy absorption of 3-D braided composite tube

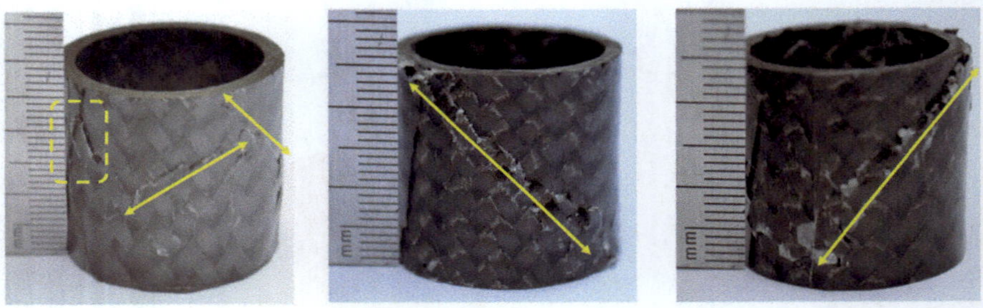

Fig. 7.10 Damage morphologies of 3-D braided composite tubes under quasi-static compression at different temperatures

with local shear failure. The significant shear failure of the tube occurs along the braided yarn direction at −50 and −100 °C. The lower the temperature is, the more serious the shear failure is. At room temperature and low temperature, the higher the strain rate is, the more serious the failure mode is. The reason is that the greater the impact pressure is, the higher the strain rate is, and the larger the impact load is.

At 0.3 MPa, the specimens were locally damaged at room temperature. Fiber breakage and resin cracking were the main failure modes, and the surface of the damaged specimens was rough. As the temperature decreased, the area of resin cracking and peeling off on the surface of the specimen increased significantly. The

Fig. 7.11 Damage morphologies of 3-D braided composite tubes under high strain rate compression at different temperatures

surface of the damaged area became smooth, which is a typical brittle failure behavior. No significant fiber breakage was observed at −50 and −100 °C.

At 0.5 MPa, the failure modes were fiber pullout, fiber breakage, and resin cracking at room temperature. At −50 °C, the resin was peeling off over a large area. The fiber tows became loose and gradually expanded outward. The specimen expanded near the loaded position and exhibited a buckling failure mode. More pieces of resin were peeling off from the surface of the specimen than at room temperature. At −100 °C, the buckling damage of the specimen was more significant, and the area of resin peeling off was larger.

At 0.7 MPa, the failure mode was crushing damage, with the specimens almost crushed into pieces at room temperature and low temperature. However, the damage degree of specimens at three temperatures was 23 < −50 < −100 °C. The reason is that as the temperature decreases, the resin becomes brittle, which increases the interfacial bond strength between the fibers and the resin. In addition, the sliding friction coefficient and shear strength are also enhanced due to the shrinkage of the resin in low temperature. Therefore, it can be found that the fiber tows are almost completely debonded from the resin at room temperature. Except for a few fragments, the remaining fiber tows are single broken. At −50 °C, the specimens split into blocks except for a few single fibers debonding. At −100 °C, there were still a few fiber tows bonded to the resin at the bottom of the specimen, maintaining good completeness.

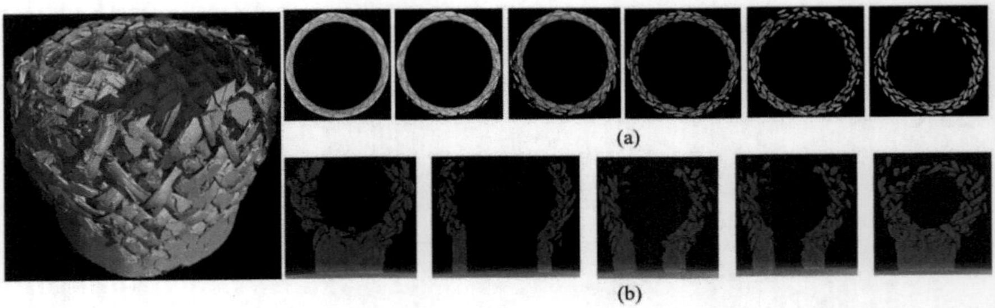

Fig. 7.12 μ-CT results of the damage morphology of 3-D braided composite tube along different orientations at −50 °C. (a) Length direction_Z-axis, (b) thickness directional_X-axis

7.5.2 μ-CT Observations

The damage morphologies of 3-D braided composite tube were scanned by X-ray CT after the tests in −50 °C and 722/s, and the scanning results along the length direction of Z-axis and thickness direction of X-axis are shown in Fig. 7.12. It can be seen from Fig. 7.12a that there are cracks in the interface between the braided yarns and the resin inside the specimen and the resin block peels off. The closer to the loading position of the specimen, the more serious the debonding phenomenon between the braided yarns and the resin. From Fig. 7.12b, we can clearly see the bending deformation of braided structure caused by resin cracking and peeling off.

Reference

1. Xianyan Wu, Qian Zhang, Bohong Gu, Baozhong Sun: Influence of temperature and strain rate on the longitudinal compressive crashworthiness of 3-D braided composite tubes and finite element analysis. Int. J. Damage Mech. **25**(4), 574–607 (2016)

Chapter 8
Multiscale Structure Mechanisms on Transverse Impact Damages in Beams

This chapter discusses the effects of braiding parameters and loading rates on the damages in 3-D braided composite beams under multiple transverse impacts. The influence of braiding structure and loading rate on the internal damage distribution will be analyzed from μ-CT observations and FEA. The energy absorption and damage mechanism of 3-D braided composites under multiple transverse impacts will also be revealed.

8.1 Braiding Angles

8.1.1 Braided Beams

Figures 8.1 and 8.2 show the transverse impact deformation of braided beam at 0.6 MPa. The spatial distribution of the yarns varies depending on the braiding angle. As the braiding angle increases, the fiber volume fraction, the interlacing points of yarns per unit length, and the thickness of the composite increase [1].

For the composite beam with the braiding angle of 20°, the mechanical properties are weakest along the impact direction compared to composites with large braiding angles. The stresses propagate rapidly along the impact direction, causing the greatest transverse deformation of the specimen. The composite beam fails completely and transverse fracture occurs at the impact point. The impact surface is subjected to compressive stresses, whereas the back is subjected to tensile stresses. In the first impact cycle, the resin on the impact surface is subjected to the impact load and cracks immediately thereafter. As shown in Fig. 4.12, the energy absorption of the first impact cycle is the largest as a percentage of the total energy absorption. Then the stress propagates diffusely from the impact point, with a greater propagation velocity along the impact direction than along the longitudinal direction. From the FEA results, it can be seen that the yarns near the back of the specimen start to break in the second impact cycle. There is no stress concentration in the damage

M. Hu, B. Gu, *Impact Damages of Braided Composites*, Engineering Materials,
https://doi.org/10.1007/978-981-16-5703-0_8

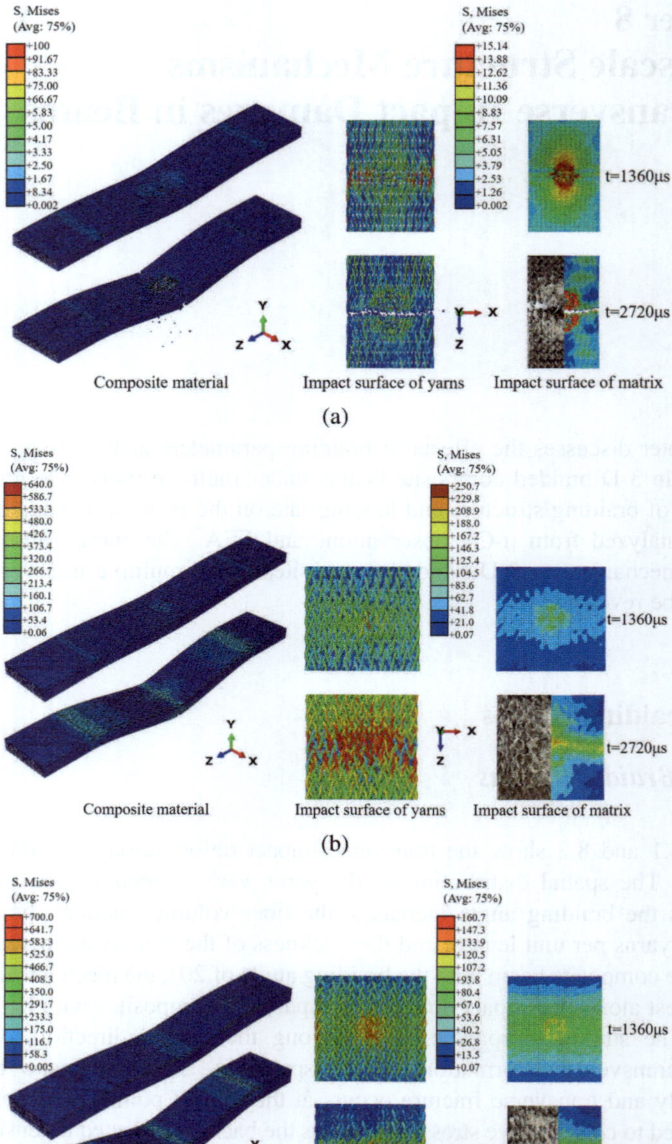

Fig. 8.1 Damage propagation of 3-D four-directional braided composites for the first four impact cycles at 0.6 MPa. (**a**) 20°, (**b**) 30°, (**c**) 40°

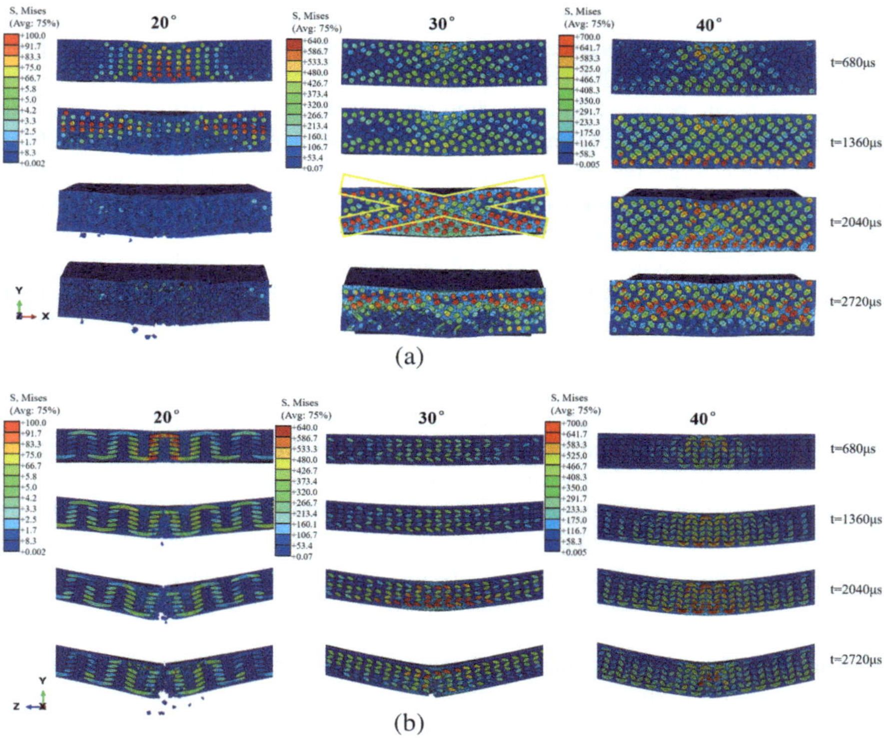

Fig. 8.2 Damage propagation of the sections of 3-D four-directional braided composites for the first four impact cycles at 0.6 MPa. (**a**) Transverse section, (**b**) longitudinal section

area. In the third impact cycle, the resin fractures completely, which can also be observed in the high-speed photographs, while the yarn continues to fracture. In the fourth impact cycle, the entire composite fails completely. As can be seen in Fig. 4.12, the energy absorption gradually decreases as the impact cycle continues. The accumulation of total energy absorption leads to fracture of the specimen.

For the composite beam with the braiding angle of 30°, the stiffness increases, and the deformation resistance of the beam increases relatively. The yarn does not fracture until the fourth impact cycle. The stress of the preform is concentrated near the impact point in the first impact cycle and gradually propagates to the entire transverse surface yarns of the beam as the number of impact cycles increases. The stress distribution in the cross section is X-shaped, and the tensile stress and compressive stress in the middle part of the resin and yarns are relatively balanced, with lower stress and less damage. When the back of the specimen fractures, the impact surface is still in high stress concentration, and the stress distribution in the cross section shows that the braided yarns near the impact surface are under greater stress. The relative increase of the compactness of the braided structure results in an increase in the impact resistance of the composite beam. The total energy absorption is higher than that of the 20° composite beam.

The composite beam with the braiding angle of 40° has the greatest resistance to deformation. From Figs. 8.1c and 8.2c, it can be seen that the maximum tensile stress concentration near the back of the composite beam leads to resin cracking and some braided yarns fracture. Compared to the composite beam with the small braiding angle, the braiding angle increases, the number of yarn interlacing points and the thickness of the specimen increase, the stiffness of the composite increases, the transverse resistance to stress propagation increases, and the resistance to damage increases. When the composite beam fractured, the stress is mainly concentrated near the damage site in the impact direction. This is different from the composite beam with the braiding angle of 30°, where the structure difference causes the stress wave to propagate at various velocities in different directions. The stress propagates rapidly along the impact direction, and the stress on the surface of the composite beam with the braiding angle of 30° propagates along the longitudinal direction when the specimen fractures. However, for the composite beam with the braiding angle of 40°, the stress wave propagation is blocked along the impact direction, and the stress propagation is attenuated. The results show that the propagation velocity of the stress wave in the composite beam decreases along the impact direction, when the longitudinal propagation of the surface stress is dissipated. The final fracture morphology of the 40° specimen is in bending, and the final fracture morphology of the 20° specimen is almost straight, which also indicates the increased toughness of the specimen with large braiding angle. The main failure mode observed from the macroscopic images includes interfacial debonding, resin cracking on the back, and small portion of fiber breakage. It is clear from Fig. 8.2 that the stress on the yarn is greater than the stress on the resin, so that the braided preform is the main load-bearing component, absorbing most of the impact energy.

From Fig. 8.3, we have chosen the yarns located in the midplane and the surface yarns of the composite beam for a special case to illustrate the damage process of the yarns under multiple impacts. The figure shows the stress distributions of the composite beam with the braiding angle of 20° are extracted every 20 μs during the first impact cycle at 0.6 MPa. From the surface yarns, it can be clearly seen that the stress propagates from the central yarn to both sides, that is, the maximum stress arises from the impact point and gradually propagates to the periphery. The three yarns in the middle break first, but the break is not at the impact point. This is due to the effect of the preform structure, which makes the stress propagate along the direction of the least resistance, and the weak part of the yarn breaks first. The double action of impact force and deformation is caused by the impact load, and the loose braiding preform structure is not enough to resist impact damage; the yarn breaks quickly after resin cracking. From the yarns of midplane, we find that the top surface yarn is first subjected to the impact load and the stress propagates gradually between the different yarns along the impact direction. The yarns break in many places in the direction of impact, mainly concentrated on the impact surface and the back surface. The damage near the impact surface is more than that near the back

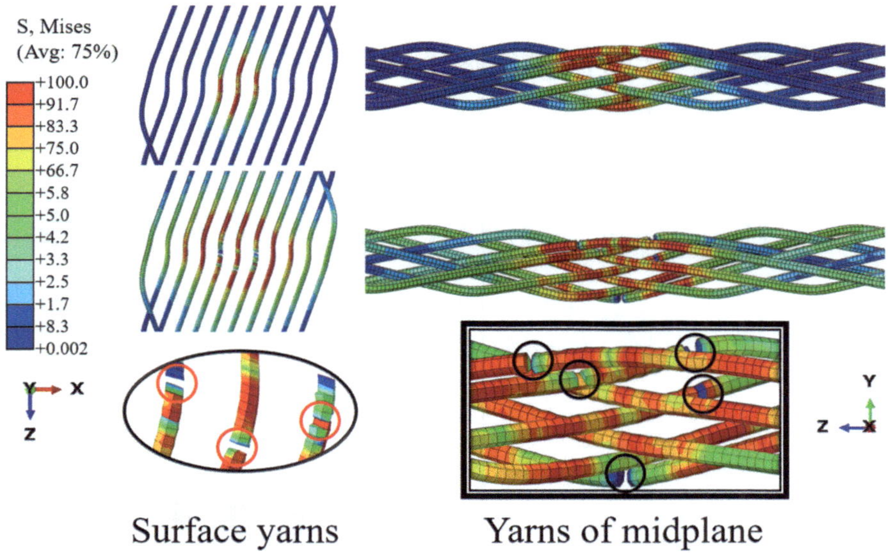

Fig. 8.3 Damage evolution of braided yarns of 3-D four-directional braided composite beam with the braiding angle of 20° at 0.6 MPa

surface. The closer to the impact area, the more severe the damage. This is mainly due to the fact that the direct propagation of stress wave in the composite beam is hindered by the yarn-yarn interface and yarn-resin interface. The propagation of stress waves in the same component is relatively easier than the propagation between different components. During propagation, the stress propagates along the path of least resistance, resulting in inconsistent fracture directions. It also shows that the fracture path is not a straight line, and in combination with Fig. 8.2, the cracking path of the specimen is also found to be curved.

Figures 8.4 and 8.5 show the comparison of μ-CT images of cross sections of the damaged region of the composite beam after impact loading and the finite element results with thermo-mechanical coupling.

The braiding angle has a significant effect on the failure mode and crack propagation path. As the braiding angle increases, the yarn interlacing points per unit volume of the preform increase, and the structure becomes more compact. The comparison between the cross and longitudinal sections through the impact point clearly shows that the 30° specimen is more deformed and more severely damaged compared to the 40° specimen. In the longitudinal section, fiber tow splitting is observed near the impact surface, and the crack extends along the gaps between the fiber tows in the impact direction (y-direction), forming a Y-shaped damage notch. In the longitudinal sections at 25% and 75% away from the impact point as observed in Fig. 8.5a, fiber tow splitting changes to fiber tow debonding, partial fibers of the surface fiber tow break, and structural integrity increases. The stress difference between the braiding preform and resin is relatively small, indicating that both the

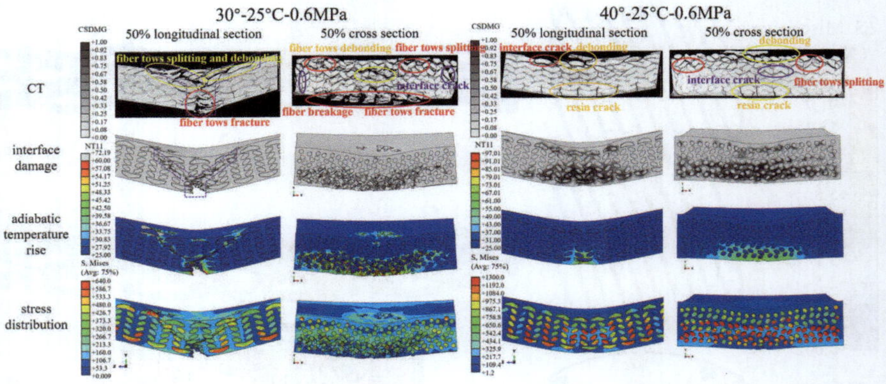

Fig. 8.4 μ-CT images of the cross and longitudinal sections of 3-D four-directional braided composites with the braiding angles of 30° and 40° at 0.6 MPa corresponding to the distributions of interfacial damage, adiabatic temperature rise, and stress distribution

preform and epoxy resin of the 30° specimen play an important role in the load bearing. The μ-CT images of cross sections show the splitting and debonding between fiber tows and resin cracking almost throughout the impact direction and that the fiber tows far away from the impact point gradually change from splitting to debonding. Part of fibers on the surface breaks. The stress wave propagates along different directions, leading to different failure modes. It can be seen from the stress distributions that the load-bearing capacity of the preform and the resin is significantly reduced after the interface failure and the difference of stress distribution is relatively reduced. The temperature rise of resin is mainly concentrated at the crack failure. Most of the plastic work is converted into heat energy when the composite is cracked. The maximum temperature can reach 72 °C, which is less than the glass transition temperature of resin [2] and has no significant effect on the mechanical properties of the composite. Part of the impact energy is consumed in the form of the deformation and damage of the braided composites, and the other part is consumed in the form of heat energy.

Since there is no heat exchange in the adiabatic shear state, the local temperature rise of the composite is significant. The temperature of the epoxy resin increases, and the temperature distribution and stress distribution interact to form a coupling effect. The meso-structure of the braided composites with the braiding angle of 40° is denser than that of the 30° composites, and relatively less fiber damage is observed in the 40° specimens. The μ-CT images show that the resin on the impact surface debonds or even peels off and the surface fiber tows split near the impact surface. The interfacial cracks are concentrated in the resin-rich region and propagate along the direction of least impedance. The resin on the back cracks along the impact direction due to tensile stress, and the compact braided structure hinders the continuous propagation of cracks. From Fig. 8.5b, it is observed that there are still fiber tows splitting and resin peeling off near the impact surface at the longitudinal sections at 25% and 75% away from the impact point. The damage on the back is significantly reduced, and the depth of resin cracking is also reduced. The crack

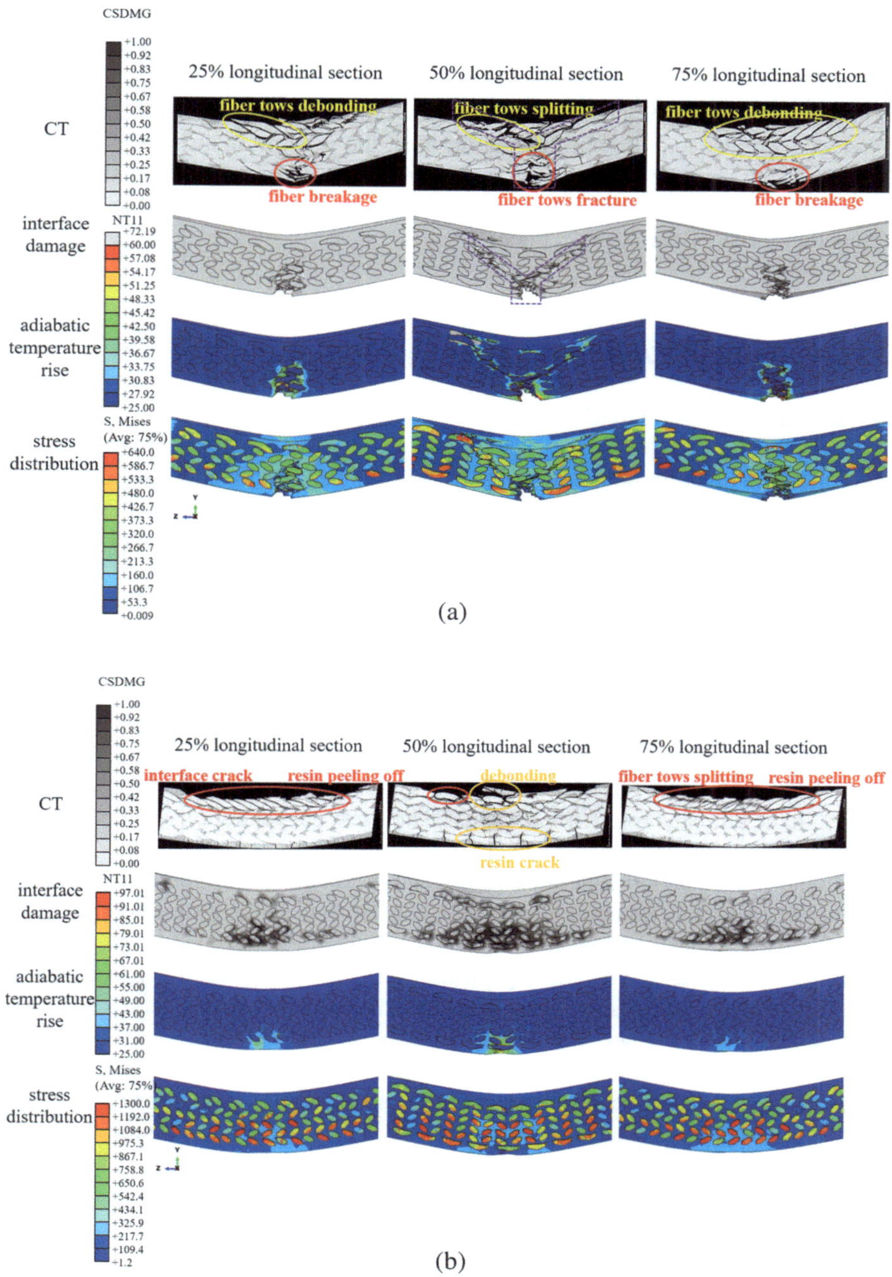

Fig. 8.5 μ-CT images of the longitudinal sections of 3-D four-directional braided composites at 0.6 MPa corresponding to the distributions of interfacial damage, adiabatic temperature rise, and stress distribution. (**a**) 30°, (**b**) 40°

extension to the yarn on the back surface is hindered and stops further expansion. The stress of yarn near the back is greater than those near the impact surface, and the interfacial damage caused by tensile stresses is more severe than that caused by compressive stresses. When the braiding angle increases, the stress of the preform is obviously greater than that of the matrix; thus, the preform mainly bears the load. The resin temperature rise is mainly concentrated on the back along the impact direction, with a maximum temperature of 97 °C, which is greater than the local temperature rise of the 30° specimen.

8.1.2 Braided Tubes

8.1.2.1 Impact Deformation History

The effect of braiding angle on the transverse impact deformation history of 3-D braided composite tubes is explored by using the impact pressure of 0.6 MPa as an example. Figures 8.6 and 8.7 show the deformation processes of 15° and 30° tubes during four impacts, respectively.

For the matrix, the stress is concentrated in the impact zone and the clamping edge. For the 15° tubes, there is an obvious elliptical shear band at the edge of the impact zone from the second impact cycle, and then the shear band extends outward, while in the third impact cycle, the 15° tube shows transverse fracture at the impact point. In contrast, for the 45° tube, there is no obvious damage, and the high stress region extends outward in the impact zone in the second impact cycle. The high stress level is also maintained in the middle region between the impact and clamping zones. In the third and fourth impact cycles, the stress concentration region continues to extend outward, and the average stress level gradually decreases [3].

For the braided preform, the stress propagates fastest in the 15° tube. The yarns from the impact zone to the clamping edge are subjected to high stress in the second and third impact cycles. In the third impact cycle, the yarns are brittle fractured in the middle of the impact zone and accompanied by shear damage at the edge of the impact zone. In the fourth impact cycle, the damage intensifies, the tube fails, and the stress decreases significantly. The yarns are most tightly arranged for the 45° tube, which has strong resistance to deformation and damage. In the fourth impact cycle, the tube still maintains a high stress level. In addition, the alternation of yarn stress in different rows in the impact zone becomes more obvious as the braiding angle increases.

8.1.2.2 Transverse Impact Stress Propagation and Distribution

As shown in Fig. 8.8, three nodes NA, NB, and NC are selected along the impact zone to the clamping zone on the preform and matrix, respectively, and the stress-time curves of nodes are extracted. Figure 8.9a1–a3 and b1–b3 show the stress-time curves of the corresponding three nodes on the preform and matrix, respectively. Node NA is in the center of the impacted region, and the stress of preform and matrix

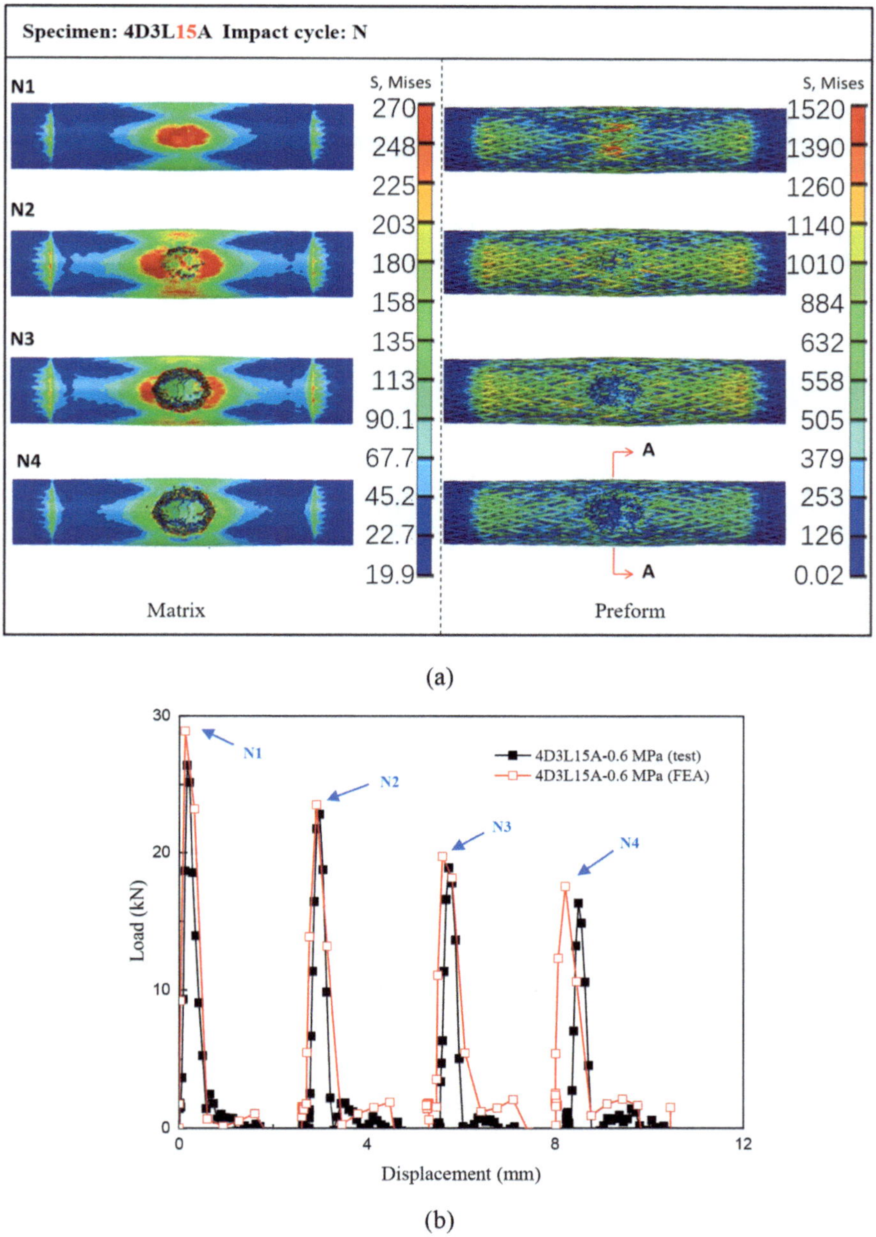

Fig. 8.6 Deformation history of 4D3L15A specimen at 0.6 MPa. (**a**) Deformation history of 15° specimens, (**b**) load-displacement curves corresponding to deformation history

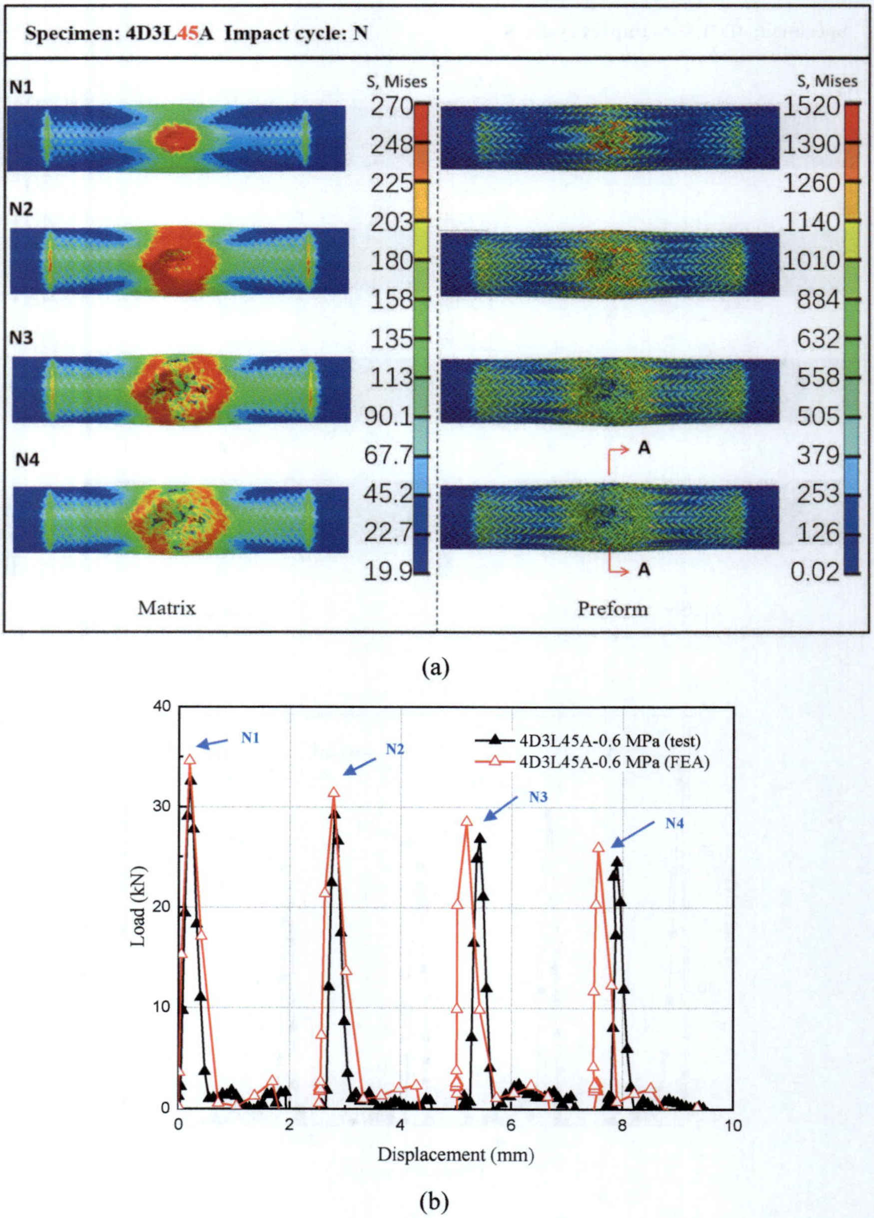

Fig. 8.7 Deformation history of 4D3L45A specimen at 0.6 MPa. (a) Deformation history of 45° specimen, (b) load-displacement curves corresponding to deformation history

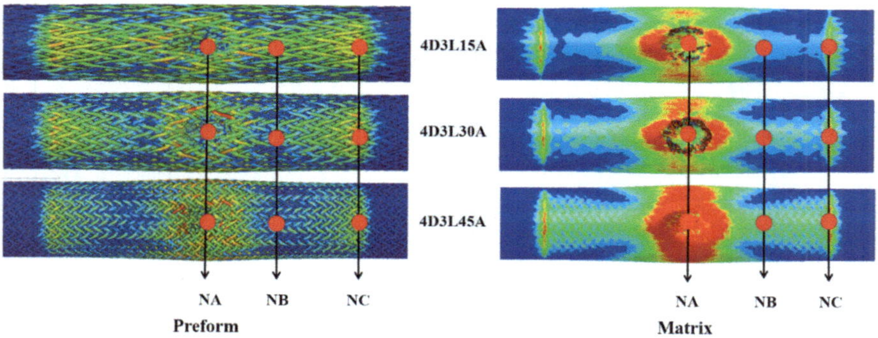

Fig. 8.8 Location of the three nodes NA, NB, and NC on the preforms with different braiding angles and matrices, respectively

with different braiding angles reaches the maximum until failure. Node NB is the middle point between the impact zone and the clamping zone. In the preform, the overall stress level of this node decreases as the braiding angle increases. This is due to the fact that as the braiding angle increases, the yarn orientation deviates from the axial direction on the one hand, and the density of yarn interlacing points increases along the braiding direction on the other hand, which are not conducive to stress propagation along the axial direction. During the propagation of the stress wave along the yarn, when it encounters the overlapping yarns or the matrix, part of the stress wave is reflected, and the compression wave is reflected by the free surface to form a tensile wave. Multiple incidences and reflections cause cracking at the interface between yarns and matrix. The propagation process of stress wave is continuously attenuated. The more interlacing points, the faster attenuation. At the fixed end face, the reflected wave is the positive image of the incident wave, and the stress is doubled. The braiding angle is an important parameter affecting the density of yarn interlacing points from the impact zone to the clamping zone. As the braiding angle increases, the greater the density of yarn interlacing points, the faster the stress wave attenuates. On the contrary, as the braiding angle increases, the nodal stress is enhanced in the matrix, and especially the 45° tube is enhanced significantly. The composite tubes with large braiding angle have better resistance to deformation and failure. The stress waves continue to propagate in the homogeneous matrix with enhanced stress before the impact zone fails completely, while the specimens with small braiding angle are damaged earlier and more severely in the impact zone.

8.1.2.3 Test Validation of the FEA Model

Figure 8.10 shows the impact damage morphologies of specimens with different braiding angles under three impact pressures, and the FEA results are in good agreement with the test results. The damage degree of specimens with different braiding angles increases with the increase of impact pressure. The damage morphology of the 15° specimen is mainly the fiber tow fracture at the impact point. Due

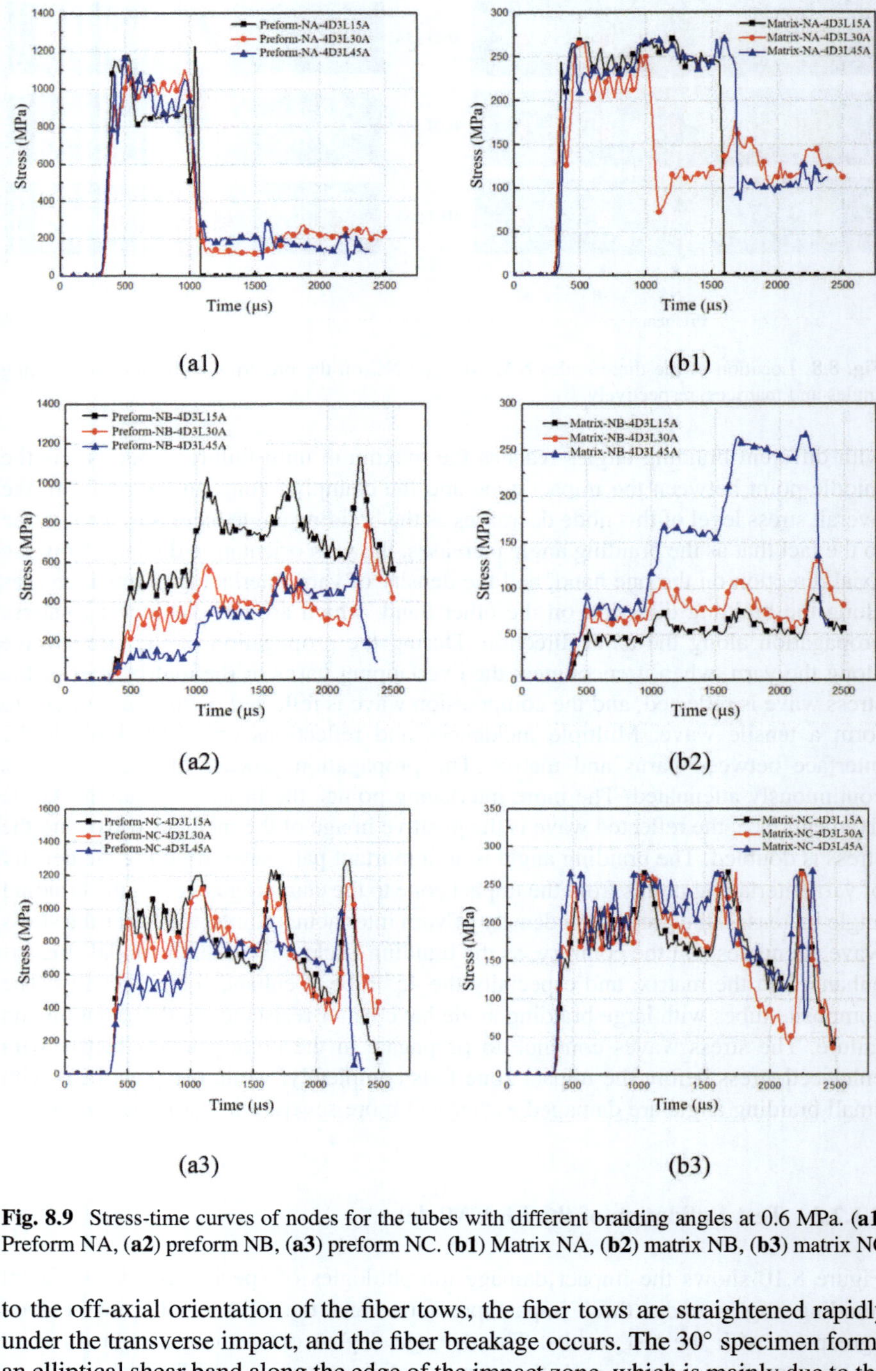

Fig. 8.9 Stress-time curves of nodes for the tubes with different braiding angles at 0.6 MPa. (**a1**) Preform NA, (**a2**) preform NB, (**a3**) preform NC. (**b1**) Matrix NA, (**b2**) matrix NB, (**b3**) matrix NC

to the off-axial orientation of the fiber tows, the fiber tows are straightened rapidly under the transverse impact, and the fiber breakage occurs. The 30° specimen forms an elliptical shear band along the edge of the impact zone, which is mainly due to the

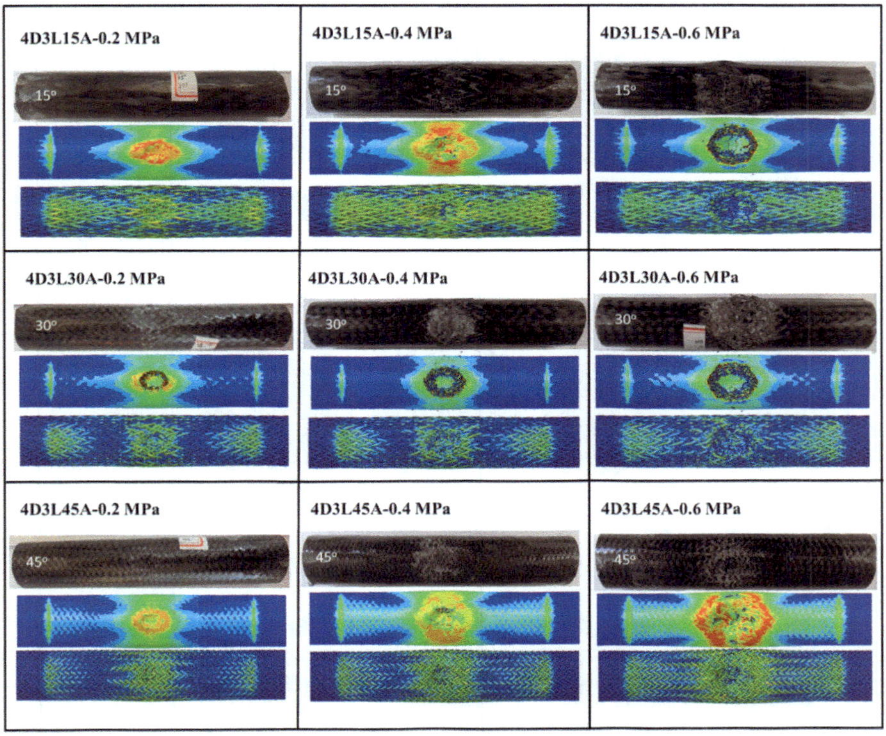

Fig. 8.10 Damage morphologies of composite tubes with different braiding angles under various impact pressures

local filling shear failure. As the braiding angle increases, the yarn deviates from the axial direction, and the density of the interlacing point increases. The stress is mainly transferred along the axial direction by the interlacing point. The deformation is blocked from both sides of the impact surface, forming a shear band at the edge of the impact zone. For the 45° specimen, the yarns are tightly arranged. The density of interlacing points is the largest. The integrity of the tube and the resistance to impact deformation and failure are the best. The main failure mode is the matrix fragmentation between the yarns and the interface cracking.

Figure 8.11 shows the cross sections of specimens with three braiding angles at different positions to compare the test results with the FEA results. For the 15° specimens, at positions a and b, shear damage has occurred at both ends except for the damage inside the central region. At positions c and d, severe damage has occurred at both the central regions and both ends of the section, while for the 30° and 45° structures, at positions a and b, the damage is concentrated in the central region with no obvious damage at both ends. The 30° specimen shows shear damage at both ends with less damage in the central region at positions c and d, while 45° specimen shows that the damage from the middle region extends to both ends with less overall damage at positions c and d.

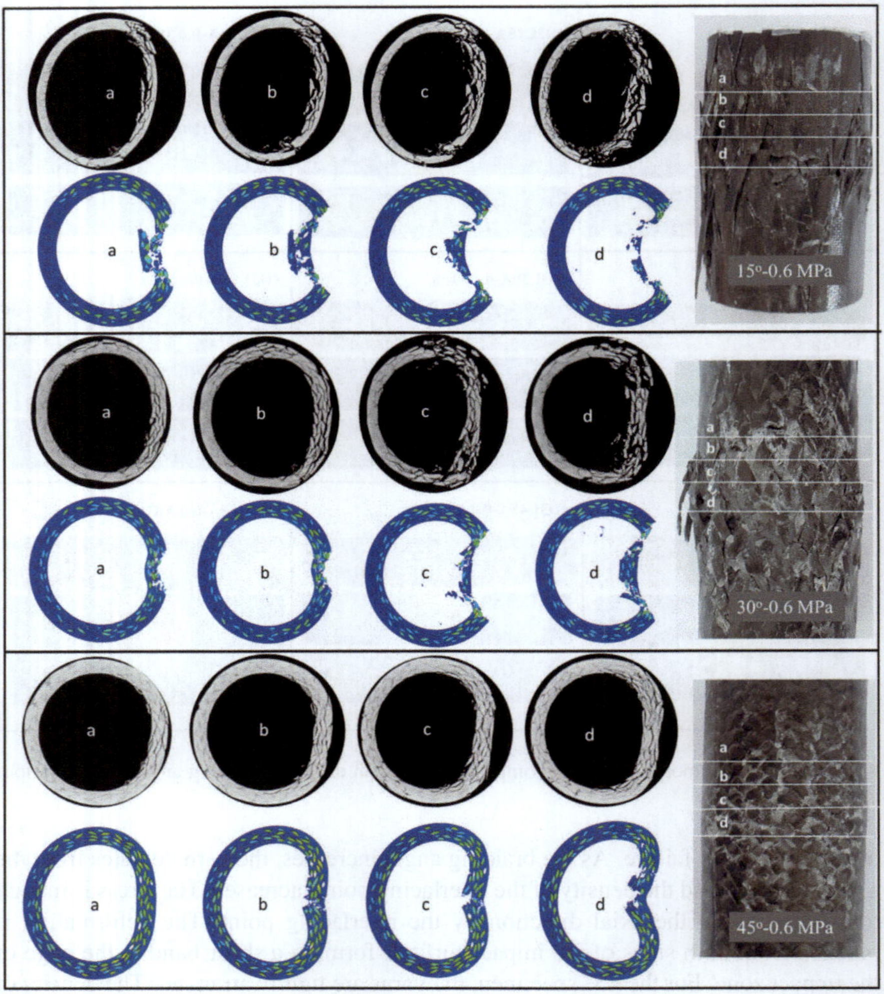

Fig. 8.11 Damage sections of composite tubes with different braiding angles at 0.6 MPa

8.2 Braided Layers

8.2.1 Impact Deformation History

As shown in Figs. 8.12 and 8.13, the effect of the number of braided layers on the transverse impact deformation history of 3-D braided composite tube is explored with the impact pressure of 0.6 MPa as an example. For the two-layer specimen, it can be seen from the second impact cycle that the yarn fractures in the middle of the impact zone. At this time, the stress of the impact surface matrix concentrates and

Fig. 8.12 Deformation history of 4D2L30A specimen at 0.6 MPa. (**a**) Deformation history of two-layer specimen, (**b**) load-displacement curves corresponding to deformation history

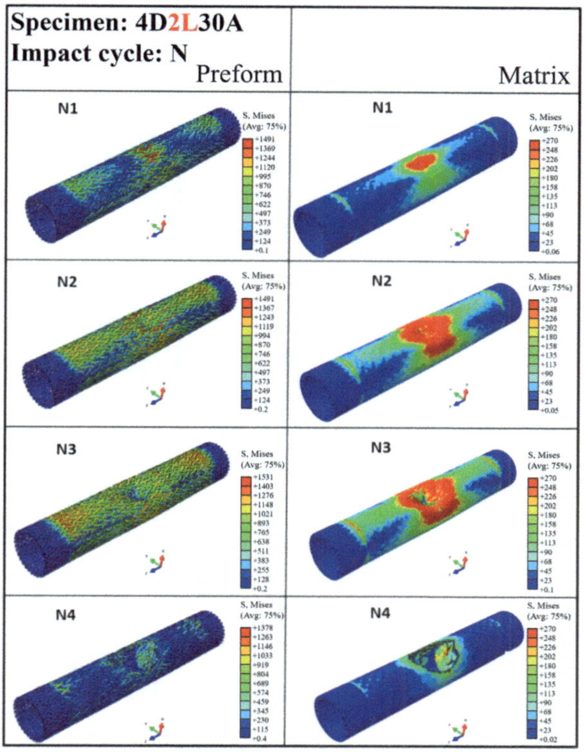

(a)

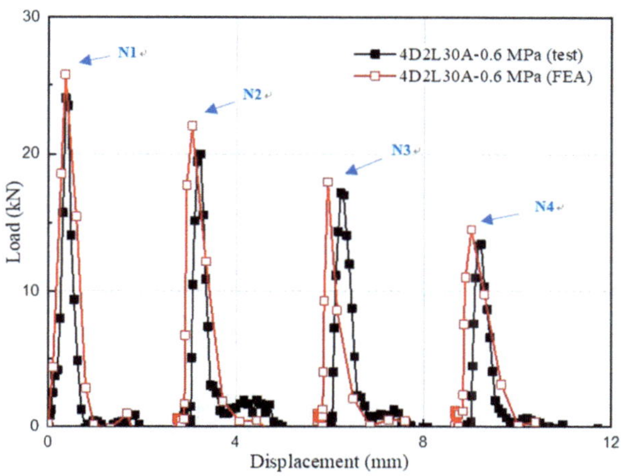

(b)

Fig. 8.13 Deformation history of 4D4L30A specimen at 0.6 MPa. (**a**) Deformation history of four-layer specimen, (**b**) load-displacement curves corresponding to deformation history

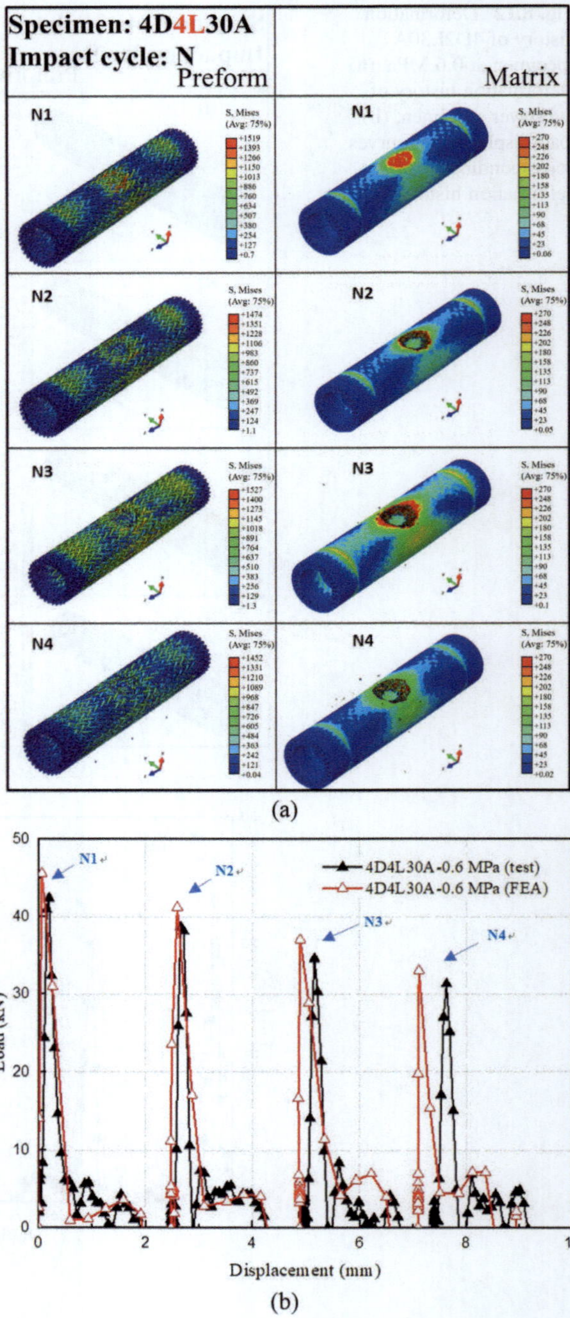

extends outward, and the flattening phenomenon of the section is obvious. In the third impact cycle, the damage of the preform and matrix in the impact zone expands rapidly from the center to the outside in a Y shape. When the fourth impact cycle is reached, the whole impact zone shears inward along the surrounding region, and the tube fails completely. The damage mechanism four-layer specimens are similar, mainly in the shear failure around the impact zone. In the fourth impact cycle, the four-layer specimen still maintains a higher stress state compared to the two-layer specimen, indicating that the four-layer braided structure has strong resistance to failure [4].

8.2.2 *Propagation and Distribution of Transverse Impact Stress*

As shown in Fig. 8.14, the preform of 3-D braided composite tube is unfolded along the back side, and the inner and outer surfaces are displayed, so that the stress distribution on the inner and outer surfaces of the entire tube can be observed. Firstly, 4D4L30A specimen (four-layer) is taken as an example to analyze the distribution and the change process of tensile and compressive stress on the preform of the tube at 0.6 MPa. In the S11 stress distribution, the red color represents the

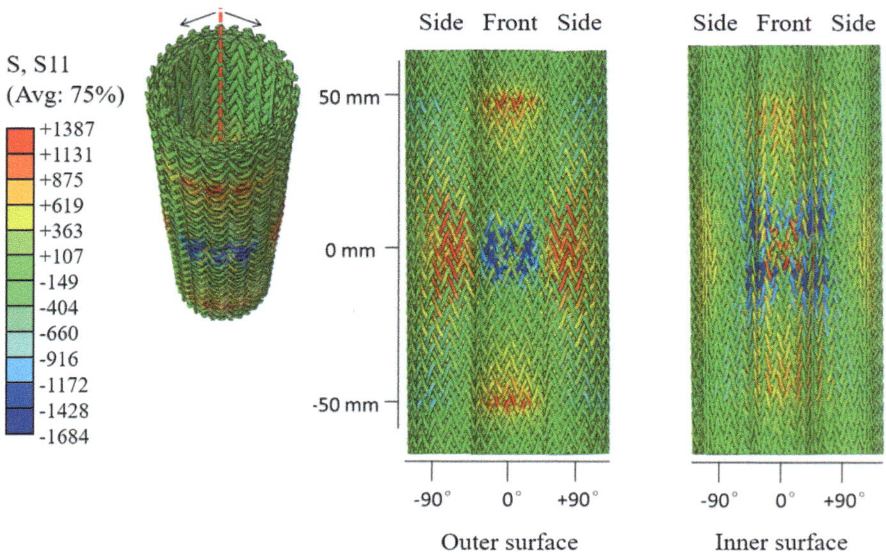

Fig. 8.14 Expanded view of the preform of 4D4L30A specimen

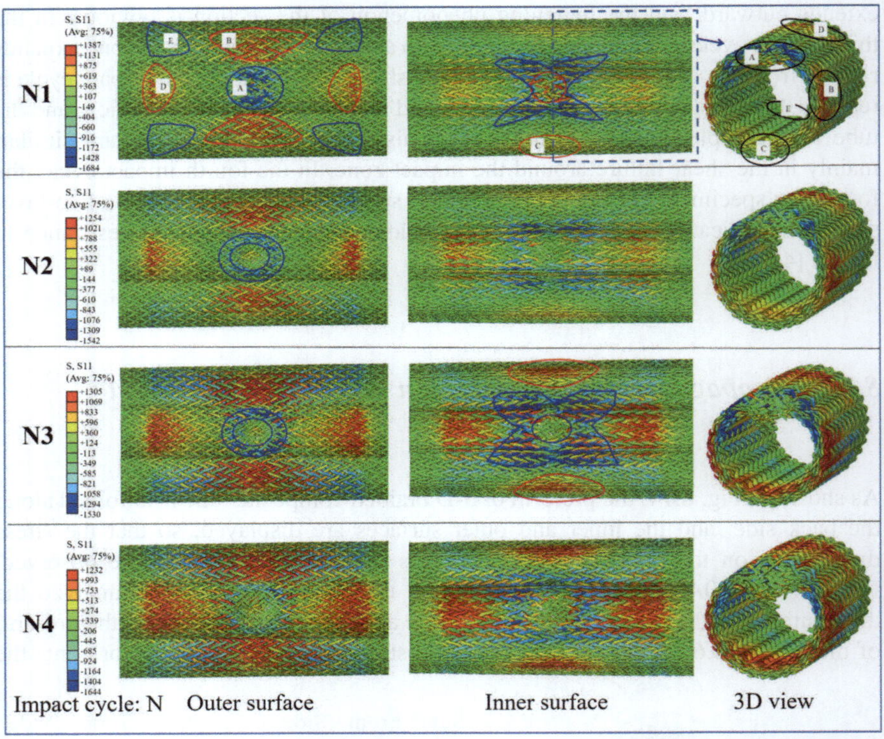

Fig. 8.15 Stress distribution in the preform of 4D4L30A specimen during the impact process at 0.6 MPa

tensile state, and the blue color represents the compressive state. It can be seen from the expanded view of the specimen that the tensile and compressive stresses in different regions are different. The following section will discuss the changes of tensile and compressive states in different regions with the increase of impact cycles.

As shown in Fig. 8.15, the distribution of tensile and compressive stress in the preform of four-layer specimen is mainly concentrated in five regions. As shown by the 3-D view, zone A is the impact contact zone, zone B is the side of the impact zone, zone C is the back of the impact zone, and zones D and E are the upper and lower parts of the clamping zone, respectively. In the first impact cycle, the outer surface of zone A is compressed, and the center of the inner surface is in tension. The yarns around it connected with the outer surface yarns are compressed, so the cross section is flattened. The outer surface of zone B and the inner surface of zone C are in tension, and zones D and E are subject to tensile and compressive stress due to overall bending deformation of the tube. In the second impact cycle, the stress in the center of zone A fades due to yarn fracture. The yarns are subjected to compressive stress at the periphery of the impact zone, and the compressive stress here continues to increase in the third impact cycle and fades in the fourth impact cycle as more

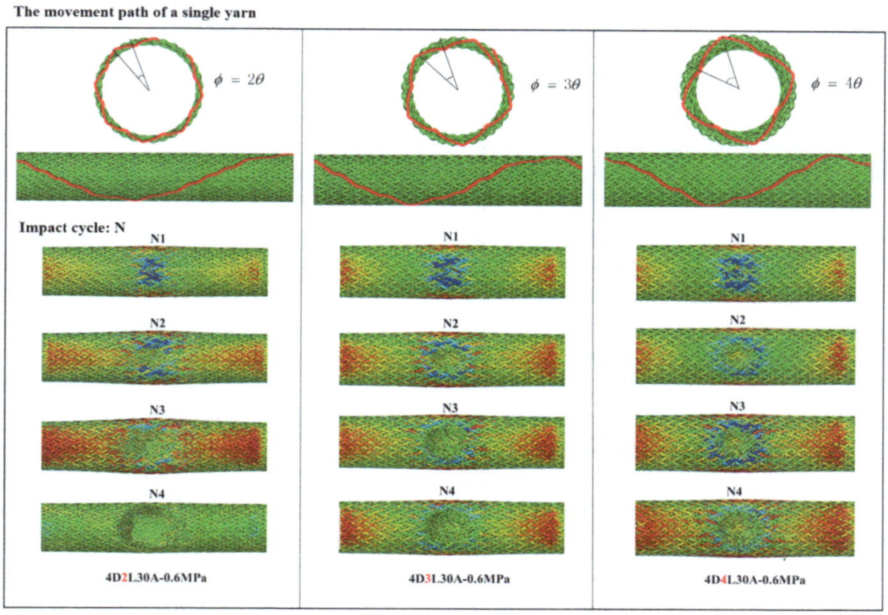

Fig. 8.16 Effect of different braided layers on the distribution of tensile and compressive stress in the tubes

yarns fracture. The tensile and compressive stresses of other regions are enhanced with continued deformation of the tube and extend to the periphery.

The main distribution regions of the tensile and compressive stress of specimens with different braided layers during impact are consistent, as shown in Fig. 8.16. The movement path of a single yarn in the tube is alternating in and out of the radial direction and spiraling around the tube wall, and the angle experienced by the yarn alternating once inside and outside the tube wall is $\varphi = m\theta$ ($\theta = 2\pi/n$, m is the number of braided layers, and n is the number of columns). With the increase of the number of braided layers, the alternation angle of yarn increases, and the number of yarn interlacing points in the direction of wall thickness increases.

In the two-layer specimen, the alternation angle of yarn inside and outside is the smallest, which is conducive to the axial propagation of the stress wave. Because the tube wall is thin, the propagation time of stress wave from the outer surface to the inner surface is short, and the number of reflection in the thickness direction increases. In the third impact cycle, compared with the other two structures, the two-layer specimen is subjected to the most obvious tensile stress from the impact zone to the clamping zone. The yarns are broken in the center of the impact surface, followed by inward fracture of the entire impact surface along the periphery and complete failure of the specimen, as shown in the fourth impact cycle. The three-layer and four-layer specimens have thick walls, and the number of interlacing points between the yarns increases, and the bending resistance and local impact resistance are improved. Shear damage around the impact zone is dominant. From the third impact cycle to the fourth impact cycle, the three-layer specimen gradually fails due to the gradual failure of the impact zone, and the tensile stress of yarns on both sides starts to decrease toward the clamping zone, while the four-layer specimen still has strong load-bearing capacity, and the tensile stress continued to extend and strengthen from the clamping zone toward the impact zone.

The damage process of representative yarns in the thickness direction of the impact zone and the stress-time curves of each yarn center node are extracted to further discuss the damage mechanism in the impact zone. Figure 8.17 show the change of tensile and compressive stress with time for the selected yarns of the two-layer specimens, with the nodes at the center of the outer surface yarn, the interior yarn, and the inner surface yarn, respectively.

As shown in Fig. 8.17, four representative yarns are selected in the impact thickness direction of the two-layer specimen. From the stress-time curve, it can be seen that when $t = 300$ μs, the stress wave propagates from the incident bar to the specimen, and the nodal stress rises rapidly, where ① and ② yarns are compressed and ③ and ④ yarns are pulled. The stress of the surface yarn is greater than that of the interior yarn. When $t = 500$ μs, most of the yarns break at the center so that the nodal stress decreases. Then, $t = 1100$ μs, the interior yarn fails completely, and the yarns on both sides are mainly fractured by inward shear damage with neat fracture.

8.2.3 *Test Validation of the Damage Mode*

As shown in Fig. 8.18, the FEA damage morphologies of 3-D braided composite tubes with three kinds of braided layers under different impact pressures are consistent with the test results. The damage of the tubes with different braided layers increases with the increase of impact pressure. The more the number of braided layers, the smaller the overall damage of the tube at the same impact pressure. The two-layer tube has no obvious impact depression under small impact pressure, while under large impact pressure, the impact zone fractures inward along the edge, showing obvious brittleness. The four-layer have similar damage mechanism, with shear damage along the edge of the impact zone.

As shown in Fig. 8.19, four CT scan sections are also selected along the axial direction in the impact zone of the tube and compared with the finite element results. In a and b sections, the three-layer and four-layer specimens are dominated by the delamination of the interface between the yarns in the middle region of the impact. While the two-layer specimen has less damage in the middle region, with shear damage on both sides of the b section and transverse cracking in the middle region. For c and d sections, the three-layer and four-layer specimens are dominated by shear damage on both sides of the impact zone. While in c section, the whole impact surface of the two-layer specimen fractures inward along both edges, and the internal damage is serious. The fibers of the impact surface in d section are completely broken. The finite element analysis results are in agreement with the test results.

Fig. 8.17 Stress-time curves and damage process of the nodes on the yarns in the thickness direction of 4D2L30A specimen

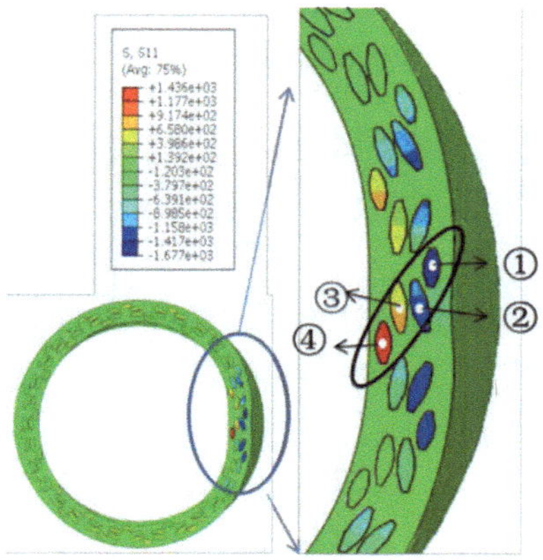

(a)

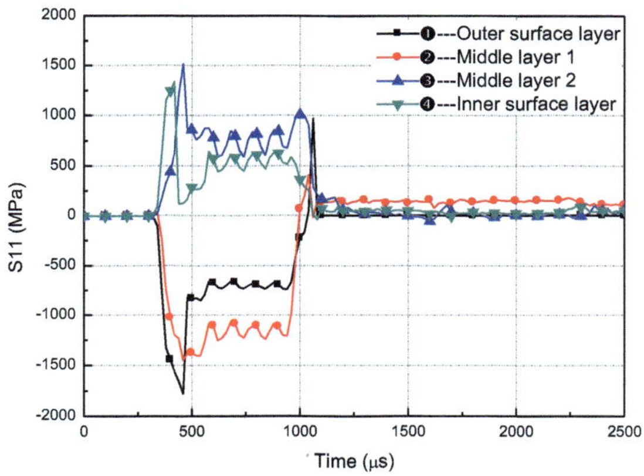

(b)

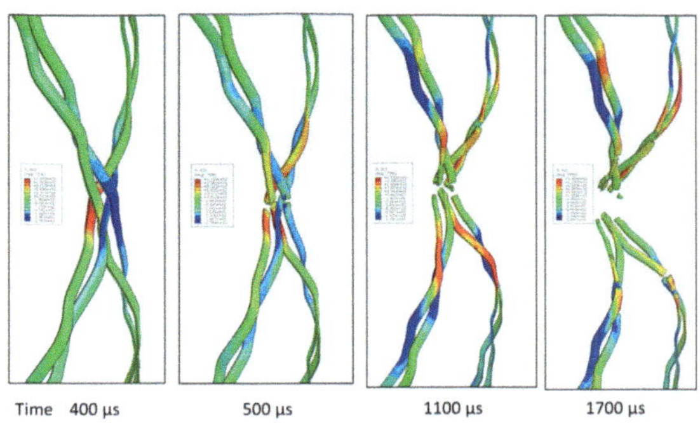

(c)

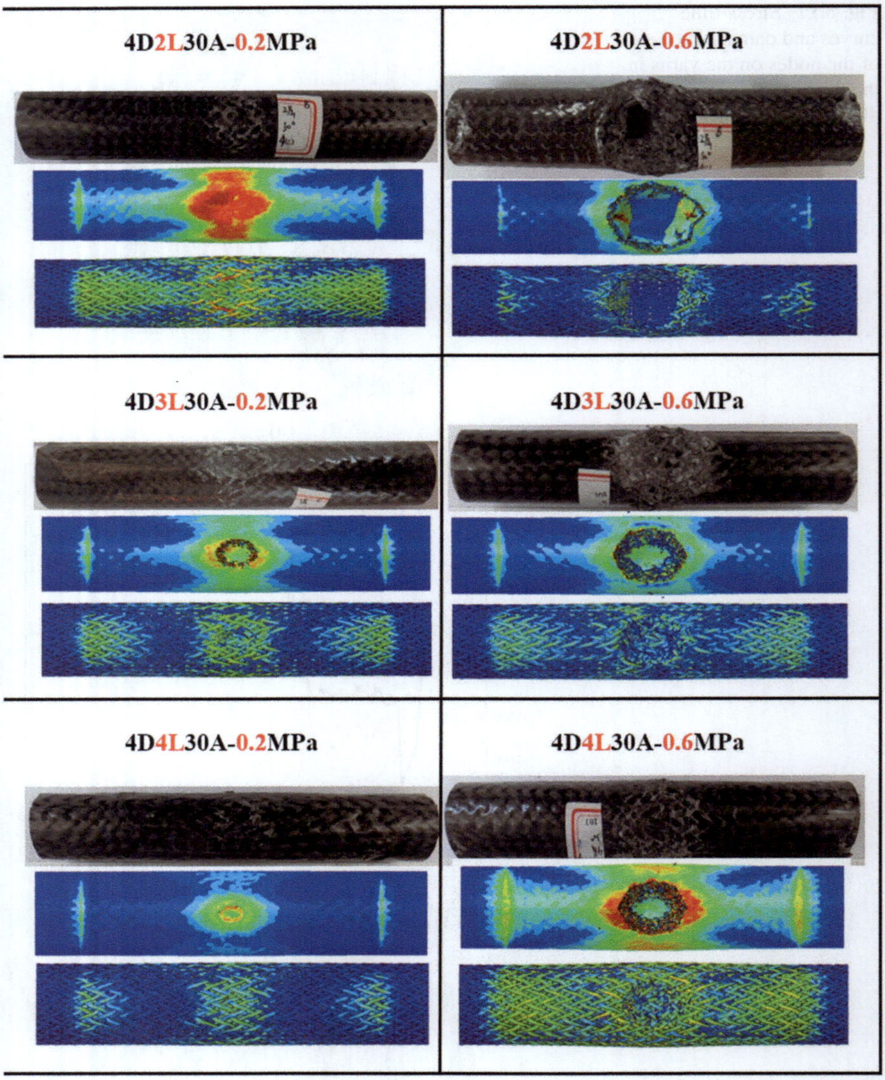

Fig. 8.18 Impact damage morphologies of 3-D braided composite tubes with different braided layers

8.3 Axial Yarns

8.3.1 Braided Beams

In order to explore the transverse impact fracture behavior of 3-D braided composite beams, the damage morphology and temperature history are discussed.

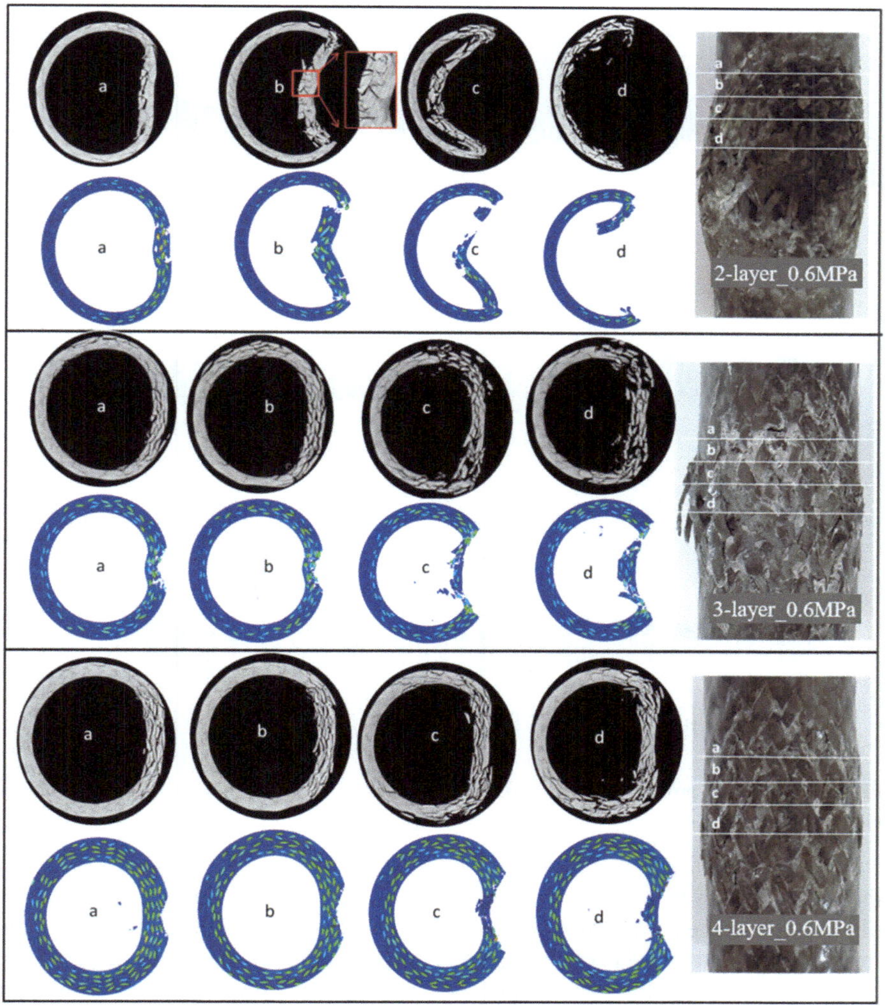

Fig. 8.19 Damage sections of composite tubes with different braided layers at 0.6 MPa

Progressive damage and adiabatic temperature rise in the longitudinal section of 3-D five-directional braided composite are depicted in Fig. 8.20. When $t = 2040$ μs, cracking damage occurs here with an X-shaped maximum stress distribution. The high stress concentration and the weak bonding of the interface between the yarn and resin accelerate the damage propagation in the local region, prompting further increase in temperature in the local region, higher than the other region. When $t = 2720$ μs, the resin cracks along the original stress concentration path in a semi-V shape. The axial yarns are twisted and deformed due to resistance to the composite beam fracture. At the same time, the local plastic strain increases significantly due to

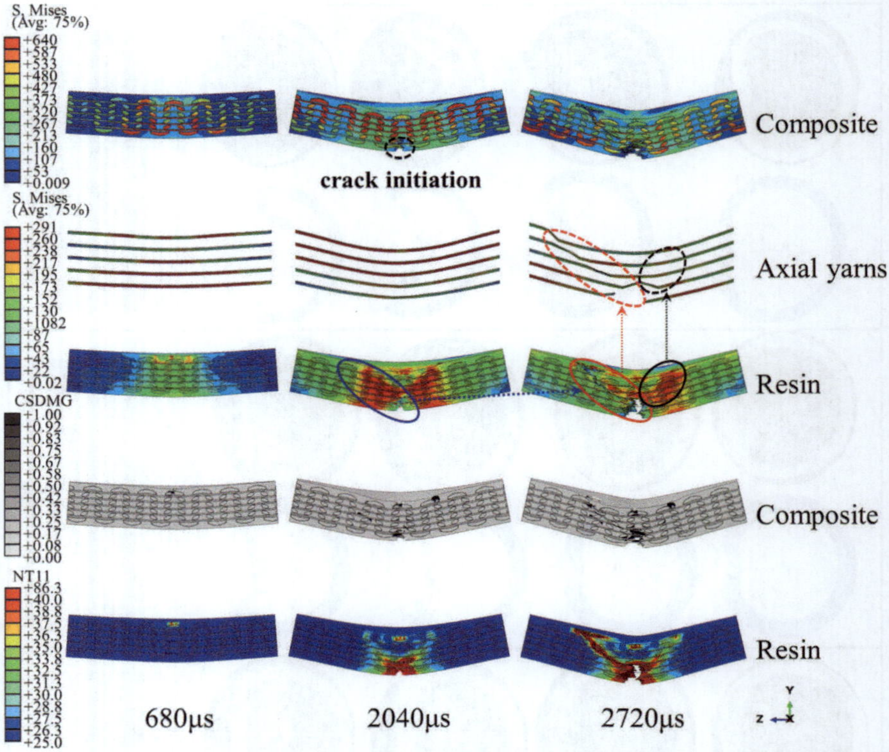

Fig. 8.20 Damage propagation and temperature evolution of 5D30 specimen for the first four impact cycles at 0.6 MPa

the weakening of material properties caused by damage and temperature rise. The plastic strain in the back, which is subjected to tensile stress, is greater than that in the front, which is subjected to compressive stress (see Fig. 7.2b). Thus, the plastic strain energy generated in the back is greater, which in turn translates into more heat energy. When the stress in the local region increases to a critical value, the resin cracks, and the fracture energy is further converted into heat energy, which continues to increase the temperature of resin. However, the contribution of the damage is much greater than that of thermal softening, because the highest temperature is about 63 °C, which is much lower than the glass transition temperature of the resin (T_g about 110 °C) [5].

Further insight into the transverse impact fracture behavior of the 5D30 braided composite can be gained by studying the stress transfer and distribution of specific yarns. We choose three yarns closest to the back impact direction, one axial yarn and two braided yarns. Figure 8.21 shows the stress distribution with time, extracting and comparing the stress along the yarns at different moments. When the stress wave first propagates to the specimen, the maximum stress of the three yarns is concentrated in

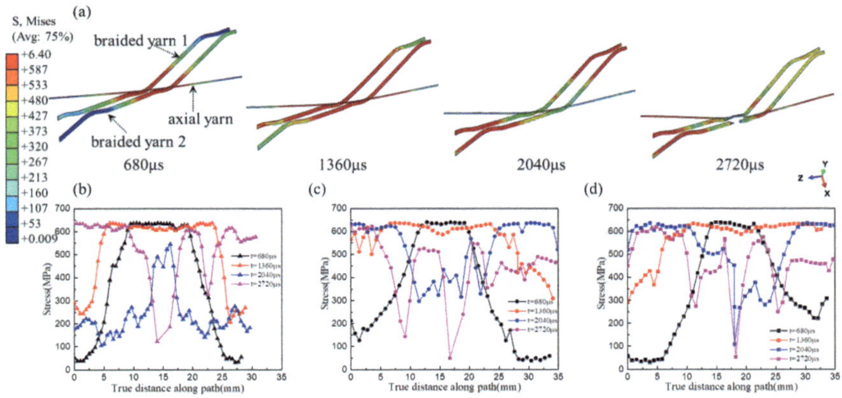

Fig. 8.21 Damage propagation for three specific yarns of 5D30 specimen. (**a**) Stress distribution, (**b**) stress variation along the axial yarn, (**c**) braided yarn 1, and (**d**) braided yarn 2 in different impact cycles

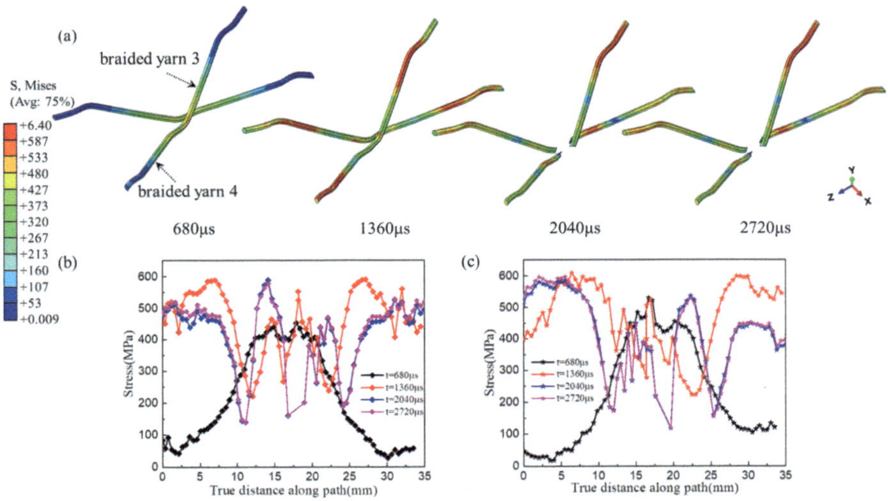

Fig. 8.22 Damage propagation for two specific yarns of 4D30 specimen. (**a**) Stress distribution, (**b**) stress variation along the braided yarn 3, and (**c**) braided yarn 4 in different impact cycles

the impact direction. As time progresses, the maximum stress gradually extends to the entire yarn. When it reaches the critical value, the yarn is damaged and broken. Compared to the 4D30 specimen in Fig. 8.22, the braided yarn of the 4D30 specimen breaks earlier, and the maximum stress is less than that of 5D30 specimen. This is due to that the axial yarn provides toughening effect, improves the stiffness and strength and delays the damage time. It can be concluded that the 3-D braided composite beam with the addition of axial yarn increases the impact damage tolerance and crack extension impedance.

8.3.2 *Braided Tubes*

8.3.2.1 Impact Deformation History

The transverse impact deformation histories of 3-D four- and five-directional braided composite tubes are obtained from the FEA results, and the deformation and damage of the preforms and the matrix of two kinds of structures for the first four impact cycles are discussed separately with the impact pressure of 0.6 MPa as an example.

Figure 8.23 shows the transverse impact deformation history and the corresponding load-displacement curves for 3-D four- and five-directional braided composite tubes. The overall damage mechanism of the five-directional specimen is similar to that of the four-directional specimen, and both are dominated by concave impact zones and edge shear damage. The overall damage degree of the four-directional specimen is greater than that of the five-directional specimen. In the second impact cycle, both structures have obvious damage to the impact zone. Due to the larger deformation of the four-directional specimen, the damage in the impact zone is smaller than that of the five-directional specimen. The stress of the matrix on the impact surface and both sides is large [6].

With the increase of impact cycle, the impact damage region of 3-D four-directional braided composite tube rapidly expands to the surrounding, while the impact damage expansion of 3-D five-directional braided composite tube is not obvious. In the third impact cycle, the stresses of preform and matrix are significantly enhanced. And in the fourth impact cycle, the specimen still remains at a high stress level.

8.3.2.2 Transverse Impact Stress Propagation and Distribution

In order to explore the effect of axial yarn on the stress propagation in the braided preform, yarns in three positions, including the braided yarn and axial yarn, are selected from the two kinds of structures along the braiding direction. The stress-time curves of nodes on the yarns are extracted, and the node locations are shown in Fig. 8.24.

Figures 8.25, 8.26, and 8.27 show the stress variation curves with time for the three nodes on the braided and axial yarns and the yarn deformation and damage process at the corresponding locations. Figure 8.25a shows the stress-time curves of the NA node on the braided and axial yarns on the impact surface of 3-D four- and five-directional braided composite tube. Figure 8.25b shows the damage process of the overlapping yarns on the impact surface of the two kinds of structures.

When the incident bar starts to impact the composite tube, the nodal stress increases rapidly, where the nodal stress of the braided yarn in the five-directional specimen is greater than the nodal stress of the corresponding location of the braided yarn in the four-directional specimen. The reason is that the yarn deformation in the

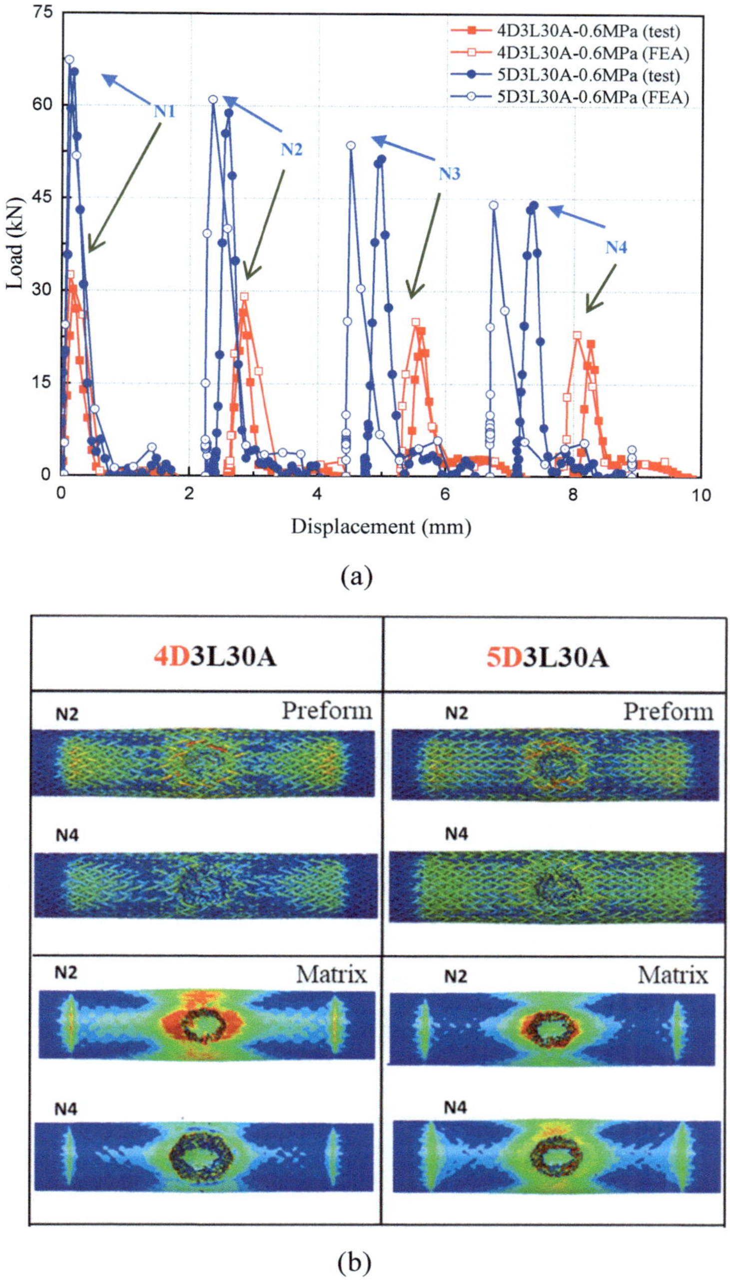

Fig. 8.23 Deformation history of 3-D four- and five-directional braided composite tubes at 0.6 MPa. (**a**) Load-displacement curves corresponding to deformation history, (**b**) deformation history

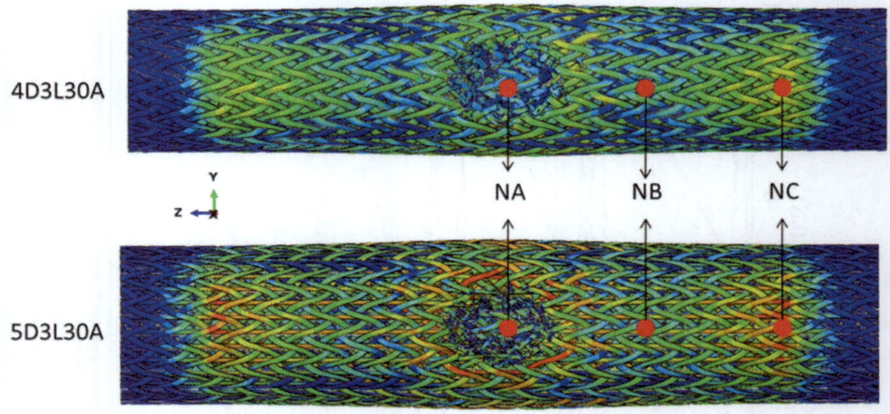

Fig. 8.24 Node locations on the four- and five-directional braided composite tubes

five-directional specimen is hindered by the addition of axial yarn, which improves the interaction force between yarns. As shown in Fig. 8.25b, when $t = 500$ μs, one side of the yarn breaks. When $t = 1100$ μs, both sides of the yarn break, and the NA nodal stress decreases rapidly, followed by increased damage on both sides of the yarn. The yarns of five-directional specimen break into more fragments.

As shown in Fig. 8.26, the nodal NB stress on the axial yarn is higher than the stress values in the braided yarn for both structures, which is particularly evident at $t = 1700$ μs. This is due to the fact that the stress propagation in the braided yarn is affected by the obstruction of the interlacing point and the inclination angle of the yarn, while in the axial yarn the stress is propagated axially along a single yarn with fast propagation and slow stress decay.

In node NC in the grip zone, as shown in Fig. 8.27, the average stress in the axial yarn node is greater than the corresponding value in the braided yarn, and for the 3-D five-way braided structure, the stress in the axial yarn is significantly greater than that in the braided yarn before $t = 1100$ μs, and later on, as the overall bending deformation of the tubing increases, both the axial and braided yarns are at higher stress levels with less difference.

8.3.2.3 Test Validation of the Damage Mode

As shown in Fig. 8.28a–c, the FEA results of impact damage morphologies of 3-D four- and five-directional braided composite tubes under three impact pressures are in good agreement with the test results. The damage degree of both structural composite tubes increases with the increase of impact pressure. The damage of 3-D four-directional braided composite tube increases with the impact pressure

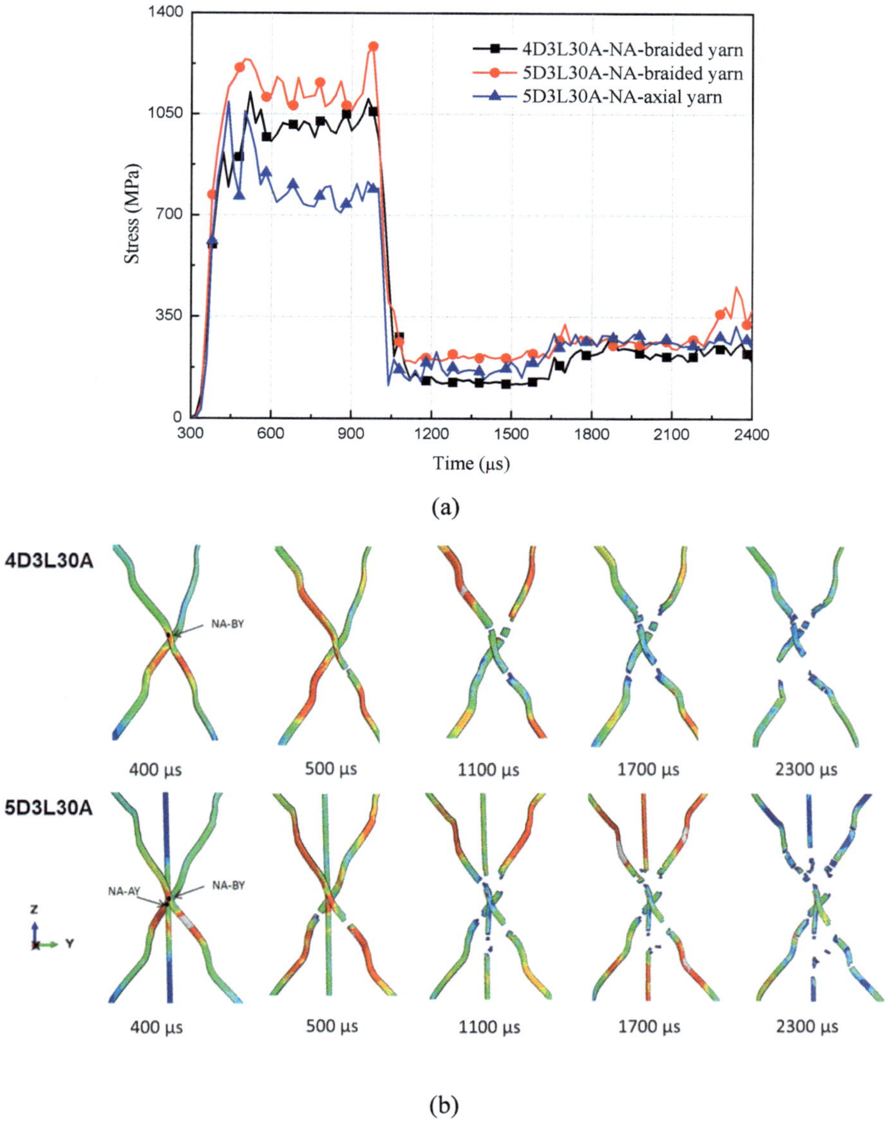

Fig. 8.25 Stress-time curves of node NA and damage histories of corresponding yarns. (**a**) Stress-time histories of node NA, (**b**) damage histories of yarns at node NA location

from matrix cracking to fiber tow fracture. Under the same impact pressure, the damage of 3-D five-directional braided composite tube is significantly less than that of 3-D four-directional braided composite tube due to the addition of the axial yarn.

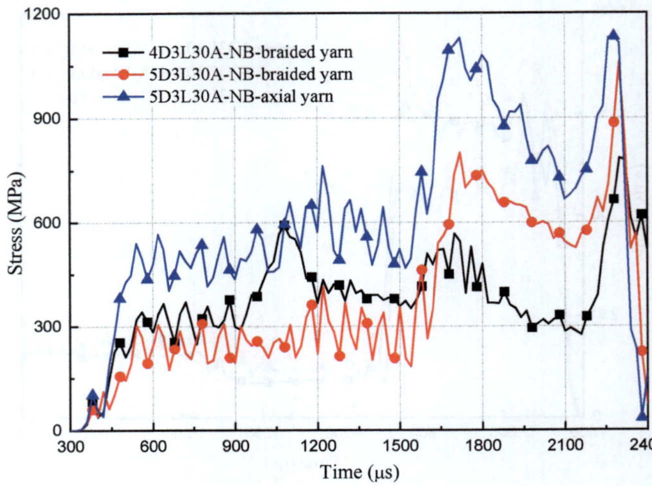

(a)

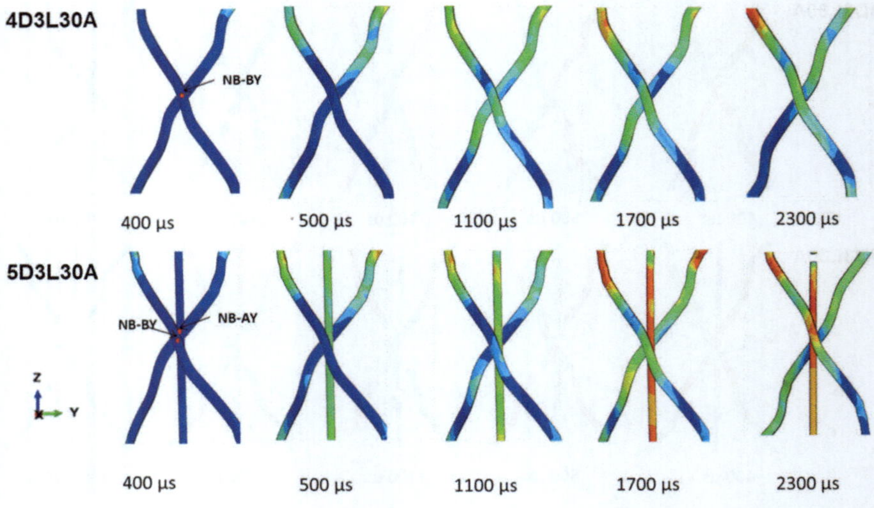

(b)

Fig. 8.26 Stress-time curves of node NB and damage histories of corresponding yarns. (**a**) Stress-time histories of node NB, (**b**) damage histories of yarns at node NB location

8.4 Impact Velocities

8.4.1 Braided Beams

In order to investigate the impact damage of 3-D braided composite beams under different impact pressures, the cross-sectional deformation processes of the matrix

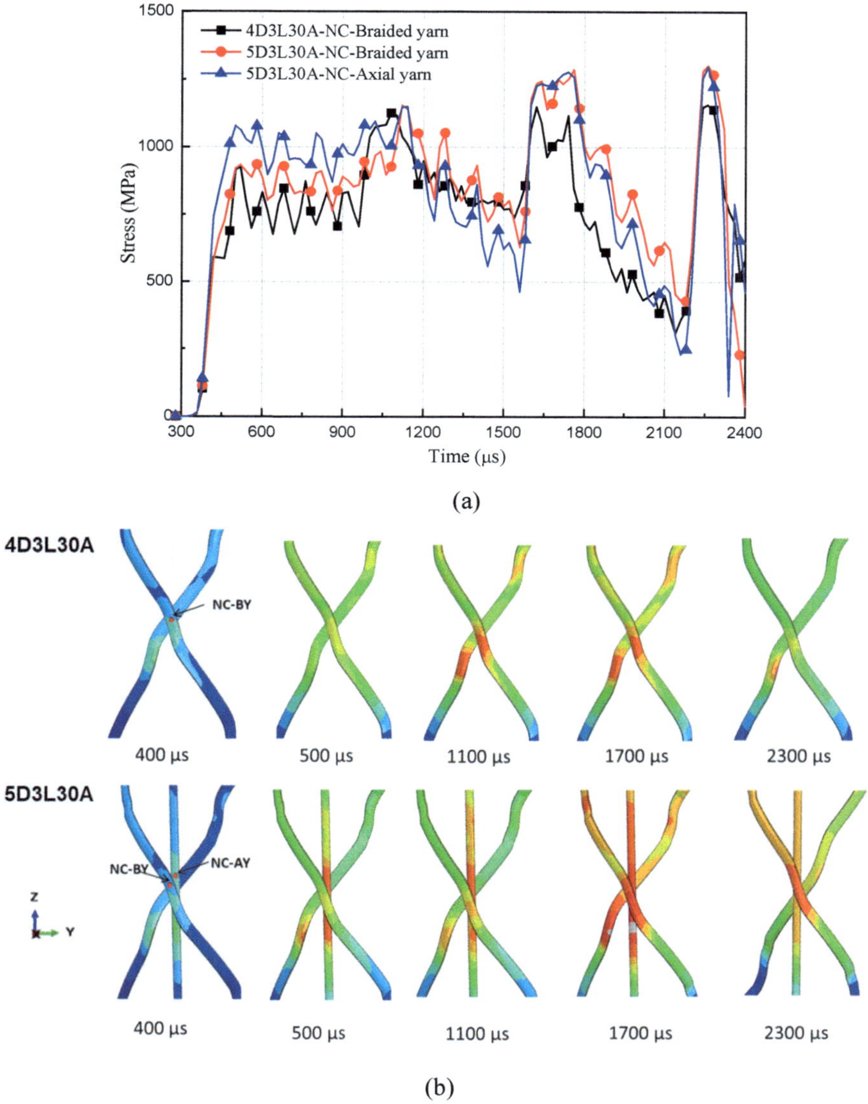

Fig. 8.27 Stress-time curves of node NC and damage histories of corresponding yarns. (**a**) Stress-time histories of node NC, (**b**) damage histories of yarns at node NC location

and preform are analyzed here (Fig. 8.29). The structure of the composite beam is determined by the different braiding angles, and the deformation of the composite beam is caused by the impact pressure. It is found that:

1. For 0.2 MPa, the distribution of high stress concentration is changed from fan shape to X shape as seen in the cross section of the matrix, and the distribution of high stress concentration in the cross section of the preform is always

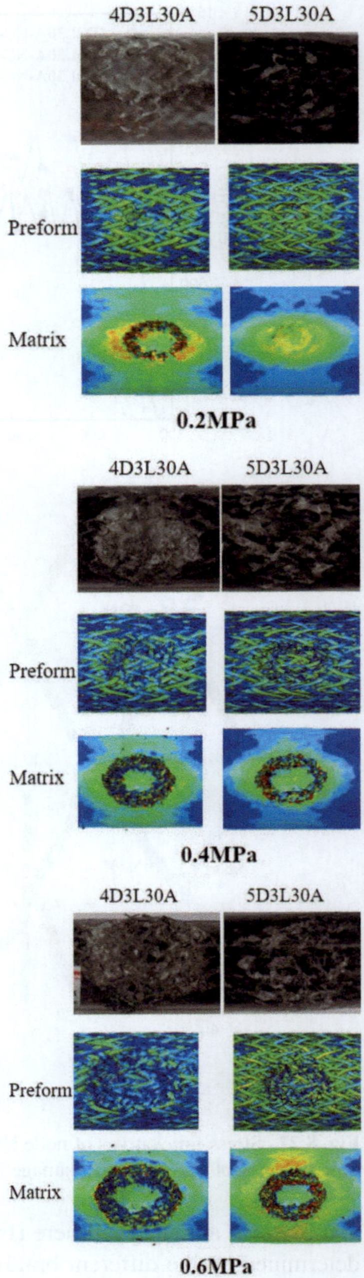

Fig. 8.28 Transverse impact damage morphologies of 3-D four- and five-directional braided composite tubes

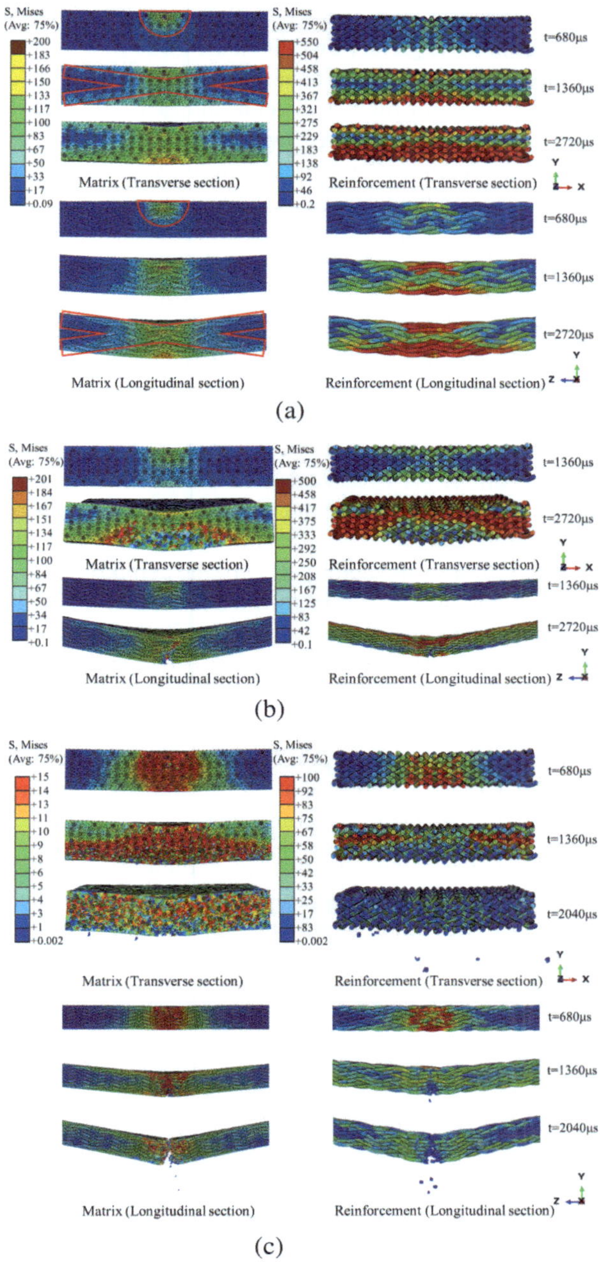

Fig. 8.29 Damage propagation of the sections of 3-D four-directional braided composite beam with the braiding angle of 20° during the first four impact cycles. (**a**) 0.2 MPa, (**b**) 0.4 MPa, (**c**) 0.6 MPa

approximate to X-shaped. The compressive stress on the impact surface at the beginning is greater than the tensile stress on the back. The stress in the middle part of yarns and resin counteracts each other, and the stress is relatively low. The distribution of high stress concentration in the longitudinal section of the resin still changes from fan shape to X shape, while the distribution of high stress concentration in the yarn is approximately I shape, but the stresses on the impact surface and the back are obviously the largest.

2. For 0.4 MPa, the yarns and resin are damaged and fractured in the fourth impact cycle. The stress drops instantly and the composite beam partially fails. It can be observed from the cross section of the resin that the stress is highest near the damage. Similarly, the yarns near the back break early under the maximum stress, after which the yarns near the broken yarns are subjected to the maximum stress and also break immediately thereafter. It can be seen from the longitudinal section that the fracture path of the composite beam is not straight but curved. This is due to the effect of the preform structure, where the maximum stress propagates along the path of least resistance. The resin and the bent part of yarns can withstand relatively little stress and preferentially fracture leading to the formation of bending cracks.
3. For 0.6 MPa, the deformation of the composite beam increases significantly, and the yarn and resin start to fracture in the second impact cycle. From the finite element results, it can be seen that in the third impact cycle, the specimen fails completely with uniform stress distribution on the fracture surface. In general, the maximum impact pressure carries the highest energy for the same braided structure specimen, which results in greater damage to the specimen. The fracture path of the beam is generated at the weakest location. In conjunction with Fig. 4.13, it is known that the accumulation of energy absorption increases the internal damage of the composite.

Figure 8.30 shows the damage evolution, adiabatic temperature rise, and stress distribution corresponding to μ-CT images of 40° specimen under different impact pressures. The impact pressure has a significant effect on the failure mode and damage morphology. The damage deformation of the specimens is significantly reduced after the impact pressure is reduced. The main failure modes are interfacial cracking and resin cracking, with no significant damage to the yarns. The interfacial damage is mainly concentrated on the surface braided yarns on the impact surface subjected to compressive stress and on the back subjected to tensile stress, and resin cracks are observed from the μ-CT images to delaminate the surface braided yarns from the interior braided yarns. The decrease of impact pressure leads to the decrease of impact energy and load peak, and the stress distribution is similar to that at 0.6 MPa. The braided yarns near the back are subjected to the higher tensile stress, and the preform is the main constituent to bear the load. The adiabatic temperature rise of resin is mainly concentrated near the impact point and the back, and the highest temperature is only 36.5 °C, which is much lower than the temperature rise at 0.6 MPa.

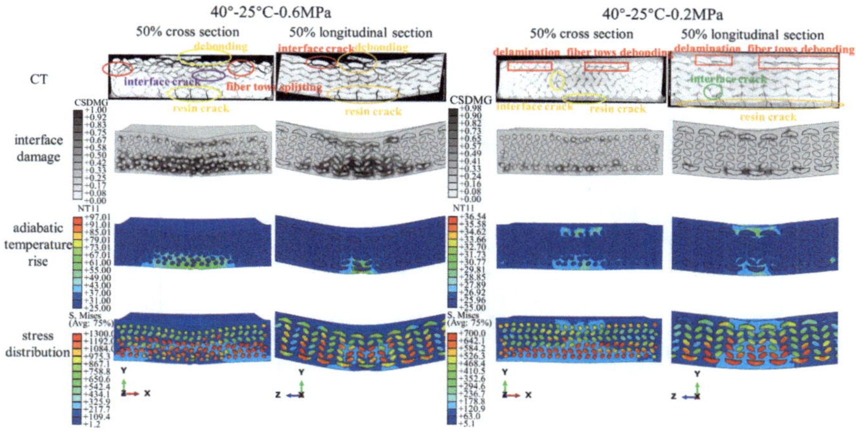

Fig. 8.30 Damage evolution, adiabatic temperature rise, and stress distribution corresponding to μ-CT images of 40° specimen under different impact pressures

8.4.2 *Braided Tubes*

8.4.2.1 Impact Deformation History

The specimen 4D3L30A is used as an example to analyze the effect of impact pressure on the transverse impact deformation history of 3-D braided composite tube. Figures 8.31, 8.32, and 8.33 show the deformation history and the corresponding load-displacement curves of the specimen 4D3L30A under three impact pressures of 0.2 MPa, 0.4 MPa, and 0.6 MPa, respectively, where the load-displacement curves include the test results and FEA results. The stress distribution and damage propagation of the preform and matrix of 3-D braided composite tube during the first four impact loading processes are discussed separately for different impact pressures.

As shown in Fig. 8.31, for the first impact cycle at 0.2 MPa, the compression stress wave is generated in the impact thickness direction of the tube. The contact surface of the composite tube and the incident bar exhibits elliptical stress concentration phenomenon. The stress decreases from the impact point to the surrounding layers, reaches the clamping zone along the axial direction, and increases again. Along the circumferential direction, the stress on both sides slightly increases due to the flattening of the section of the impact zone. Comparing the stress concentration region of the preform and resin, it is found that the resin is homogeneous and its elliptical concentration region is relatively regular in shape, while the stress of the preform is obviously different due to the different alignments of the two adjacent columns of yarns forming the braiding angle, and the stress shows the alternation of strength among the columns. During the first impact cycle, both the preform and the matrix are not significantly damaged. In the second impact cycle, the stress concentration region of the composite continues to expand outward. Several yarn fractures

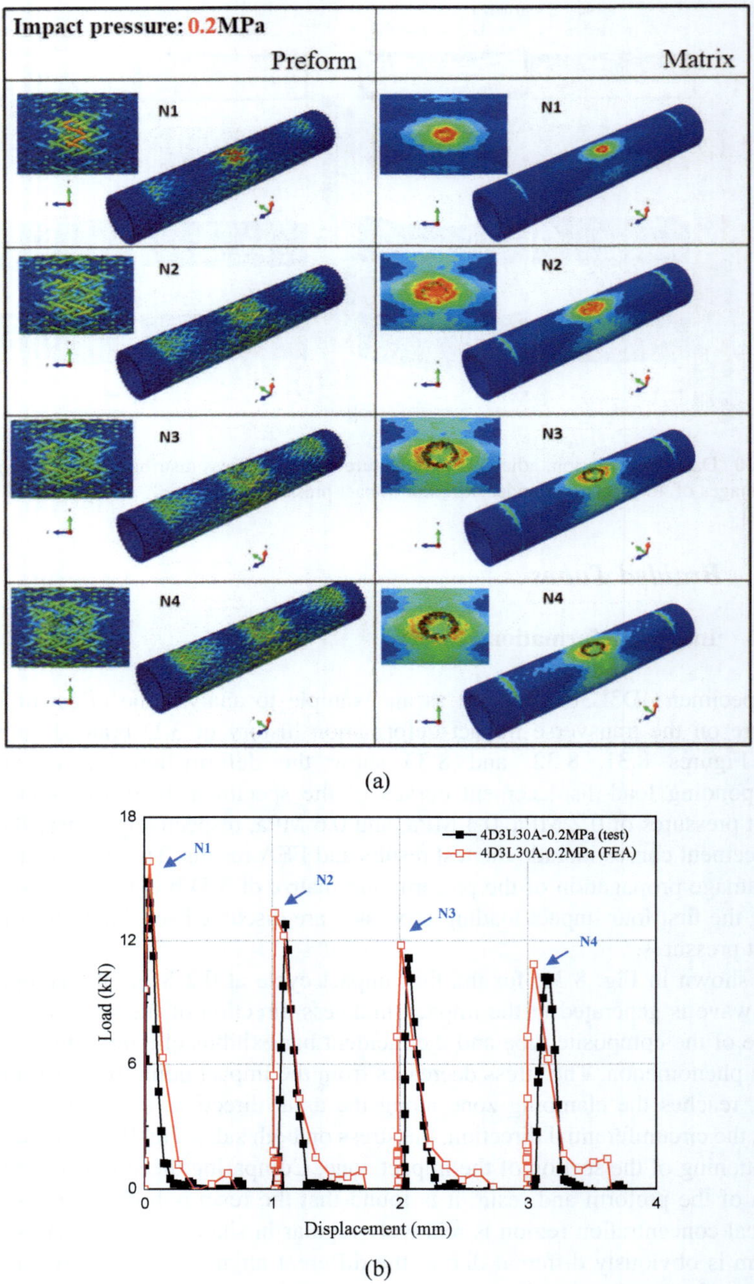

Fig. 8.31 Deformation history of specimen 4D3L30A at 0.2 MPa. (**a**) Deformation history at 0.2 MPa, (**b**) load-displacement curves corresponding to deformation history

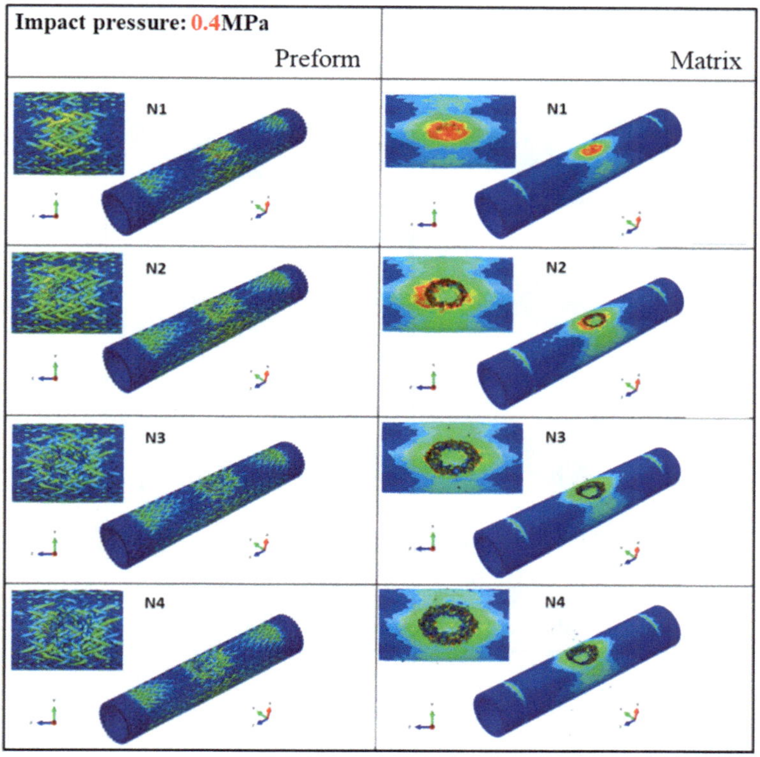

(a)

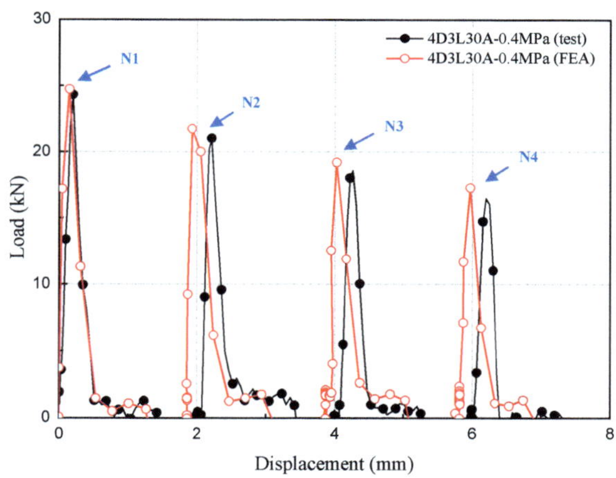

(b)

Fig. 8.32 Deformation history of specimen 4D3L30A at 0.4 MPa. (**a**) Deformation history at 0.4 MPa, (**b**) load-displacement curves corresponding to deformation history

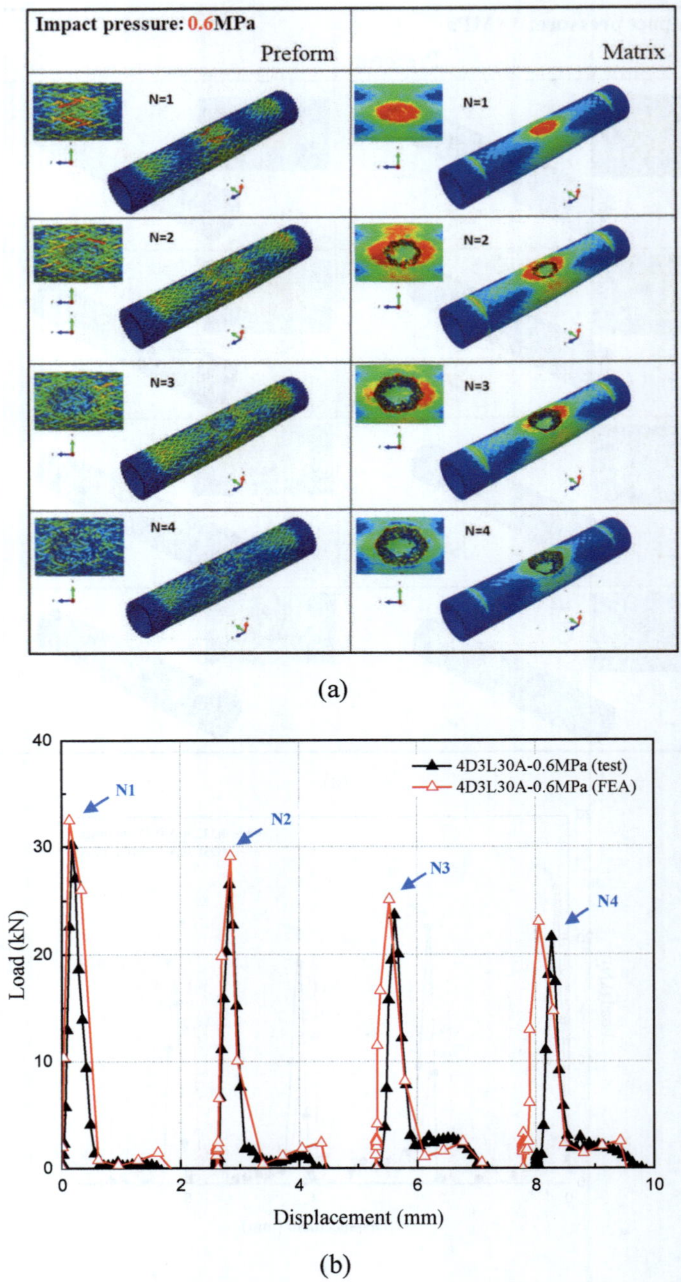

Fig. 8.33 Deformation history of specimen 4D3L30A at 0.6 MPa. (**a**) Deformation history at 0.6 MPa, (**b**) load-displacement curves corresponding to deformation history

appear in the preform, and cracking appears in the matrix, but not obvious. In the third and fourth impact cycles, the stress concentration at the contact surface of the composite and the incident bar gradually disappears, while the stress concentration around the impact surface is obvious. At this time, the tube is mainly subjected to shear stress, and shear damage occurs along the edge of the contact surface. Obvious yarns fracture and matrix fragmentation can be seen, but the overall damage degree of the tube is small.

As shown in Figs. 8.32 and 8.33, the damage processes of the first four impact cycles of the tube are similar for the two impact pressures of 0.4 and 0.6 MPa. The first impact cycle is similar to that under the impact pressure of 0.2 MPa, and no obvious damage occurs, mainly showing the phenomenon of stress concentration at the impact point and clamping zone. The area and stress of the concentration region increase as the impact pressure increases. In the second impact cycle, obvious shear damage has occurred around the impact surface. In the third and fourth impact cycles, the elliptical shear zone gradually expands to the periphery. The tube shows obvious filling shear phenomenon at 0.6 MPa. At the same time, the entire composite tube has weak bending deformation during the whole impact process.

To sum up the above analysis, the deformation damage law of tubes is basically the same under different impact pressure. In both cases, at the beginning of the impact, the stress concentrates on the contact surface and then dissipates. The stress concentrated region transfers to the surrounding of the impact surface and forms a shear band at the outer edge of the impact surface. With the increase of impact pressure, the faster the above process occurs, the more serious the material damage is. In addition, the load-displacement curves extracted from the FEA results agree well with the test results.

8.4.2.2 Propagation and Distribution of Transverse Impact Stress

Figure 8.34 shows the stress distribution of the preform and the matrix. It can be seen from the stress distribution that the bearing capacity of the preform of 3-D braided composite tube is much higher than that of matrix, indicating that the 3-D braided preform is the main load-bearing component of the composite.

The stress wave propagation in the tube is explored by using the impact pressure of 0.6 MPa as an example, and stress-time curves are extracted at several nodes on the preform and matrix, respectively. As shown in Fig. 8.34, three nodes are selected along the braiding direction, where NA and NC represent the impact zone and clamping zone and NB represents the intermediate zone. The three nodes along the circumferential direction (section A-A) are NA, NE, and NF in order.

Figures 8.35 and 8.36 show the stress-time curves of the selected nodes on the preform and the matrix and the corresponding local magnification of the stress propagation at the beginning. The stress propagation of the two components is similar. It can be seen that when the incident bar is impacted, a one-dimensional

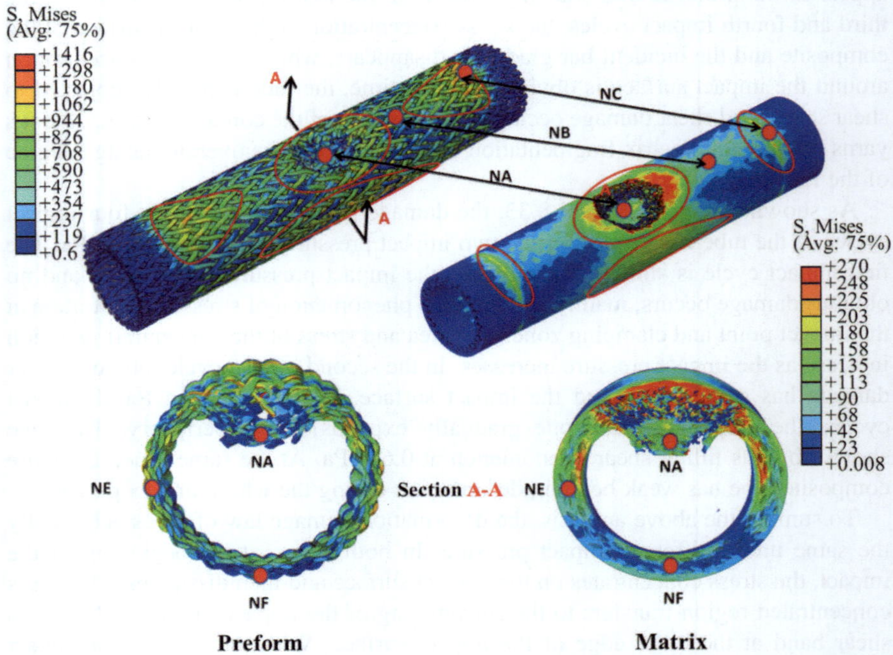

Fig. 8.34 Stress distribution and node locations of preform and matrix

stress wave is generated and propagates along the incident bar. Before the stress wave reaches the specimen, the nodal stress is zero on the specimen. When the specimen is impacted, the stress wave propagates from the impact point to the surrounding region, reaching nodes NA, NB, and NC in order along the braiding direction and nodes NA, NE, and NF in order along the circumferential direction. From the whole process of stress propagation, the stress amplitude decreases sequentially along the circumferential direction (NA-NE-NF), while it decreases first and then increases along the braiding direction (NA-NB-NC). This phenomenon can be explained by the stress wave propagation and reflection, and the amplitude of stress wave is decreasing in the process of propagation. Because the stress wave is reflected at the fixed end and the reflected wave is the positive image of the incident wave, the stress is doubled when the two waves are superimposed. In addition, the impact zone starts to destroy after the first impact. The stress of node NA decreases rapidly, while the nodes in other regions are subjected to multiple stress waves and show fluctuation.

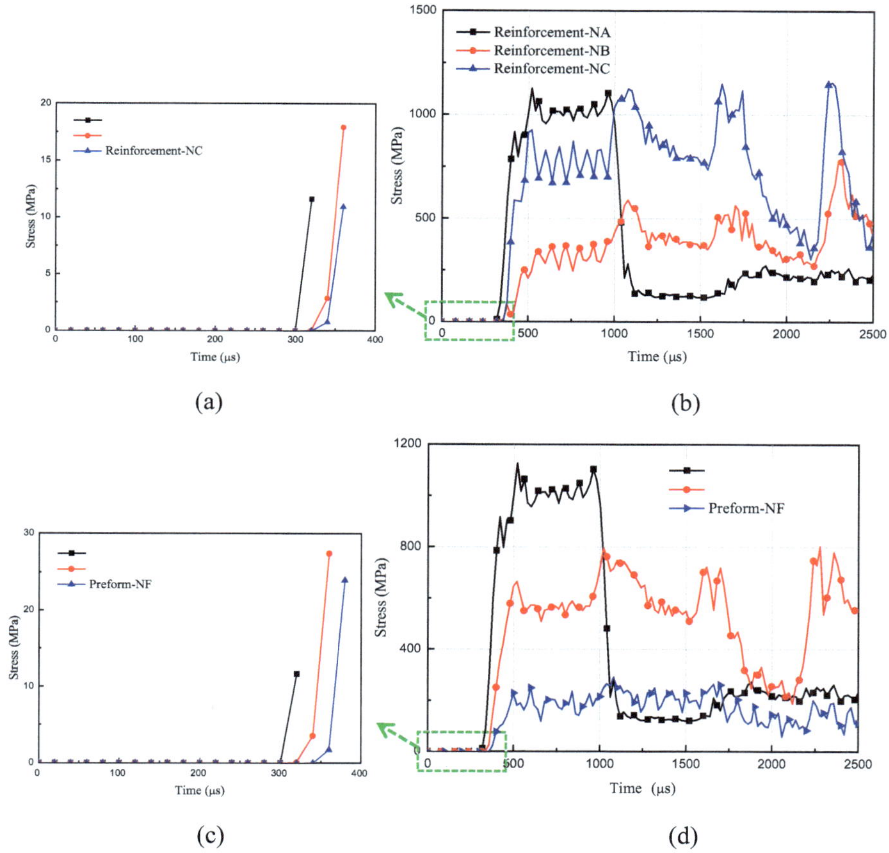

Fig. 8.35 Nodal stress-time curves on preform

8.4.2.3 FEA Model Verification

Figures 8.37, 8.38 and 8.39 show the transverse impact damage morphologies of specimen 4D3L30A under different impact pressures, including the test results and the FEA results. Four locations a–d are selected along the braiding direction from the edge of the impact zone to the center of the impact surface, and the transverse sections at each location are extracted from the μ-CT results for comparison with the finite element results. The damage morphologies of the specimens obtained by the two methods are found to be in good agreement by comparison. From a to d, the damage region expands from the central position to both sides. The central position of the cross section a, b is dominated by matrix fragmentation between fiber tows, which is shown as a gap between fiber tows in the μ-CT result and as matrix element elimination in the finite element results. The obvious shear band can be observed at

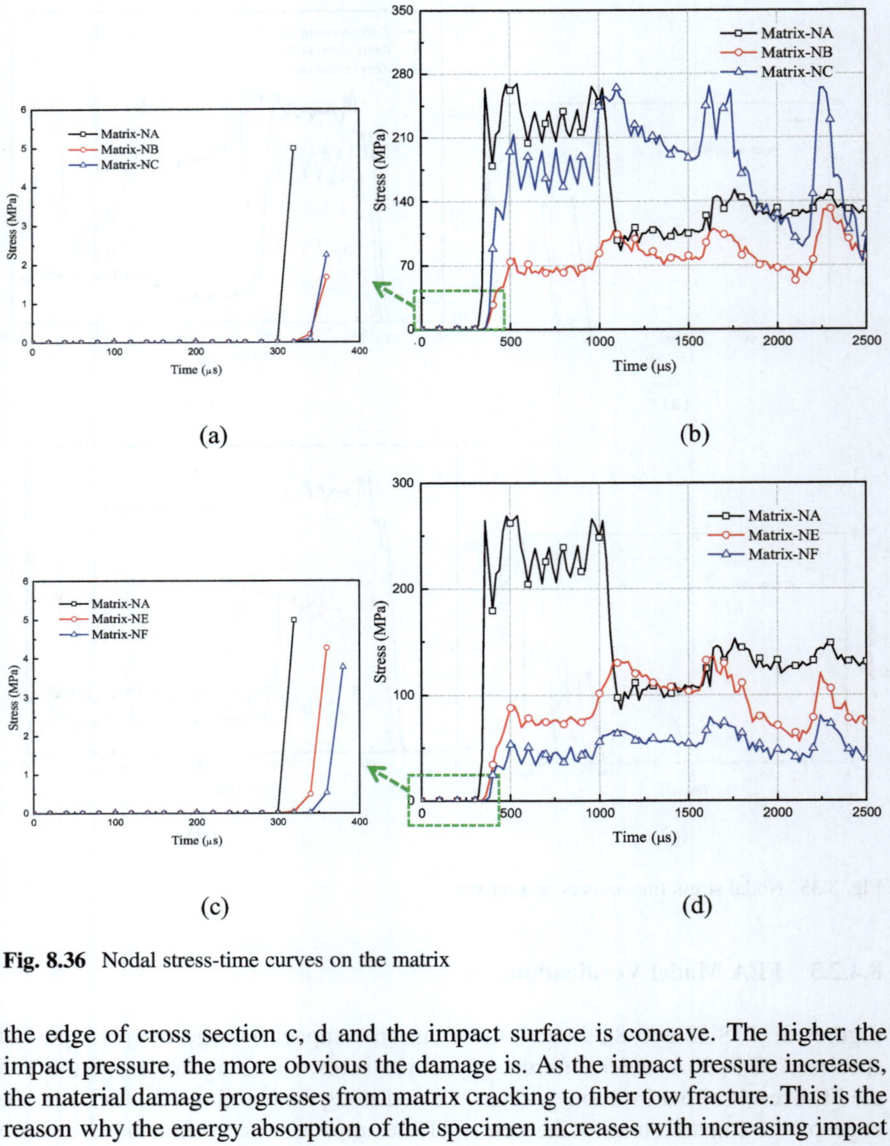

Fig. 8.36 Nodal stress-time curves on the matrix

the edge of cross section c, d and the impact surface is concave. The higher the impact pressure, the more obvious the damage is. As the impact pressure increases, the material damage progresses from matrix cracking to fiber tow fracture. This is the reason why the energy absorption of the specimen increases with increasing impact pressure.

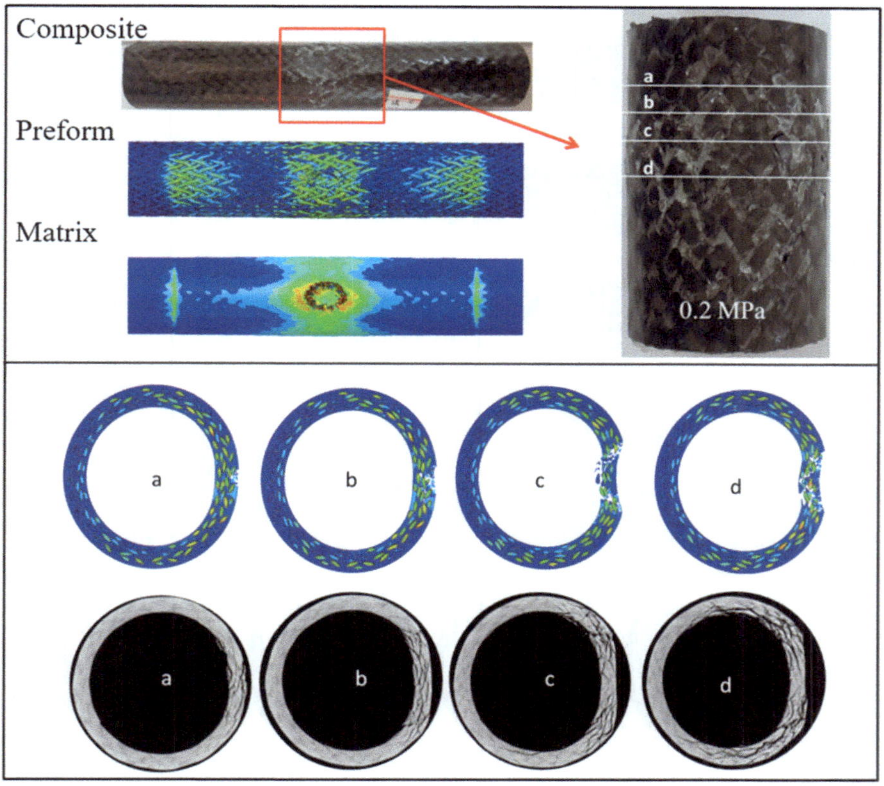

Fig. 8.37 Damage morphology of specimen 4D3L30A at 0.2 MPa

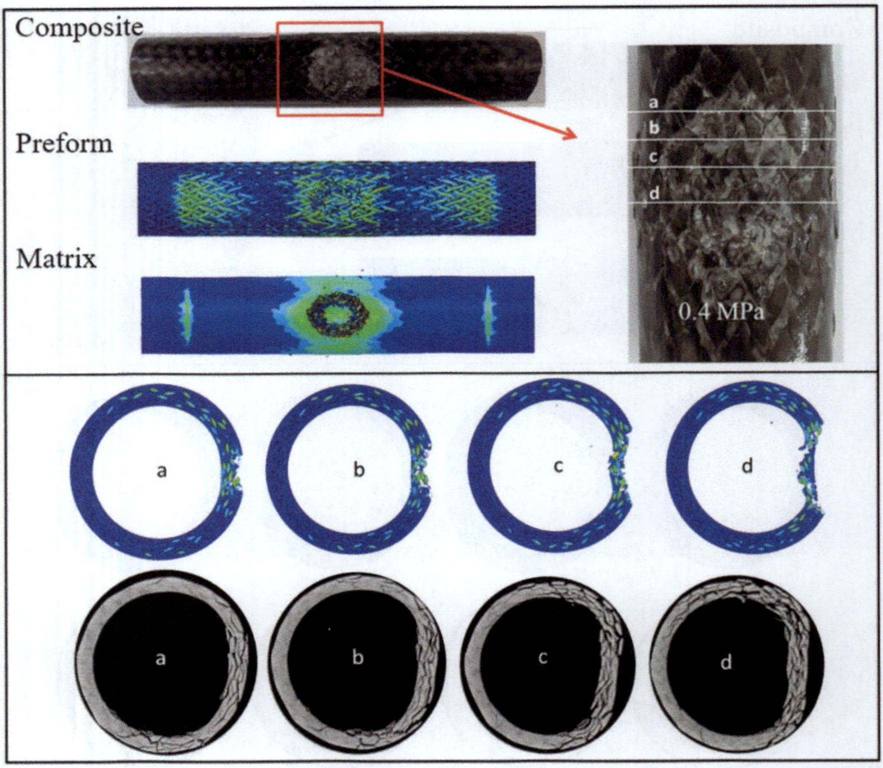

Fig. 8.38 Damage morphology of specimen 4D3L30A at 0.4 MPa

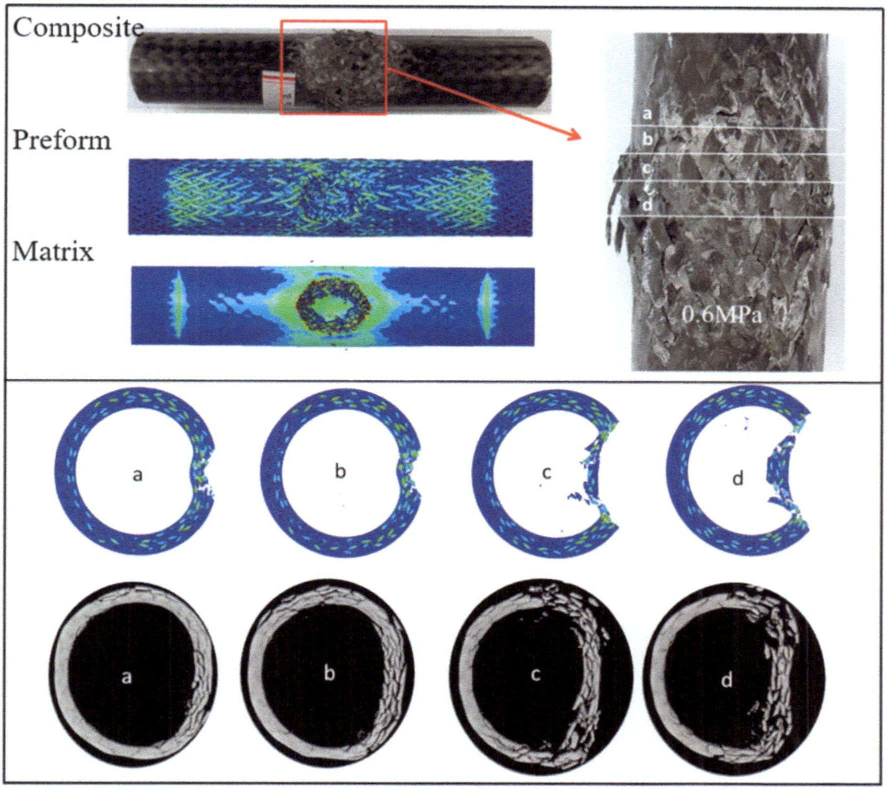

Fig. 8.39 Damage morphology of specimen 4D3L30A at 0.6 MPa

References

1. Hu Meiqi, Zhang Junjie, Sun Baozhong, Gu Bohong: Finite element modeling of multiple transverse impact damage behaviors of 3-D braided composite beams at microstructure level. Int. J. Mech. Sci. **148**, 730–744 (2018)
2. Dong, K., Zhang, J., Cao, M., Wang, M., Gu, B., Sun, B.: A mesoscale study of thermal expansion behaviors of epoxy resin and carbon fiber/epoxy unidirectional composites based on periodic temperature and displacement boundary conditions. Polym. Test. **55**, 44–60 (2016)
3. Haili Zhou, Wei Zhang, Tao Liu, Bohong Gu, Baozhong Sun: Finite element analyses on transverse impact behaviors of 3-D circular braided composite tubes with different braiding angles. Compos. A: Appl. Sci. Manuf. **79**, 52–62 (2015)
4. Haili Zhou, Dongmei Hu, Bohong Gu, Baozhong Sun: Transverse impact performance and finite element analysis of three dimensional braided composite tubes with different braiding layers. Compos. Struct. **168**, 345–359 (2017)
5. Hu Meiqi, Sun Baozhong, Gu Bohong*: Microstructure modeling multiple transverse impact damages of 3-D braided composite based on thermo-mechanical coupling approach. Compos. Part B. **214**, 108741 (2021)
6. Haili Zhou, Baozhong Sun, Bohong Gu: Responses of 3D four-directional and five-directional circular braided composite. Int. J. Crashworthiness. **21**, 353–366 (2016)

Chapter 9
Multiscale Structure Mechanisms on Axial Compressive Impact Damages

This chapter discusses the effects of braiding parameters and strain rates on the damage distribution and propagation of 3-D braided composite tubes during axial compressive impact. The damage mechanism of 3-D braided composite tubes under axial compressive impacts will be revealed both from tests and FEA.

9.1 Braiding Angles

9.1.1 Axial Impact Compression History

The impact compression damage histories of the two-phase constituent structures (preform and matrix) of 3-D four-directional braided composite tubes with three different braiding angles (25°, 35°, and 45°) at 676/s, 650/s, and 627/s are given in Figs. 9.1, 9.2, and 9.3, respectively. From Fig. 9.1a–d, it can be seen that the stress wave propagates along the length of the specimen with different propagation velocities on the braided preform and the resin matrix, which is significantly faster on the braided preform than the resin. Due to the fact that the propagation velocity of stress wave on the material depends on the stiffness and density of the material, the stiffness and density of carbon fiber tows are different from that of the epoxy resin. As the stress wave propagates, the whole composite begins to bear the load. Because of the different properties of the carbon fiber tow and the epoxy resin, the two are subjected to different loads. As seen in Fig. 9.1e–h, the braided preform is subjected to significantly higher loads than the resin matrix. When the composite reaches the maximum compressive stress, the whole braided preform reaches the maximum stress, while the stress at the contact part between the resin and the surface braided yarn is the minimum, and the stress at other parts also reaches the maximum. The initial damage to the braided yarn occurs at 230 μs in Fig. 9.1f, and the initial damage to the matrix occurs at 260 μs in Fig. 9.1g.

M. Hu, B. Gu, *Impact Damages of Braided Composites*, Engineering Materials,
https://doi.org/10.1007/978-981-16-5703-0_9

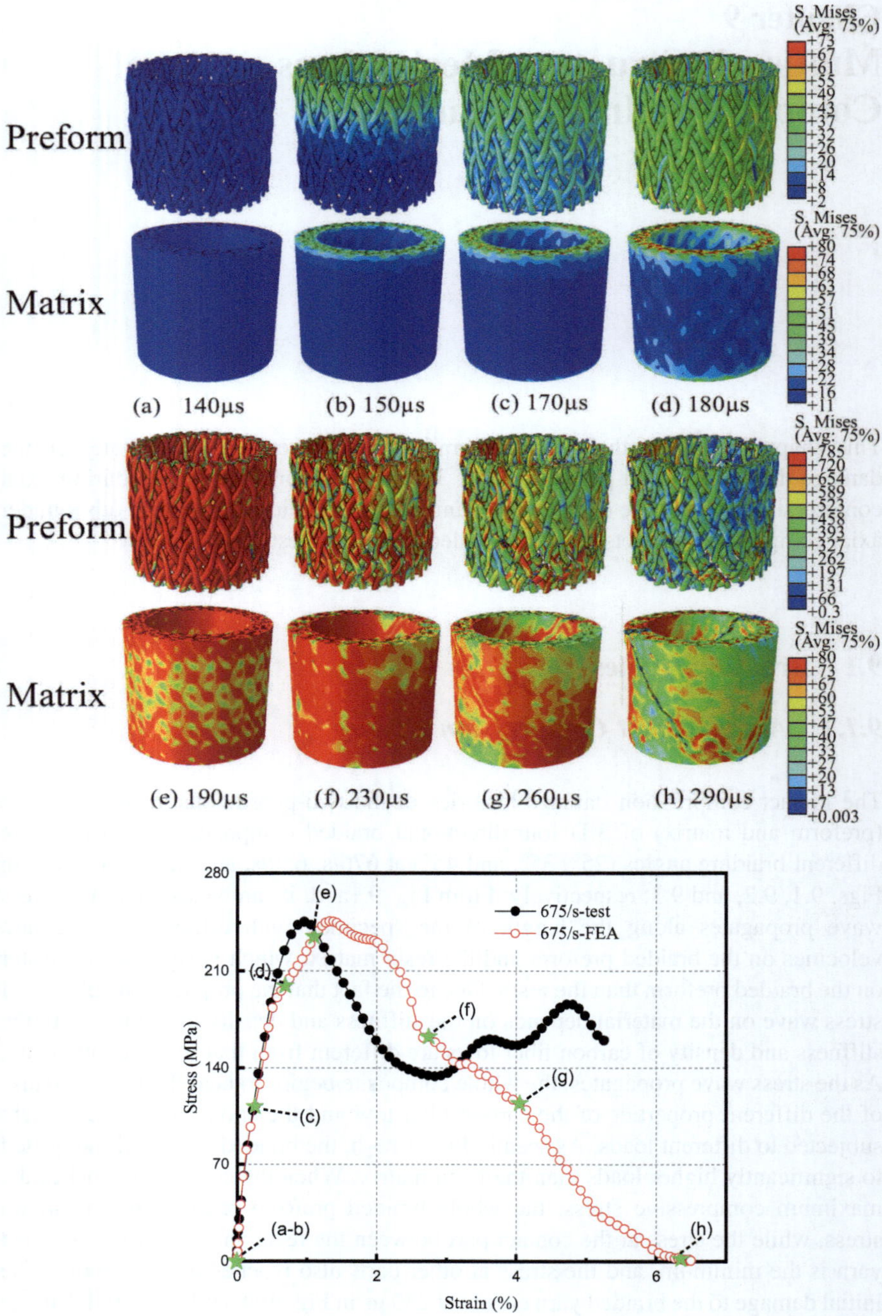

Fig. 9.1 Axial compression history of the three-layer three-dimensional and four-directional braided composite tube with the braiding angle of 25° at 676/s

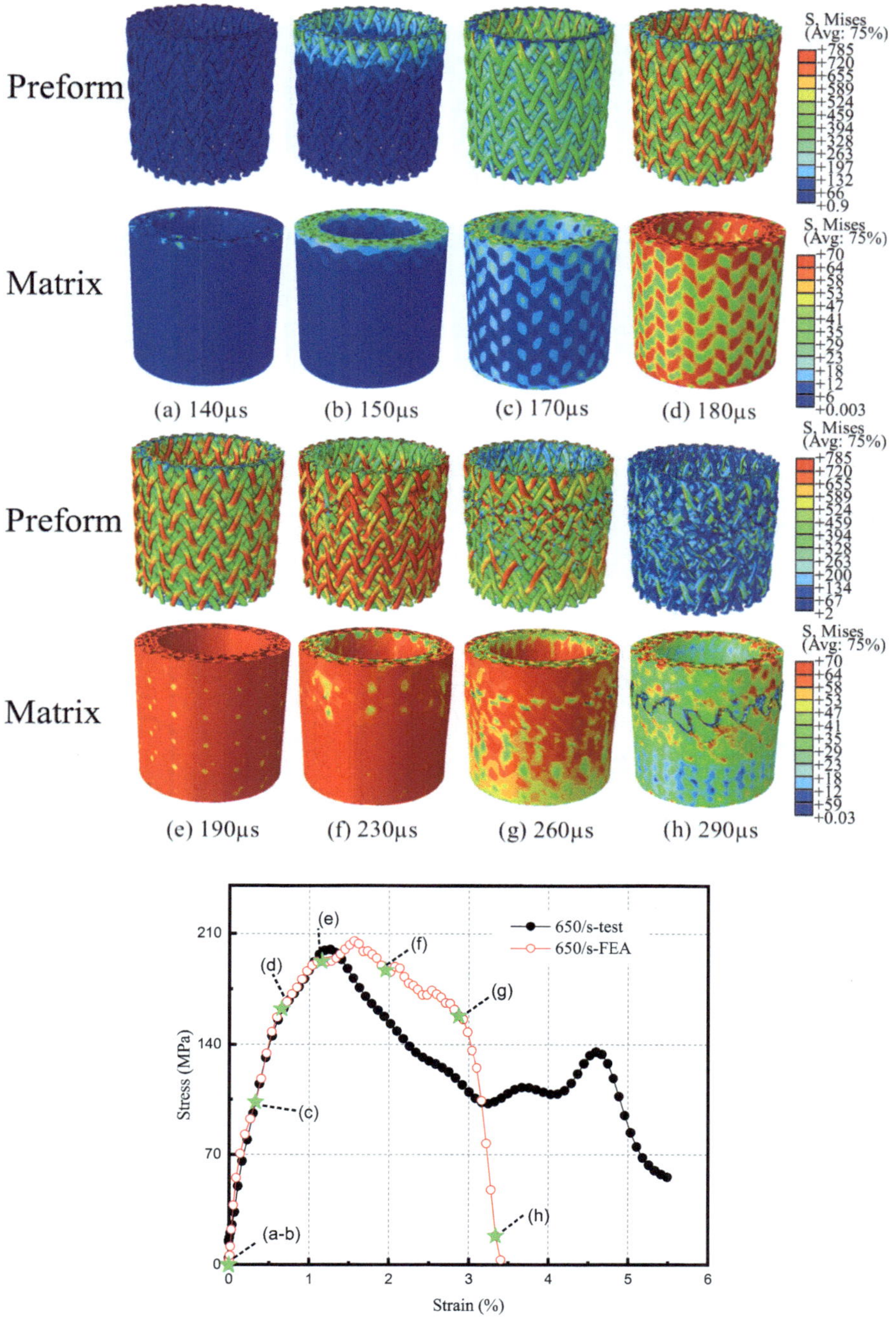

Fig. 9.2 Axial compression history of the 3-layer three-dimensional and four-directional braided composite tube with the braiding angle of 35° at 650/s

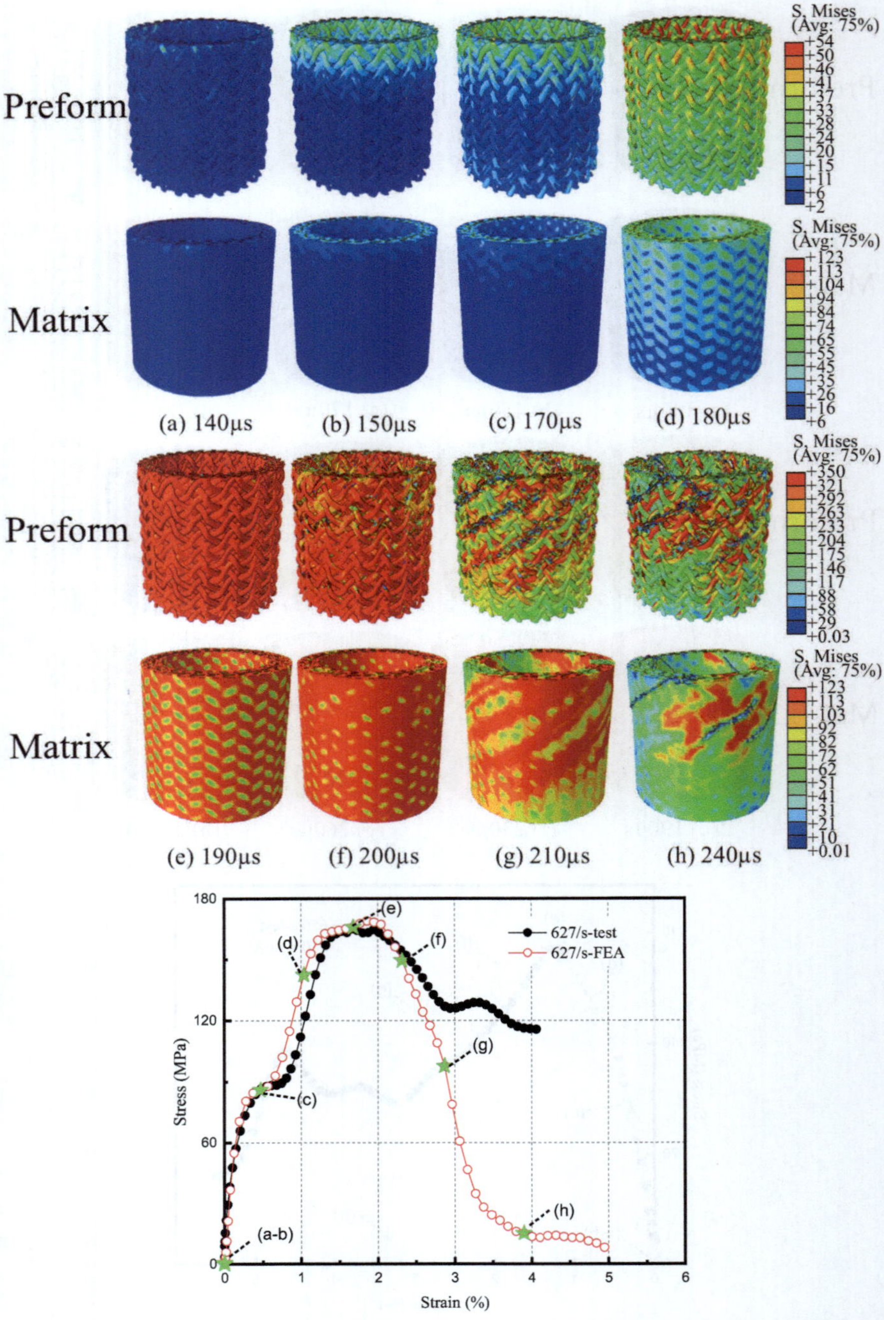

Fig. 9.3 Axial compression history of the three-layer three-dimensional and four-directional braided composite tube with the braiding angle of 45° at 627/s

From Fig. 9.2a–d, it can be seen that for the 35° specimen, the stress wave does not propagate much faster through the braided preform than the resin compared to the 25° specimen, which is related to the orientation of braided yarn. From the stress-strain curves of the compression history, it can be seen that the composite tube reaches the maximum compressive stress in Fig. 9.2e, the whole braided preform reaches the maximum stress at this time, and the stress is concentrated on the inner and outer surface of the braided yarn. The overall stress of the resin is the largest in Fig. 9.2f. This is due to the fact that the initial damage of the braided yarn occurs at 230 μs as shown in Fig. 9.2f after the maximum stress is reached, while the initial damage of the resin occurs at 260 μs as shown in Fig. 9.2g. It can be seen from Fig. 9.3a–d that the propagation velocity of stress wave in the braided preform is slightly faster than that in the resin for the 45° specimen. The composite reaches the maximum compression stress at Fig. 9.3e, which is different from that of the 35° specimen. At this time, the stress of all the yarns in the whole braided preform reaches the maximum value, while the minimum stress appears at the place where the resin contacts with the surface yarns. The initial damage of yarn occurs at 200 μs as shown in Fig. 9.3f, and the stress of resin reaches the maximum. The initial damage to the resin occurs later than the initial damage to the yarn at 240 μs as shown in Fig. 9.2h.

9.1.2 Stress-Strain Curves

Figure 9.4 shows the comparison of test and FEA stress-strain curves for 3-D four-directional braided composite tubes with three different braiding angles. Both the test and FEA stress-strain curves show a nonlinear increase, and the stiffness and strength increase with the increase of the strain rate. The strain rate dependence is in good agreement with the test results. It can also be seen from Fig. 9.4 that the FEA results are in good agreement with the test results before the tube failure. There are differences between the test stress-strain curves and the FEA calculated curves in the plastic stage. This is because the carbon fiber tow in the FEA model is regarded as an elastic-plastic model and the addition of plastic parameters makes the yield stage of the calculated stress-strain curves more pronounced than the test stress-strain curves. Also when the tube reaches the maximum stress, the difference between the FEA curves and the test curves is greater. This is due to the fact that in the FEA model, when the stress reaches the maximum value, the elements are eliminated and no longer bear the load due to the failure criterion, while the material can still bear a certain load after failure during tests.

The comparison of the material stiffness and strength is extracted from the test and FEA stress-strain curves, as shown in Fig. 9.5. The FEA calculated stiffness values are in good agreement with the experimentally obtained stiffness values, while the FEA calculated strength values are slightly larger than the test values. Although the meso-structure model is closer to the real structure of the composite than the unit cell model, there are still errors in predicting the material properties.

Fig. 9.4 Test and FEA stress-strain curves for three-layer three-dimensional and four-directional braided composite tubes with three different braiding angles. (**a**) 25°, (**b**) 35°, (**c**) 45°

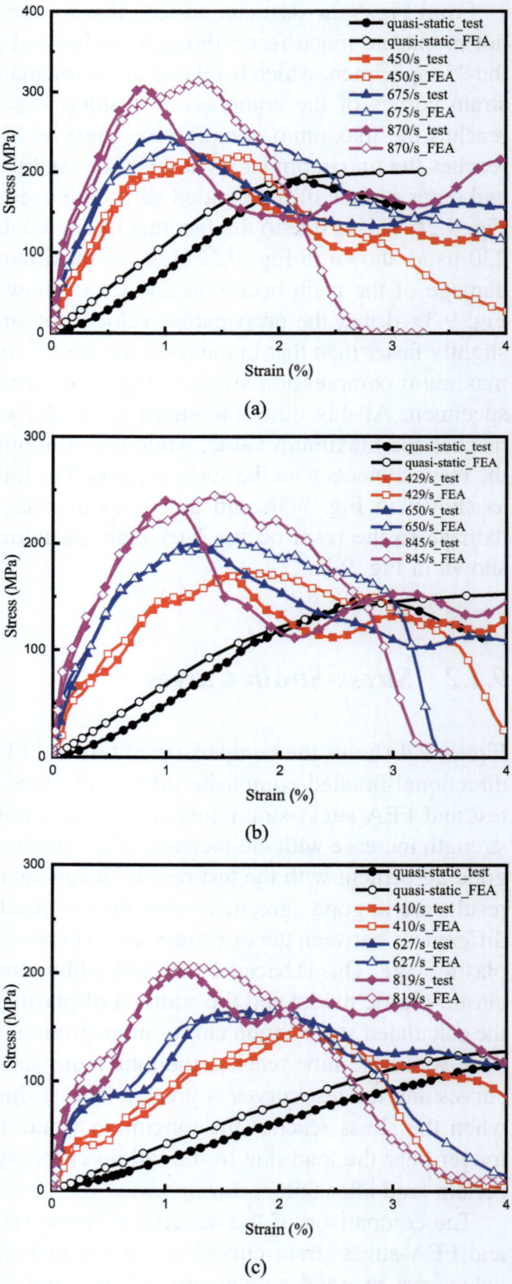

Fig. 9.5 Comparison of test and FEA results of strength and stiffness of three-layer three-dimensional and four-directional braided composite tubes with three different braiding angles. (**a**) 25°, (**b**) 35°, (**c**) 45°

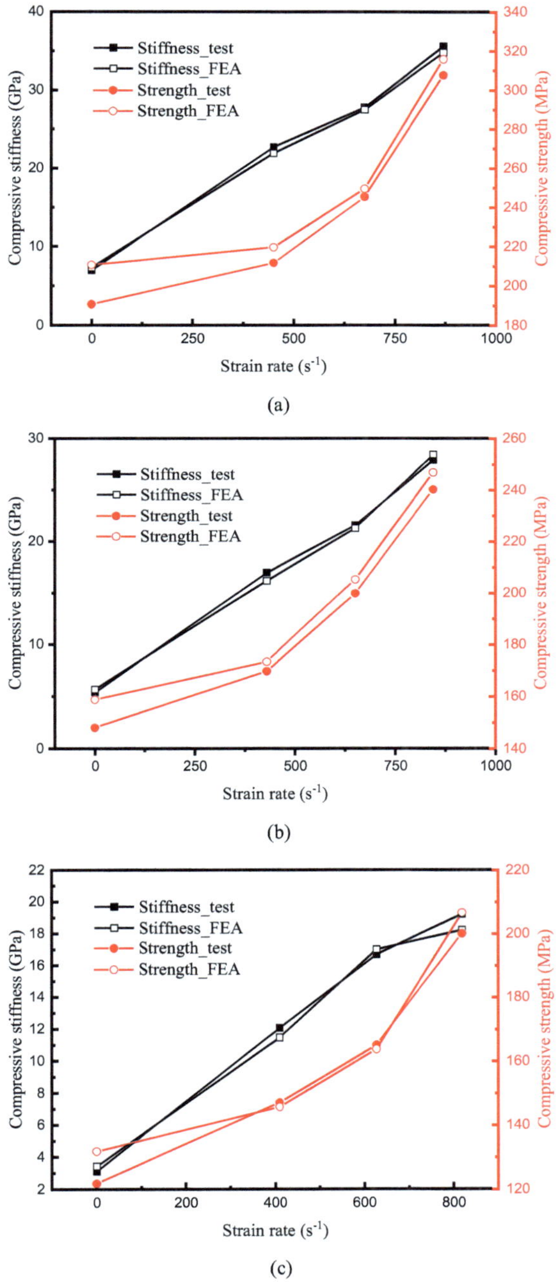

This is due to the neglect of the morphology of fibers within the yarns and the interface between the yarn and the resin in the geometric model. In the actual compression loading, the internal fibers do not break at the same time, while the yarn-resin interfaces debond at different parts.

9.1.3 Damage Morphologies

Figure 9.6 shows the test and FEA damage morphologies of 3-D four-directional braided composite tubes with three different braiding angles after axial impact compression at various strain rates, and the comparison shows that the two are in good agreement. The test damage morphologies of the specimens are as described in Chap. 7. When the braiding angle is small (25°), the tube shows shear damage bands, fiber pullout, and fiber breakage under the strain rates of 450/s and 675/s. Under the strain rate of 870/s, the whole specimen is destroyed. Due to the shear failure, some specimens are broken into petals, and some yarns are completely debonded from the resin. The finite element model shows the same shear damage and fiber breakage as in the experiment under the strain rates of 450/s and 675/s. As the strain rate increases, the damage morphologies of the specimens become more severe. At 870/s, the specimen does not show the complete failure as in the experiment, but shows an obvious crushing failure mode with a large failure area and split flap cracking [1].

When the braiding angle increases to 35°, the specimens have no shear damage and maintain good integrity under three strain rates. With the increase of the strain rate, the compression deformation of tube increases due to massive resin damage and yarn buckling. The damage morphologies obtained from the FEA model under three strain rates are similar to the experiment, showing fiber breakage and resin cracking and exhibiting significant strain rate effects. When the braiding angle increases to 45°, the buckling deformation of tube is not obvious, and the failure mode is mainly the resin damage under impact compression loading. At 410/s, resin damage and a small amount of fiber breakage occur at the front end of the specimen contacting with the incident bar. The difference is that there are resin cracking and fiber breakage at the other end of the specimen in the finite element model in addition to the failure modes. Under the strain rate of 627/s, the damage morphologies of experiment and finite element model are in good agreement, with small shear bands appearing along the braided yarn direction in addition to resin damage. Under the strain rate of 819/s, the resin damage extends from the front end in contact with the incident bar to the end, accompanied by small shear bands. The damage is relatively serious. Two clear shear bands can be seen in the finite element model. It can be seen that the braiding angle as an important structural parameter of the preform affects the damage mode and failure mechanism of 3-D braided composite tube under different strain rates.

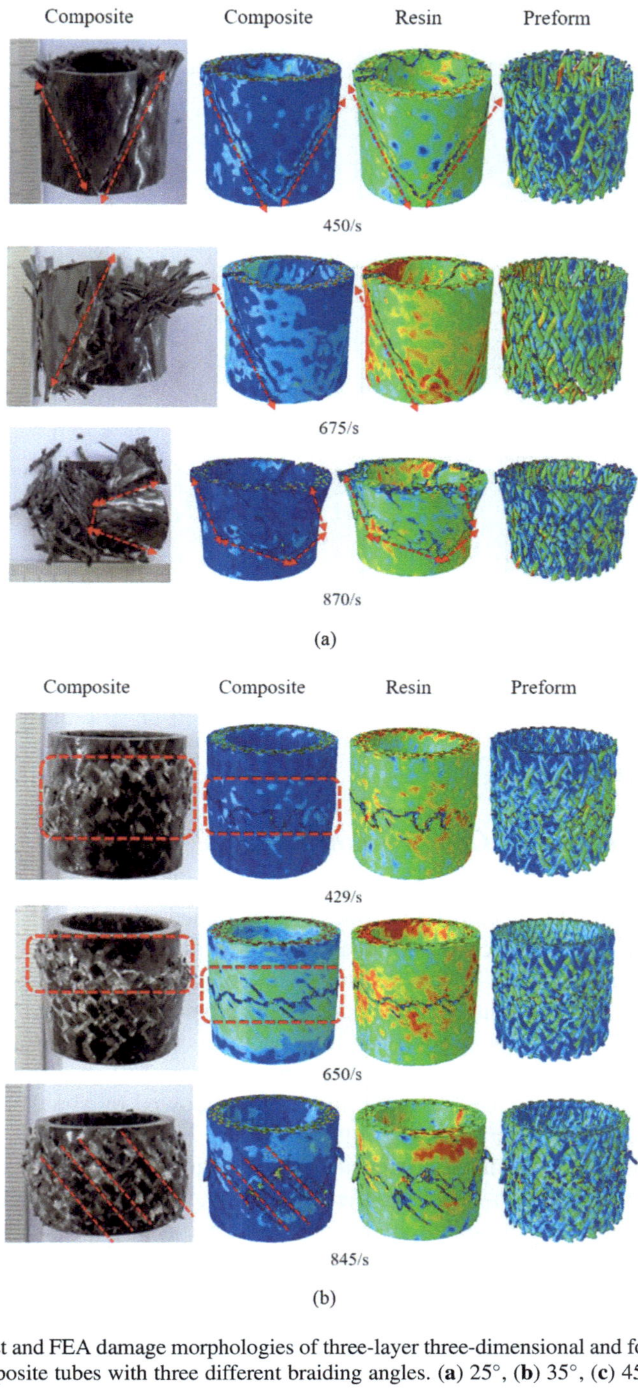

Fig. 9.6 Test and FEA damage morphologies of three-layer three-dimensional and four-directional braided composite tubes with three different braiding angles. (**a**) 25°, (**b**) 35°, (**c**) 45°

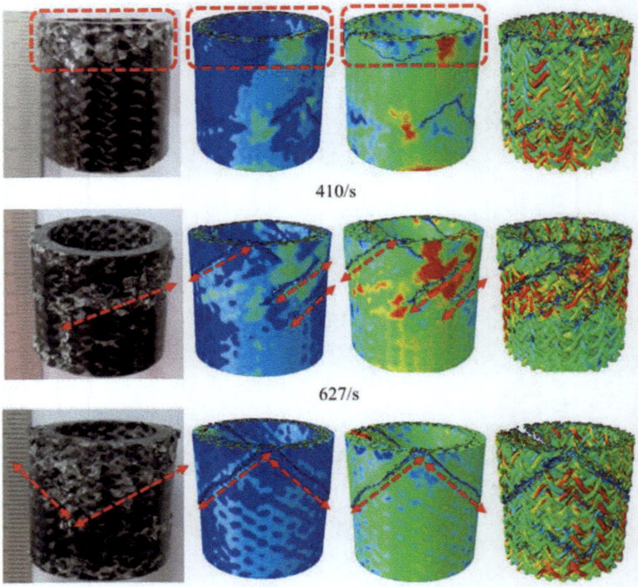

Fig. 9.6 (continued)

9.2 Braided Layers

9.2.1 Axial Impact Compression History

The impact compression damage histories of the braided preform and matrix in 3-D four-directional braided composite tubes with three different braided layers (two-, three-, and four-layer) at 442/s, 429/s, and 416/s are given in Figs. 9.7, 9.8, and 9.9, respectively. For the two-layer braided composite tube, it can be seen from Fig. 9.7b that the stress wave propagates onto the specimen with little variability in the propagation velocity over the braided yarn and resin at 130 μs. The stress of the composite tube reaches the maximum at 210 μs, followed by simultaneous initial damage to the braided yarn and resin at 220 μs [2].

For the three-layer braided composite tube, the stress wave propagation law is similar to that of the two-layer specimen as can be seen in Fig. 9.8. As the stress wave propagates, the stresses on the braided yarns and resin gradually increase. As can be seen in Fig. 9.8d, e, the stresses on the inner and outer surface yarns are greater than those on the interior yarns, and the stresses on the resin in contact with the inner and outer surface yarns are greater than those in the surrounding area. Significant yielding of the composite tube occurs at 200 μs, and the stress of resin

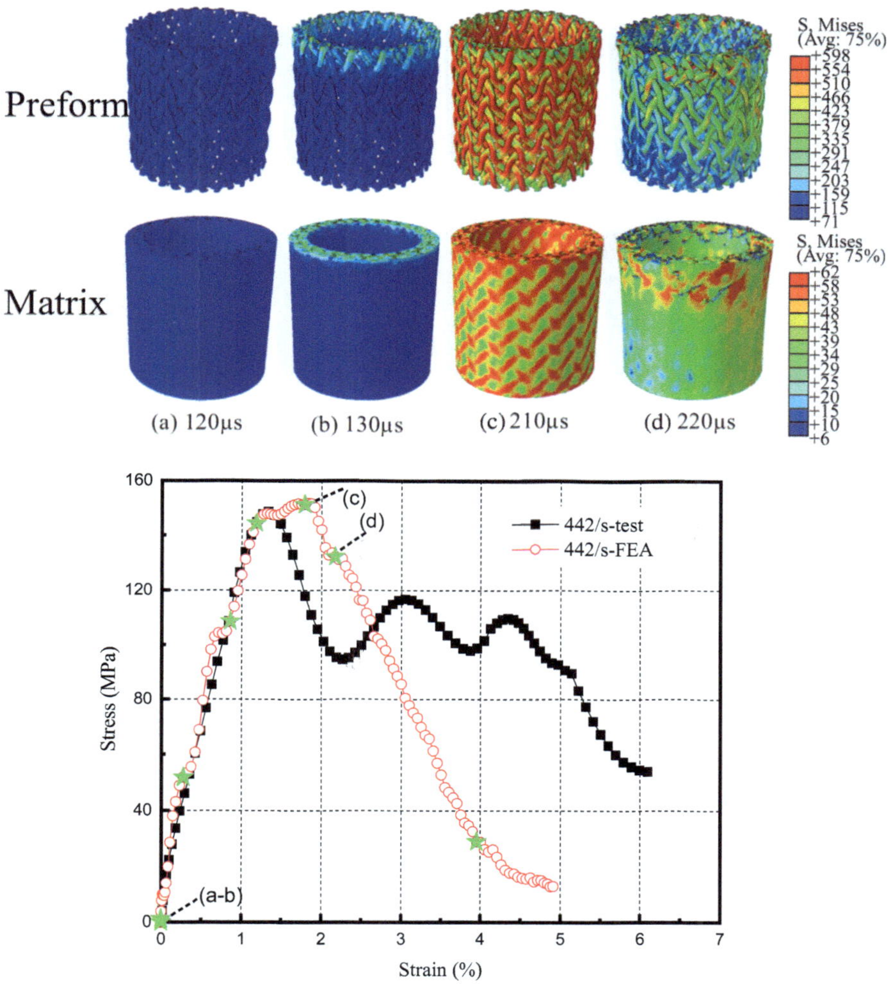

Fig. 9.7 Axial compression process of two-layer 3-D four-directional braided composite tube with the braiding angle of 35° under the strain of 442/s

reaches the maximum. At 230 μs, the whole composite reaches the maximum stress, and the initial damage to the braided yarn occurs, while the initial damage to the resin occurs at 260 μs.

For the four-layer composite tube (Fig. 9.9), the stress wave propagation characteristics and stress distribution law are similar to those of the three-layer specimen, with the material also yielding significantly at 200 μs. At this time, the stress of resin reaches the maximum. The difference is that due to the increase in the number of

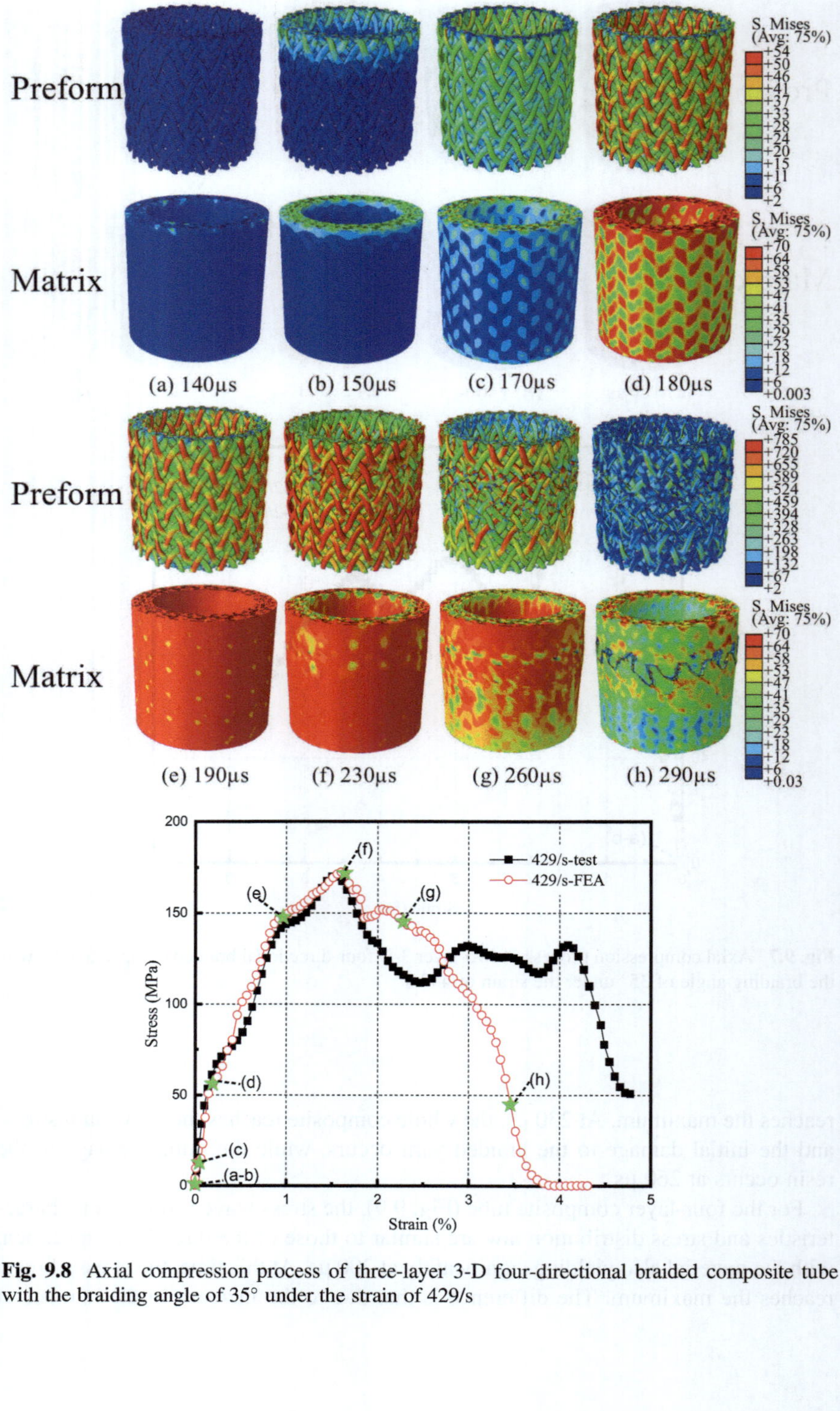

Fig. 9.8 Axial compression process of three-layer 3-D four-directional braided composite tube with the braiding angle of 35° under the strain of 429/s

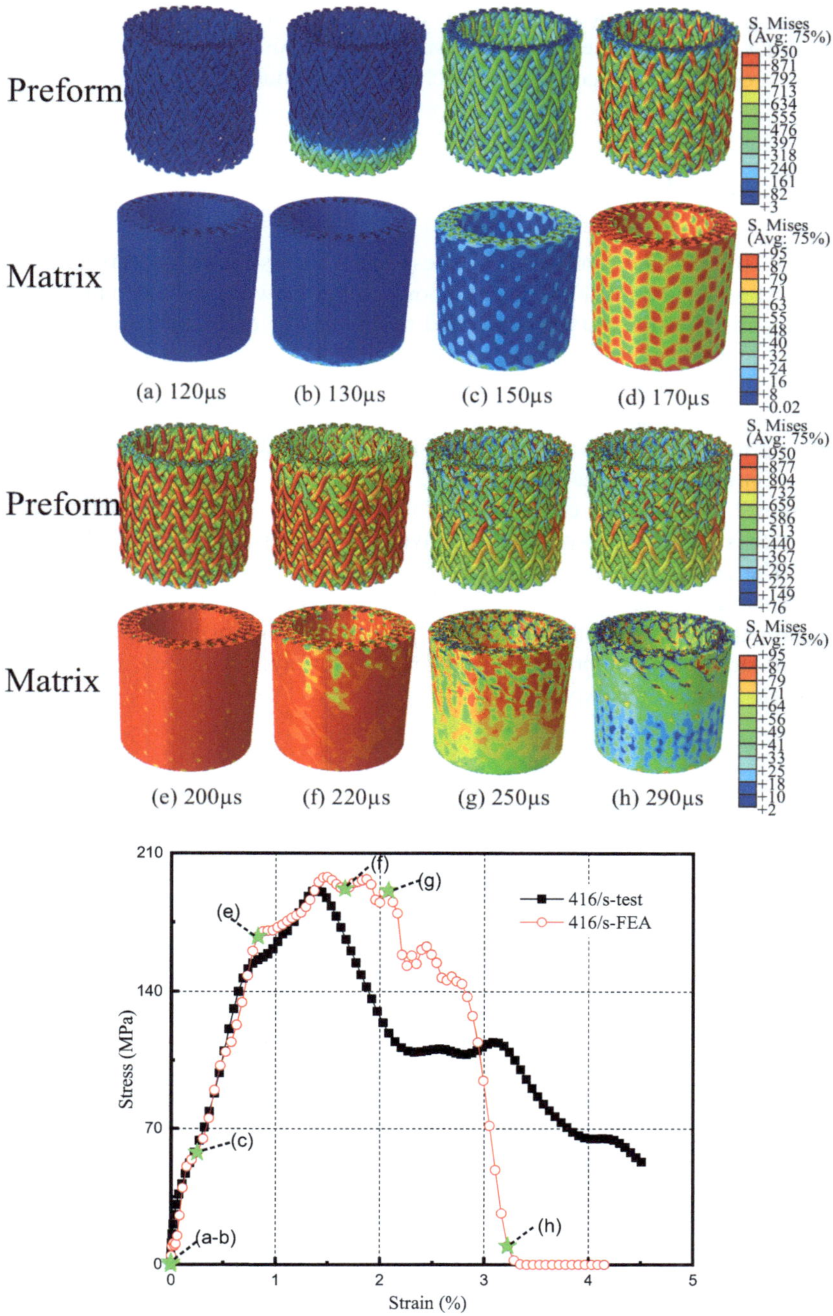

Fig. 9.9 Axial compression process of four-layer 3-D four-directional braided composite tube with the braiding angle of 35° under the strain of 416/s

braided layers, the braided preform has an increased ability to withstand the load. The initial damage to the yarn and resin occurs almost simultaneously when the overall stress of the composite tube reaches the maximum at 220 μs.

9.2.2 *Stress-Strain Curve*

Figure 9.10 shows the comparison of stress-strain curves for the two-layer and four-layer 3-D four-directional braided composite tubes between the test and FEA, and the comparison of the test and simulated results for the three-layer specimens with the same braiding angle have been given in Fig. 9.4b. It can be seen from the curves that it is similar to the composite tubes with different braiding angles. The overall trend of the curves shows that the test and the FEA stress-strain curves both exhibit nonlinear increasing trend and good consistency. The FEA stress-strain curves also have a significant strain rate dependence.

Figure 9.11 shows the comparison of test and finite element calculated strength and stiffness for the two-layer and four-layer 3-D four-directional braided composite tubes. The calculated stiffness of the FEA model for the two-layer and four-layer specimens agrees well with the test values. For the two-layer specimen, the finite element calculated strength is 2.2–7.3% greater than that of the experiment under different strain rate compression. And for the four-layer specimen, it is 2.0–9.0% greater than that of the experiment.

9.2.3 *Damage Morphologies*

Figure 9.12a, b shows the test and FEA simulation damage morphologies of the two-layer and four-layer 3-D four-directional braided composite tubes after axial impact compression under different strain rates. Similar to the above three-layer braided composite tube, the simulated damage morphologies also have a significant strain rate effect. The damage morphologies of the two-layer specimen at 676/s and 884/s are different from that of the three-layer and four-layer specimens. There are a large number of yarn-resin interfaces debonding at 676/s due to resin damage. At 884/s, the entire specimen is pulverized and no longer retains structural integrity due to the increase in impact pressure. While only the cracking and peeling off of the resin block and the fiber breakage can be seen in the finite element model, the debonding of the fiber-resin interface cannot be observed. The specimen still maintains structural integrity at 884/s. There is discrepancy between the experiment and the simulation. This is due to the fact that the interface between the fiber tow and resin is not considered in the finite element meso-structure modeling, but the interface between fiber tow and resin is regarded as perfectly bonded. In addition, the finite element model does not consider the defects and microcracks caused during the braiding process and actual preparation of the composite tube. Compared

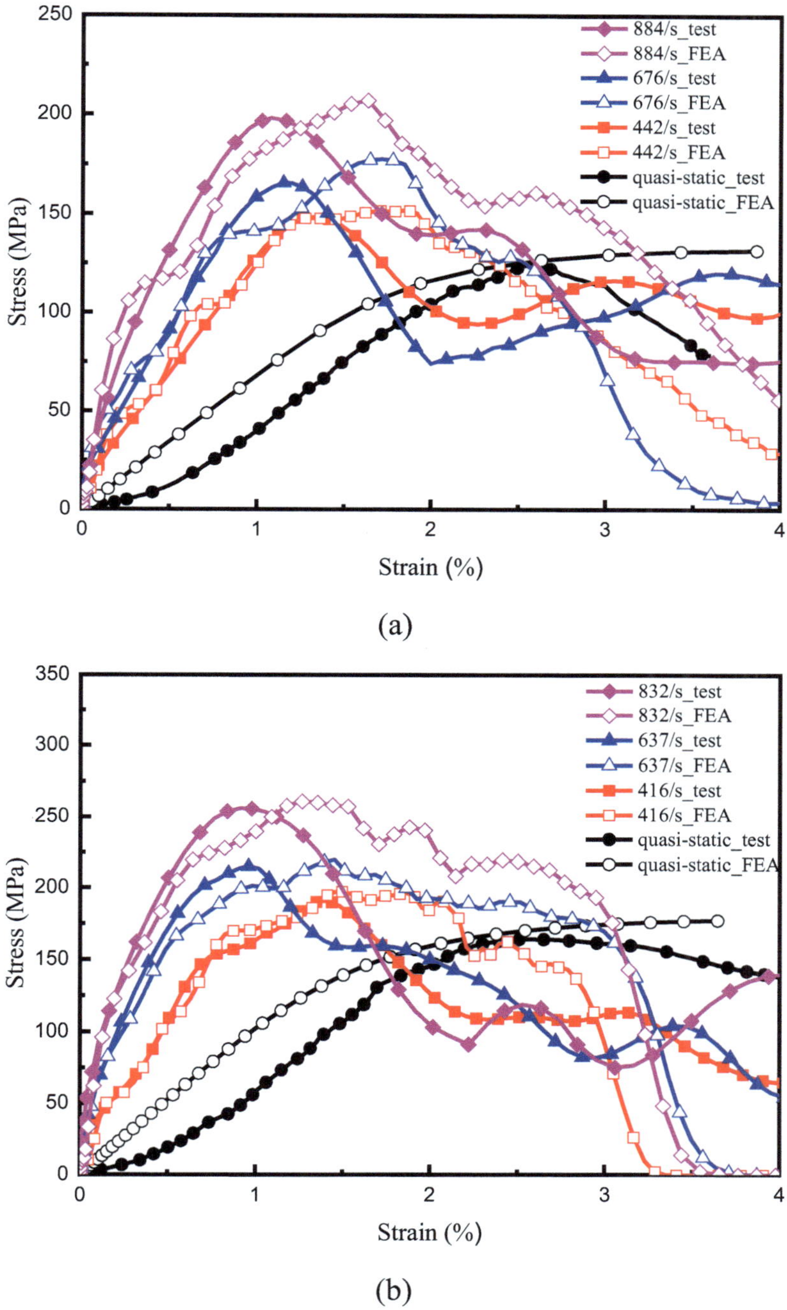

Fig. 9.10 Test and FEA stress-strain curves for 3-D four-directional braided composite tubes with the braiding angle of 35° and different braided layers. (**a**) Two-layer, (**b**) four-layer

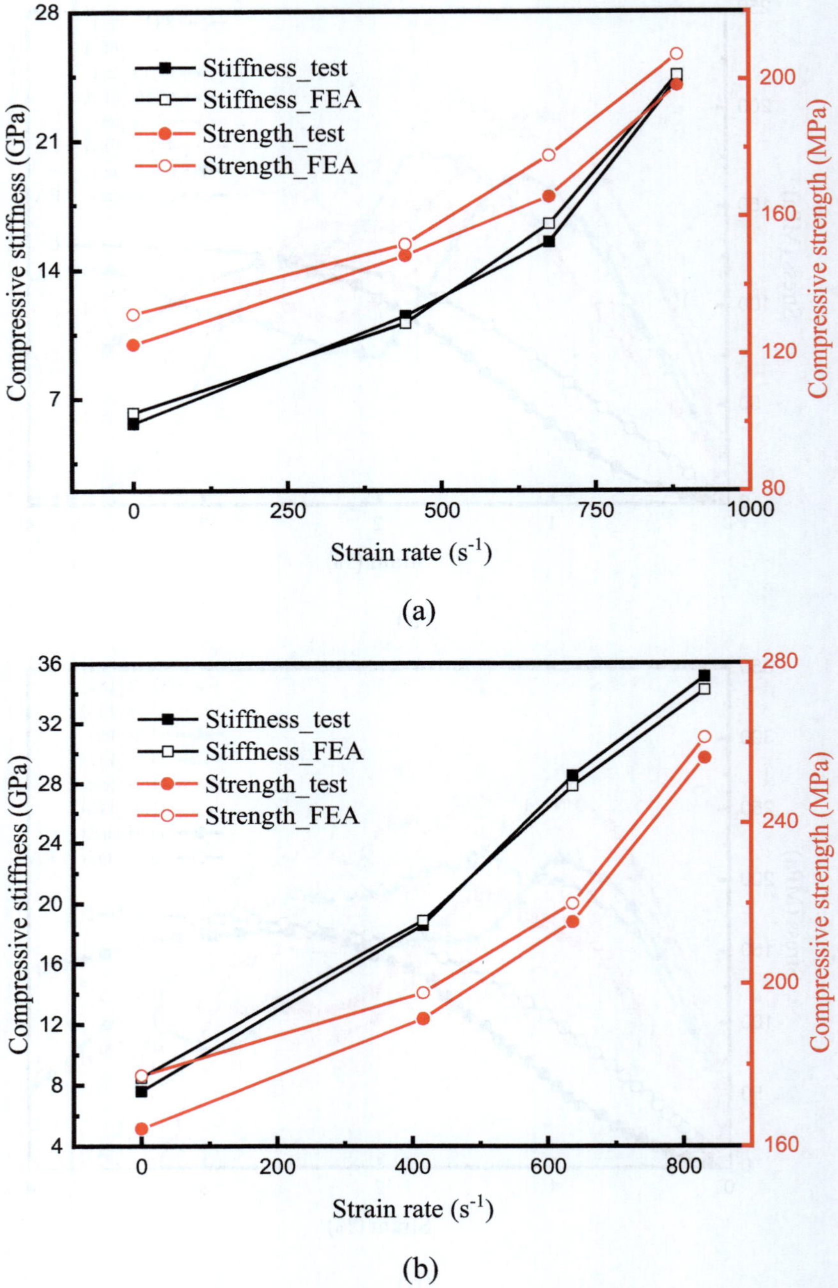

Fig. 9.11 Comparison of strength and stiffness for 3-D four-directional braided composite tubes with the braiding angle of 35° and different braided layers between test and FEA. **(a)** Two-layer, **(b)** four-layer

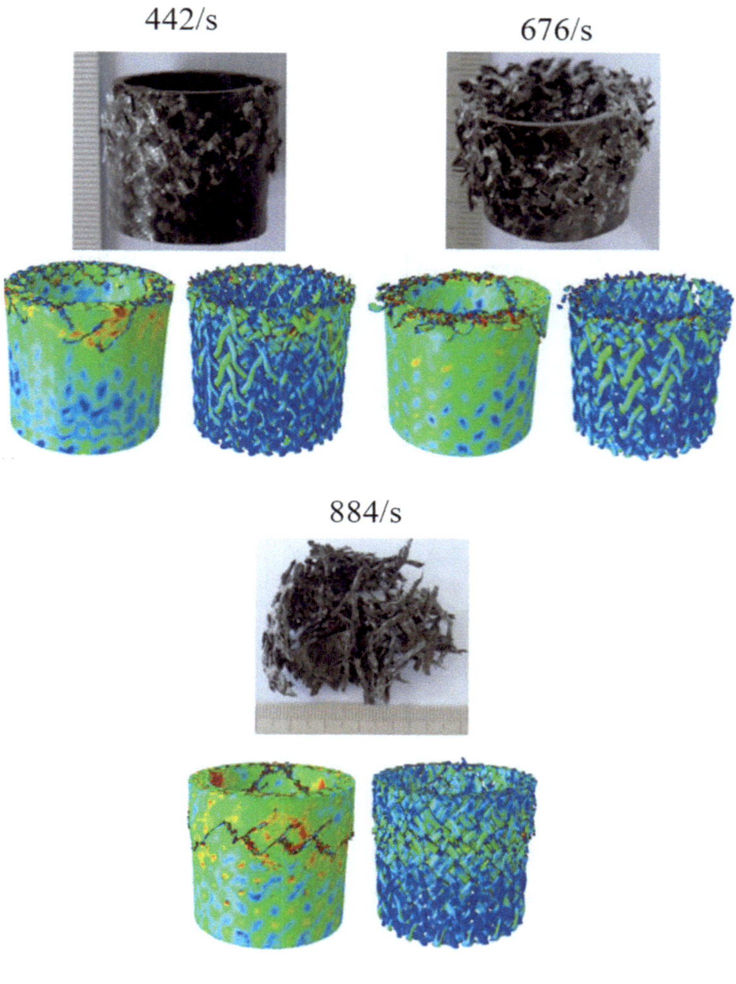

(a)

Fig. 9.12 Test and FEA damage morphologies for 3-D four-directional braided composite tubes with the braiding angle of 35° and different braided layers. (**a**) Two-layer, (**b**) four-layer

with the two-layer tube, the four-layer tube in Fig. 9.12b has higher axial compressive and shear stiffness. Only compressive buckling and lower shear damages occur under axial impact.

For the four-layer braided composite tube, the impact resistance is significantly improved due to the increase in the number of braided layers. The damage morphologies of specimens are less distinctly different than for the two-layer and three-layer specimens under three strain rates. The failure mode of the specimen is dominated by surface resin damage under impact compression loading with three strain rates. The

Fig. 9.12 (continued)

FEA damage morphologies are similar to the experiment. It can be seen that the number of braided layers also affects the failure mode and damage mechanism of the whole composite tube with the same braiding angle.

9.3 Axial Yarns

9.3.1 Quasi-static Axial Compression History

Figure 9.13a, b shows the quasi-static loaded finite element model and the finite element calculated stress-strain curve for 3-D four-directional braided composite tube, respectively. Four representative points a, b, c, and d on the curve are selected for stress analysis.

Figures 9.14 and 9.15 give the stress distributions at the four points corresponding to a, b, c, and d during the quasi-static axial compression deformation of the preform

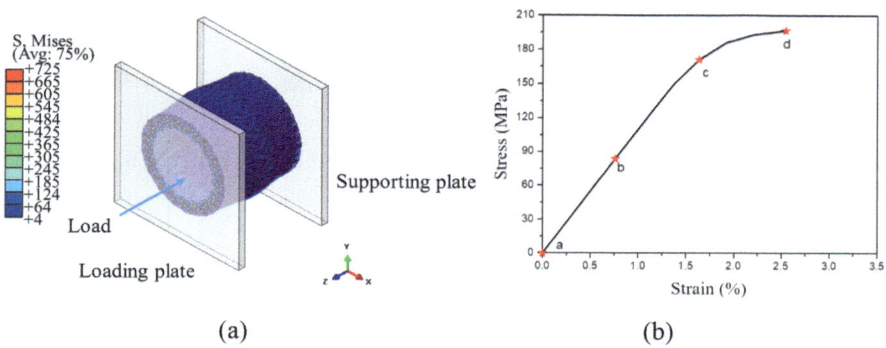

Fig. 9.13 Quasi-static axial compression of 3-D braided composite tube. (**a**) Loading method, (**b**) FEA stress-strain curve

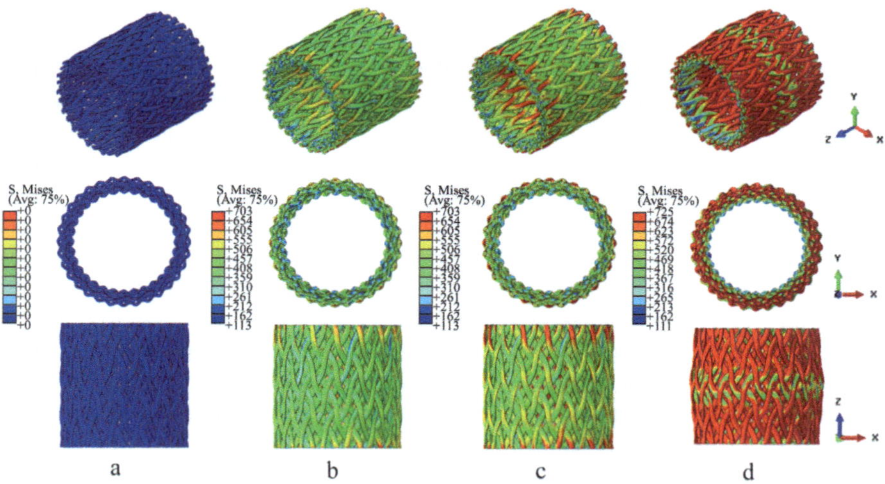

Fig. 9.14 Quasi-static compressive deformation process of the preform of 3-D four-directional braided composite tube

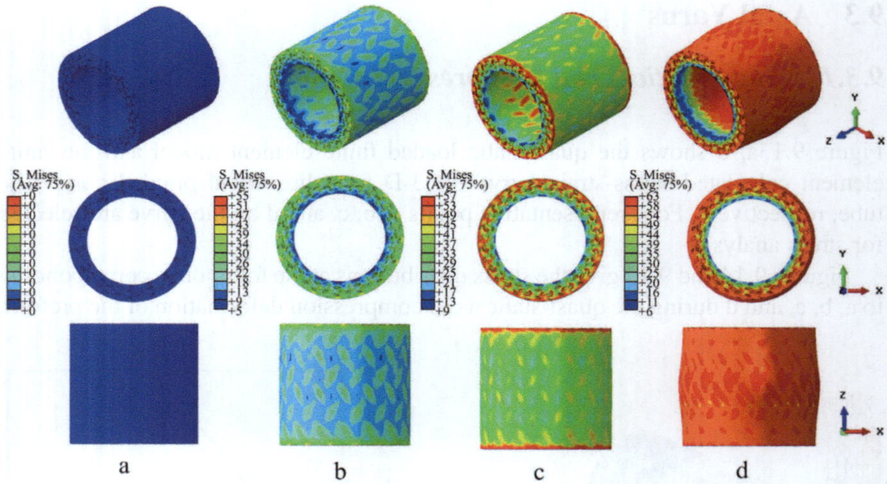

Fig. 9.15 Quasi-static compressive deformation process of the matrix of 3-D four-directional braided composite tube

and the resin of 3-D four-directional braided composite tube with the braiding angle of 25° from XYZ, YX, and ZX views, respectively. From the Mises stress distributions, it is clear that the stress (100–725 MPa) of the preform is significantly higher than those of the resin (<100 MPa) of 3-D braided composite tube. In the elastic stage b, the stress on the preform and matrix starts to distribute unevenly. The stress of the preform near the loading plate and the support plate on the inner and outer surface is large, and the stress of the middle part of the matrix contacting with the yarns and the outer circumference of the end contacting with the loading plate is the largest. At the yield point c, the uneven distribution of stress is more obvious; with the propagation of compression load, the overall stress of preform and matrix increases gradually and reaches the maximum at d. Due to the compression deformation of the yarns in the middle of the preform, the stress of the yarns on the inner surface contacting with the loading plate and the inner circumference of the end of the resin contacting with the loading plate decreases.

To further analyze the internal stress distribution of the material, 1/4, 1/2, and 3/4 μ-CT results of the tube are performed along the Z-axis in the length direction and Y-axis in the thickness direction in XYZ coordinate system to extract the stress distribution of the corresponding cross section of the tube as shown in Fig. 9.16. It again verifies that the stress of the internal preform of the composite tube is much greater than that of the matrix. The braided yarn has the most uniform stress distribution at the 1/2 cross section in the length direction, and the stress of the inner surface yarns is less than that of the outer surface yarns and the internal yarns.

The quasi-static compressive deformation process of the preform and matrix of 3-D five-directional braided composite tube is shown in Figs. 9.17 and 9.18. As can be seen from Fig. 9.17, the stress (≈1000 MPa) of the axial yarn reaches the

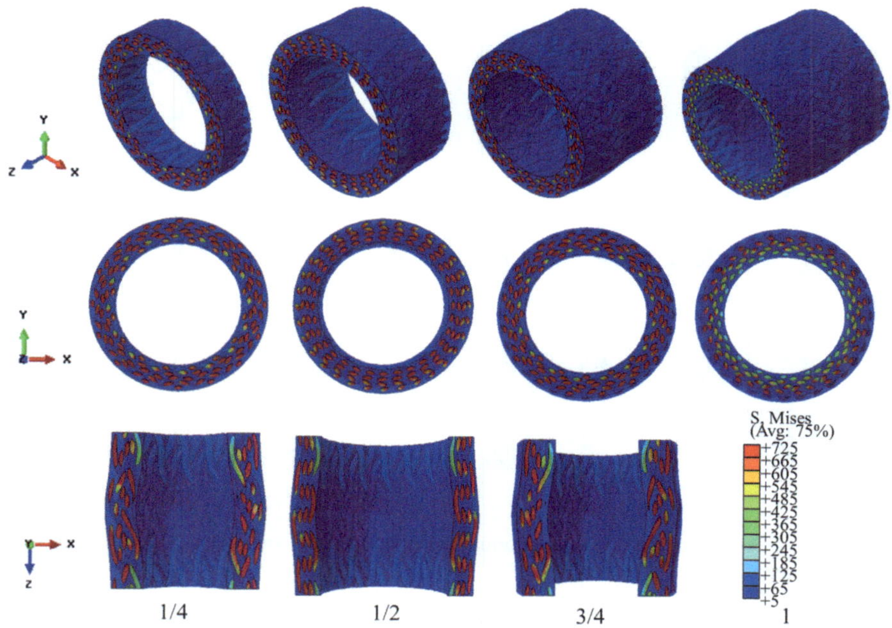

Fig. 9.16 Stress distribution in different sections of 3-D four-directional braided composite tube

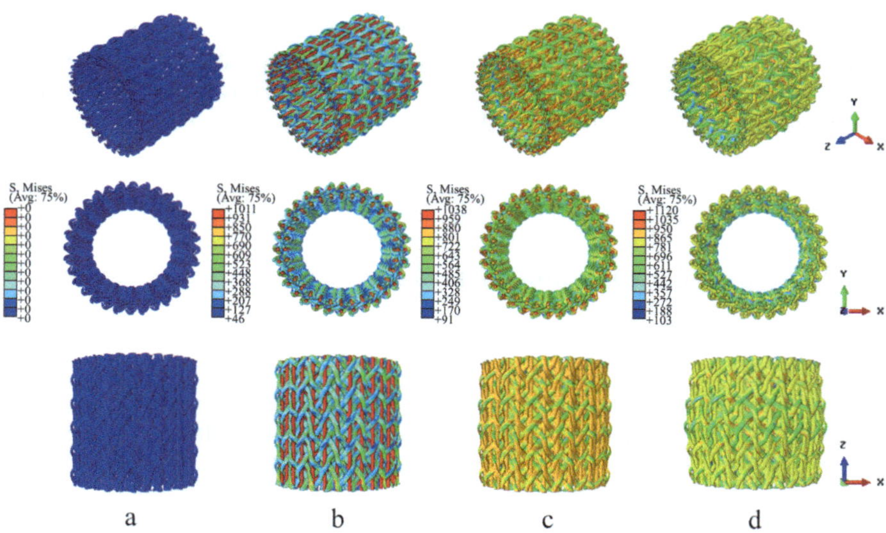

Fig. 9.17 Quasi-static compressive deformation process of the preform in 3-D five-directional braided composite tube

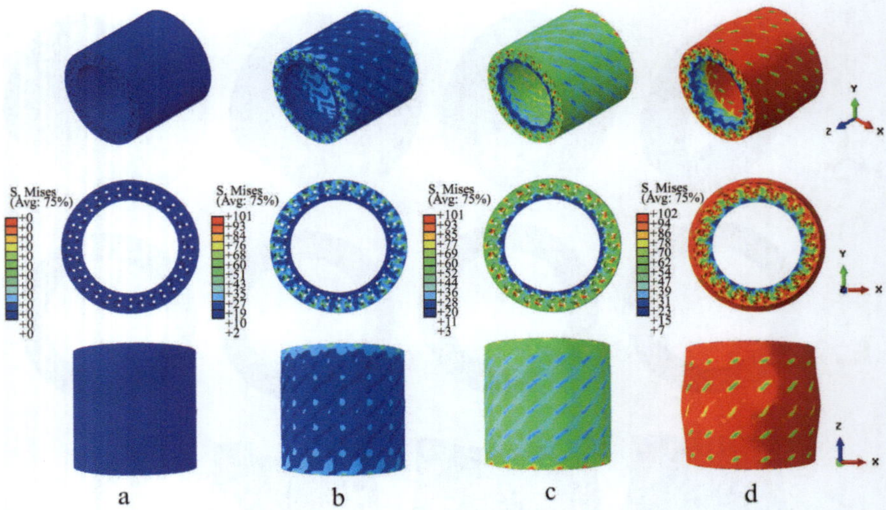

Fig. 9.18 Quasi-static compressive deformation process of the matrix in 3-D five-directional braided composite tube

maximum value and is much larger than that (≈368 MPa) of the braided yarn at the elastic stage b, indicating that the axial yarn bears the main load at this stage. The stress of the axial yarn decreases and that of the braided yarn increases at the yield point c and the peak stress d. The stress of both is finally in the range of 722–958 MPa. As seen in Fig. 9.18, the overall stress of the matrix is the largest at d. The stress on the resin-yarn contact area is the minimum along the braided yarn direction during compression. The stress distribution of 3-D five-directional braided composite tube along the length and thickness directions is shown in Fig. 9.19, and the 1/4, 1/2, and 3/4 sections meet the requirement of $\sigma_{\text{axial yarn}} > \sigma_{\text{braided yarn}} > \sigma_{\text{resin}}$.

9.3.2 Impact Compression Behaviors

9.3.2.1 Axial Impact Compression History

Figure 9.20 shows the impact compression response history of the preform and matrix of 3-D five-directional braided composite tube at 423/s. It can be shown that the stress wave propagates throughout the specimen at 150 μs as shown in Fig. 9.20c and the stress wave propagates significantly faster in the axial yarn than in the braided yarn. Also from Fig. 9.20c, d, it can be seen that the axial yarn is the main load-bearing part during the whole impact compression process and reaches the maximum stress first. The stress of the tube reaches the maximum as shown in Fig. 9.20e, when the stress of the entire braided preform and resin also reaches the maximum. Yielding of the composite tube occurs in Fig. 9.20d, with initial damage

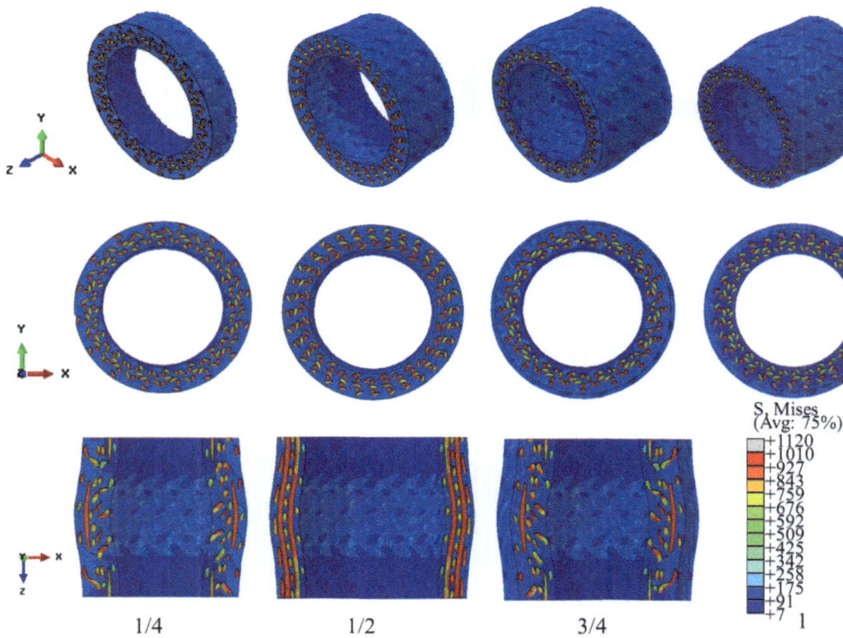

Fig. 9.19 Stress distribution in different sections of 3-D five-directional braided composite tube

to the braided yarn and axial yarn occurring at 220 μs and initial damage to the resin at 230 μs.

In order to further study the stress propagation and distribution law at different locations of the braided preform, five nodes are selected along the length and thickness direction of the braided preform in order to extract the stress-time curves for comparison. Five locations A, B, C, D, and E are selected along the length direction of the inner surface yarn, outer surface yarn, and axial yarn of the preform as shown in Fig. 9.21. The stress-time curves for the corresponding nodes are given in Fig. 9.22a–c. From Fig. 9.22a, it can be seen that the maximum stress distribution of the braided yarn on the inner surface is $\sigma_E > \sigma_D \approx \sigma_C \approx \sigma_B > \sigma_A$, which indicates that the middle part (B, C, and D) of the braided yarns on the inner surface is subjected to the major load, while the side (A) near the input end is subjected to the less load under axial compression loading. Since the damage occurs on the side of the specimen near the output end, the stress-time curve of node E exhibits significant fluctuations before reaching the maximum stress. The maximum stress distribution law of the braided yarn on the outer surface is $\sigma_E \approx \sigma_D > \sigma_C > \sigma_B > \sigma_A$, with less stress nearer to the loading end. For the axial yarn, the stress-time curves of the five nodes at different positions are more consistent. It can be found that the maximum stress of nodes on the yarns on the outer surface is larger than that on the inner surface and the stress of nodes on the axial yarn is significantly greater than that of the braided yarn on the inner and outer surface.

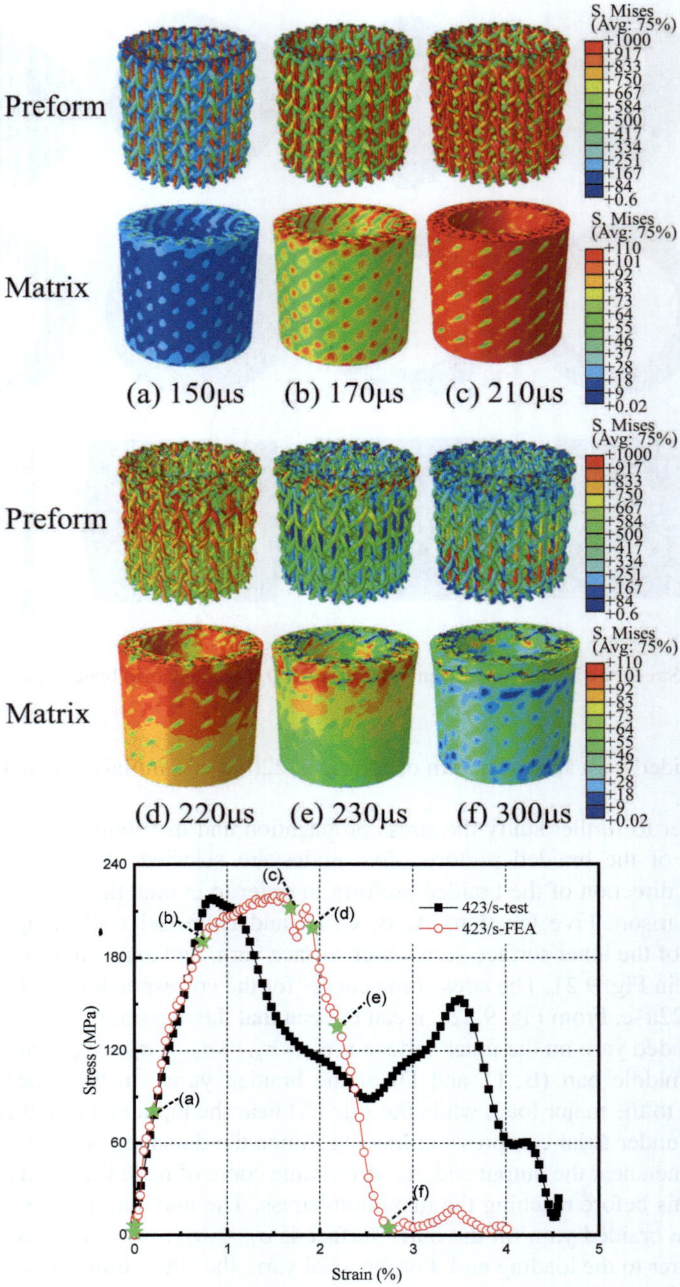

Fig. 9.20 Axial compression process of 3-D five-directional braided composite tube with the braiding angle of 35° at 423/s

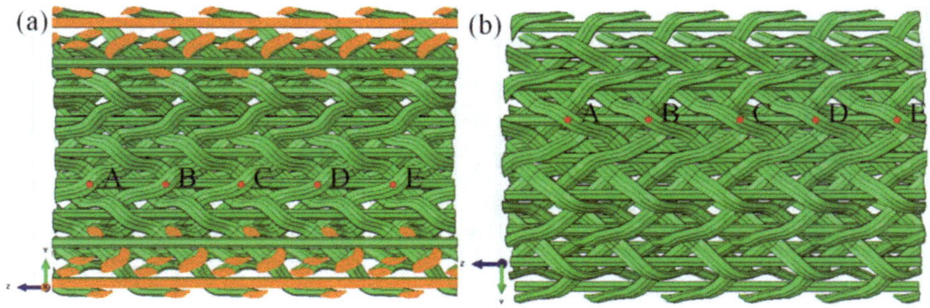

Fig. 9.21 Locations of selected nodes on the preform. (**a**) Inner surface of the braided yarns, (**b**) outer surface of the braided yarns

In order to study the nodal stress at different locations in the thickness direction of the braided preform, five nodes are selected on the cross sections of the braided yarn and axial yarn along the thickness direction of the preform, sequentially from the outer layer to the inner, and the selected locations are shown in Fig. 9.23a. The corresponding nodal stress-time curves are given in Fig. 9.23b. It can be seen that the maximum stresses are satisfied $\sigma_2 > \sigma_1 > \sigma_3$ at three nodes on the cross section of the braided yarn, which indicates that for the braided yarn the inner yarn bears more load and the outer surface yarn bears more load than the inner surface yarn. Also before reaching the maximum stress, the nodal stresses (4 and 5) on the axial yarn are both greater than the three nodal stresses (1, 2, and 3) on the braided yarn. The stress-time curves of the two are almost coincident, which is consistent with the abovementioned tendency of extracting nodal stresses along the length direction. In addition, it can also be seen from Fig. 9.23b that the nodal stresses (4 and 5) on the axial yarn take significantly less time to reach the maximum stress than the nodal stresses (1, 2, and 3) on the braided yarn, which again verifies that the stress wave propagates faster on the axial yarn than on the braided yarn.

Figure 9.24 shows the locations of the three nodes (A, B, and C) selected at the left, middle, and right positions of the resin along the axial and thickness directions of the specimen, respectively. The corresponding stress-time curves are given in Fig. 9.25. Comparison of Fig. 9.25a, b shows that the stress-time curves of the nodes selected at different positions along the axial and thickness directions of the resin have similar laws and the maximum stress is almost equal, which indicates that the stress propagation in the resin is uniform and the resin plays a role in propagating the stress in the impact compression of the composites. The nodal stresses on the resin are much less than those on the yarn, so the damage behavior of the braided composite tube in axial compression is mainly controlled by the mechanical behavior of the yarn.

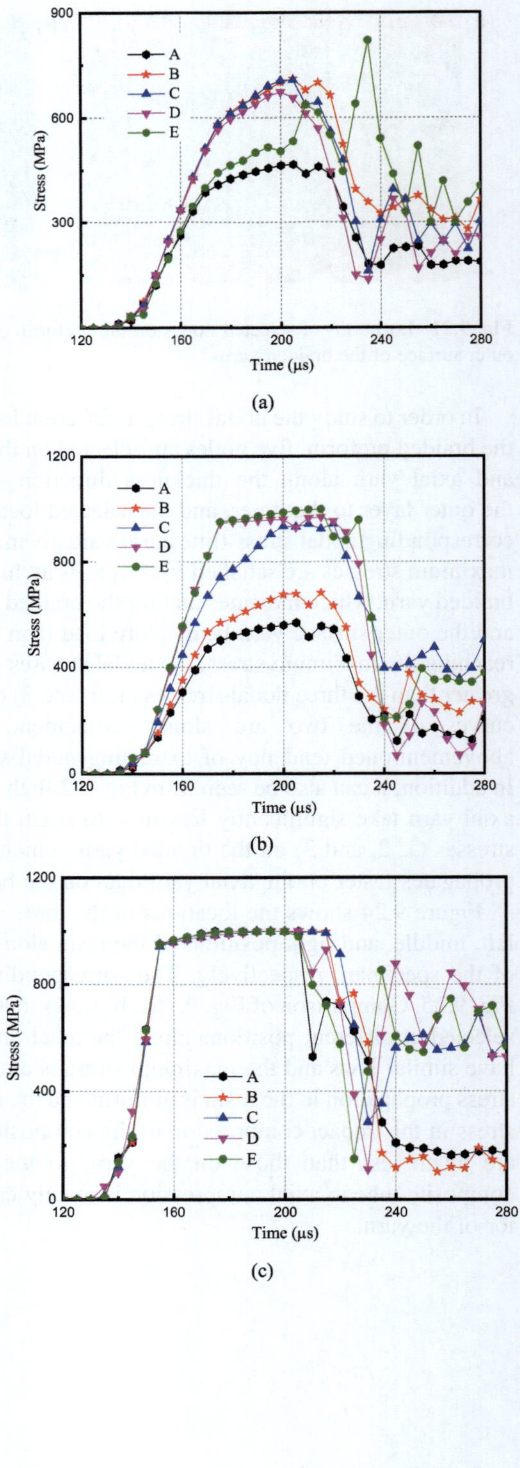

Fig. 9.22 Stress-time curves for different nodes selected along the axial direction of the braided preform. (**a**) Braided yarns on the inner surface, (**b**) braided yarns on the outer surface, (**c**) axial yarns

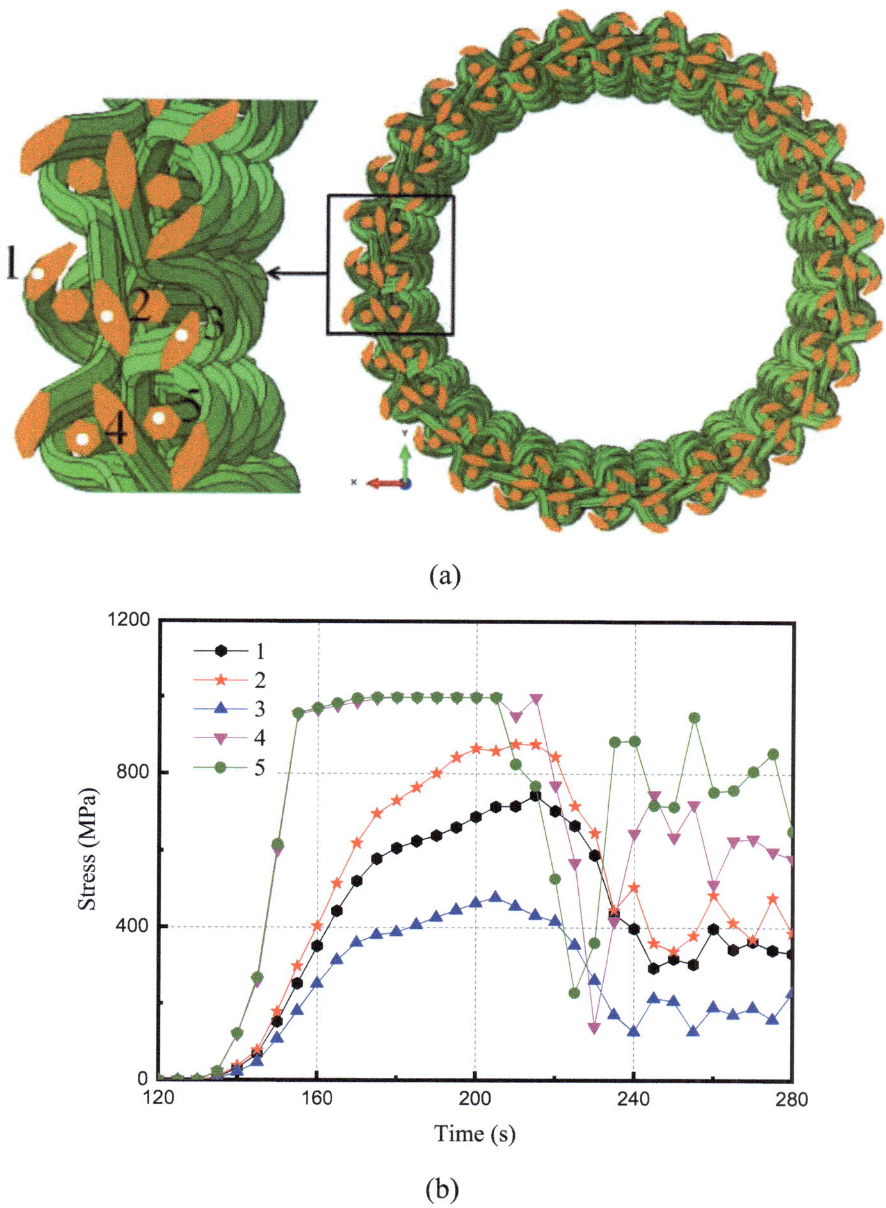

Fig. 9.23 Selected nodal stress-strain curves along the thickness direction of the preform. (**a**) Selected nodes along the thickness direction of the preform, (**b**) nodal stress-time curves

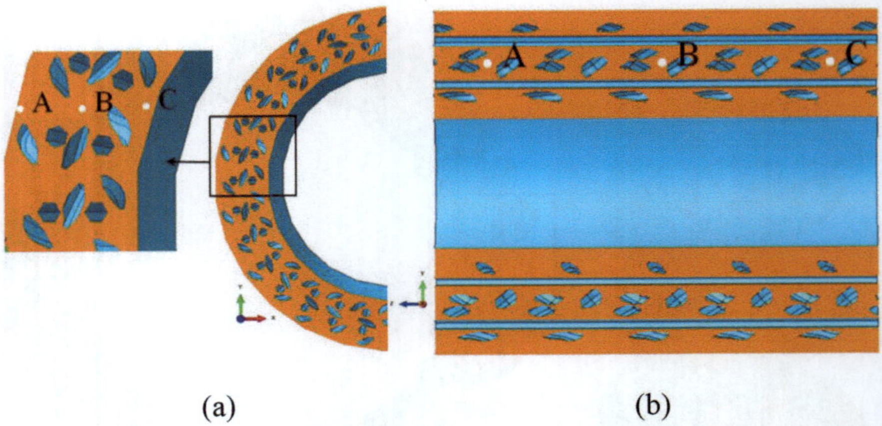

Fig. 9.24 Special nodes selected from the resin. (**a**) Along the thickness direction, (**b**) along the axial direction

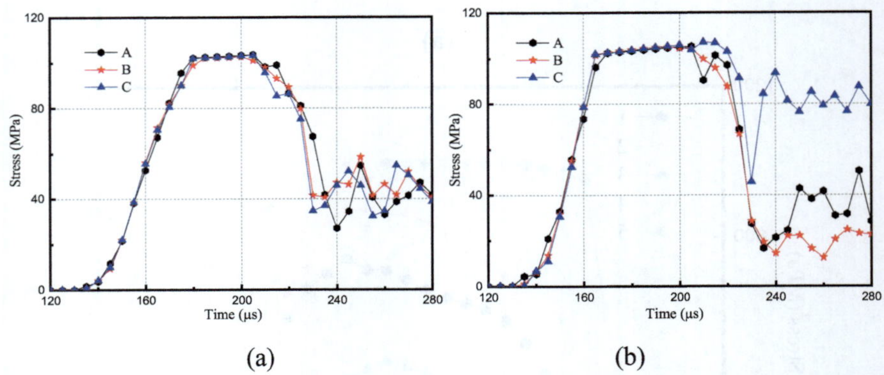

Fig. 9.25 Stress-time curves of nodes on the resin. (**a**) Along the thickness direction, (**b**) along the axial direction

9.3.2.2 Stress-Strain Curves

Figure 9.26 shows the comparison of the test and FEA stress-strain curves for 3-D five-directional braided composite tube. The comparison of strength and stiffness between the test and FEA is shown in Fig. 9.27. Similar to the results discussed above for the 3-D four-directional braided composite tube, the FEA stiffness of the composite tube agrees well with the test values, and the FEA strength of the composite tube is 0.6–8.5% higher than the test values for similar reasons. The comparison results again validate the feasibility of the 3-D meso-structure model for predicting the compression properties of 3-D braided composite tubes.

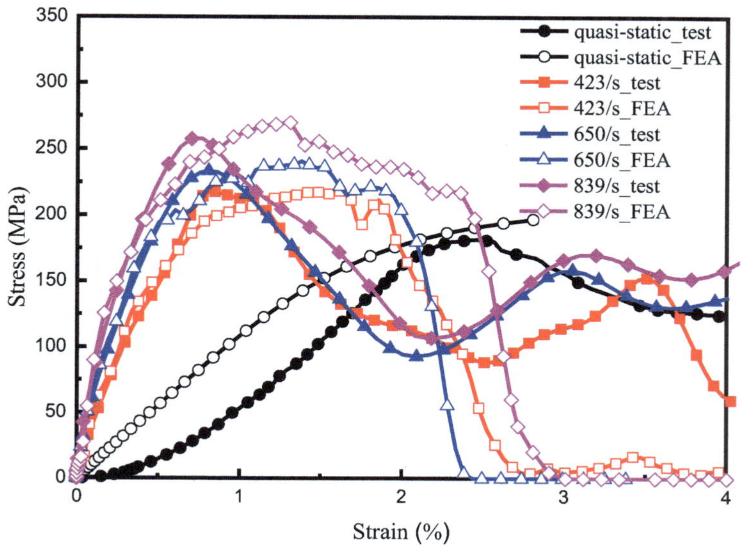

Fig. 9.26 Test and FEA stress-strain curves for 3-D five-directional braided composite tube with the braiding angle of 35°

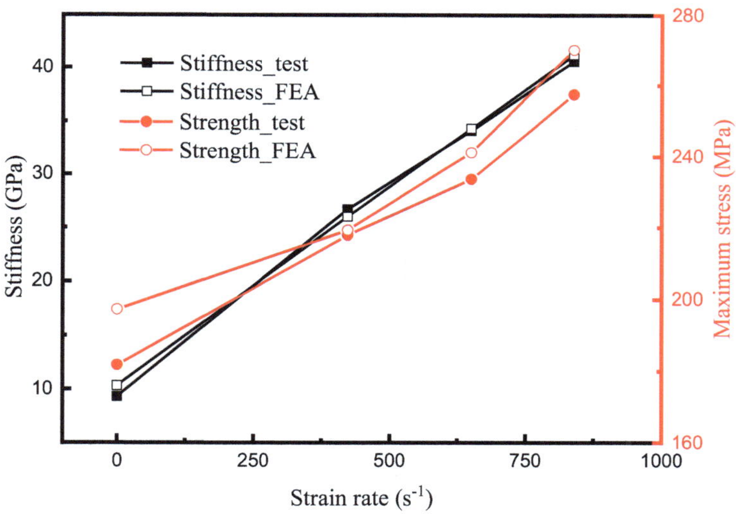

Fig. 9.27 Comparison of test and FEA results of stiffness and maximum stress for 3-D five-directional braided composite tube with the braiding angle of 35°

Fig. 9.28 Test and FEA damage morphologies of the two-layer 3-D five-directional braided composite tube with the braiding angle of 35°

9.3.2.3 Damage Morphologies

Figure 9.28 shows the test and FEA damage morphologies of the two-layer 3-D five-directional braided composite tube after axial impact compression under different strain rates. Compared with the above damage morphologies of the two-layer 3-D four-directional braided composite tube, it can be found that the axial impact resistance of 3-D five-directional braided composite tube is significantly improved.

At the strain rates of 423/s and 650/s, the damage morphology is mainly resin failure at the contact end with the loading plate. Under the high strain rate compression loading of 839/s, the specimen still maintains structural integrity with a large amount of interface debonding between the yarns and resin. The FEA results are in good agreement with the test results under different strain rates.

References

1. Xianyan Wu, Qian Zhang, Wei Zhang, Bohong Gu, Baozhong Sun: Axial compressive deformation and damage of four-step 3-D circular braided composite tubes under various strain rates. J. Text. Inst. **107**(12), 1584–1600 (2016)
2. Xianyan Wu, Bohong Gu, Baozhong Sun: Comparisons of axial compression behaviors between four-directional and five-directional braided composite tubes under high strain rate loading. J. Compos. Mater. **50**(28), 3905–3924 (2016)

Chapter 10
Impact Strength with Thermo-mechanical Coupling Effect

This chapter discusses the structural and temperature effects of stress distribution in 3-D braided composites under impact loading through the results of finite element analysis. The influences of environment temperature on the internal damage distribution in combination with μ-CT technique are analyzed to reveal the thermo-mechanical coupling damage and energy absorption mechanisms.

10.1 Structural Effect at Room Temperature

Figure 10.1a shows the stress distribution, interfacial damage, and adiabatic temperature rise in the longitudinal section of the composite through the impact point along the Z-axis. The composite maintains structural integrity at the impact pressure of 0.2 MPa. It can be seen that the preform is the main load-bearing structure of the composite beam and plays a dominant role in the plastic deformation of the whole beam. The continuous inelastic deformation leads to an increase in adiabatic thermal energy because the heat generated by the local violent plastic deformation cannot be dissipated immediately. The adiabatic temperature rise is concentrated at the impact point and the back. The interface is completely damaged when the value of the damage variable D (see Chap. 2, Sect. 2.3) reaches 1. Local interface damage is present in the 4D30 specimen, while there is almost no damage in the 5D30 specimen.

When the impact pressure increases to 0.4 MPa, the resin of 4D30 specimen produces semi-V-shaped shear fracture, and the braided yarns are twisted and deformed, resulting in relative position slip. The adiabatic temperature rise and interfacial damage are concentrated in the region with geometrical defects, i.e., the semi-V-shaped fracture and the impact surface, which undergoes local violent plastic deformation. However, when the resin of the 5D30 specimen cracks from the back of the impact point, the presence of the axial yarn prevents further crack extension and protects the structural integrity of the preform, and the presence of the axial yarn

M. Hu, B. Gu, *Impact Damages of Braided Composites*, Engineering Materials,
https://doi.org/10.1007/978-981-16-5703-0_10

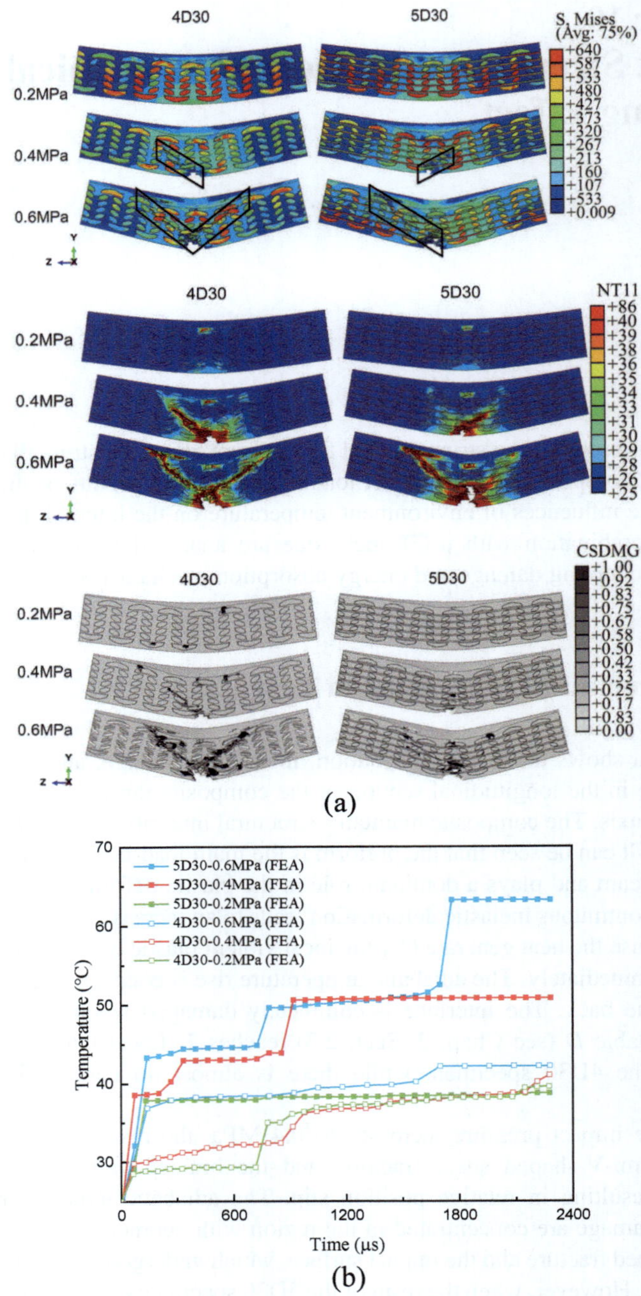

Fig. 10.1 Impact damage of 3-D four-/five-directional braided composites. (**a**) Stress distribution, interfacial damage, and adiabatic temperature rise in the longitudinal section of the composite through the impact point along the Z-axis, (**b**) temperature history of the composite at the impact point

inhibits the deformation of the braided yarn due to the transverse impact compressive stress, while the axial yarns near the back surface are subjected to tensile stress, which causes the elongation and deformation. When the back stress reaches the critical value, the resin cracks, and the heat generated raises the temperature, which leads to the thermal expansion of the resin. This in turn causes the resin and yarn to squeeze each other, generating internal stress and local deformation, forming a thermal stress coupling closed loop. The more severe the damage, the more energy is generated.

At 0.6 MPa, the 4D30 specimen shows V-shaped shear fracture in the longitudinal section and relative slip and twist deformation in many of the braided yarns. Due to the deformation caused by high local tensile stress around the yarns, the yarn-resin interface is damaged. After significant plastic deformation and eventual damage evolution, resin fragmentation of localized regions results in adjacent yarn contact. When this occurs, the strength of the composite increases locally until the shear bands merge to form the V- or semi-V-shaped failure mode, as shown in Fig. 10.1a. Thus, the 5D30 specimen has more yarns transferring the applied load through contact after the surrounding resin is broken under transverse load, resulting in higher peak load (see Fig. 4.26a). In addition, the yarns of the 5D30 specimen are deformed near the cracks, and the axial yarns are significantly bended and deformed. The outermost axial yarn on the back of the composite is broken, and the braided yarn is damaged. The presence of the axial yarns prevents further deformation of the braided yarn. It is worth mentioning that the braided structure has a significant effect on the failure path. The stress concentration leads to a local increase in damage near the crack, which leads to an increase in temperature under tensile stress. The interfacial damage is more severe for 4D30 specimen than for 5D30 specimen, while the area of interfacial damage is much larger than for 5D30 specimen. For 5D30 specimen, smaller deformation induces less interfacial stress, which reduces interfacial damage. The above results further indicate that the addition of axial yarn can improve the damage resistance of the composites.

Under the combined effect of thermal and mechanical loads, the constitutive equation of the resin is:

$$\{d\varepsilon_i^m\} = \left[S_{ij}^m\right]\left\{d\sigma_j^m\right\} + \{\alpha_i^m\}dT \tag{10.1}$$

where α is the thermal expansion coefficient of the resin at the initial temperature T_0. The internal stress of the resin $\{d\sigma_i^m\}$ is the sum of the mechanical stress caused by the mechanical load and the thermal stress caused by the temperature change, i.e.:

$$\{d\sigma_i^m\} = \{d\sigma_i^m\}^{(M)} + \{d\sigma_i^m\}^{(T)} \tag{10.2}$$

Deformation plasticity theory provides fracture mechanic applications for the plastic analysis of ductile composites. By using von Mises stress potential and related flow law, the nonlinear relation is extended to multiaxial stress state, giving a multiaxial model of:

$$E\boldsymbol{\varepsilon} = (1+\upsilon)\mathbf{S} - (1-2\upsilon)p\boldsymbol{I} + \frac{3}{2}\alpha\left(\frac{q}{\sigma_0}\right)^{n-1}\mathbf{S} \tag{10.3a}$$

$$p = -\frac{1}{3}\boldsymbol{\sigma} : \boldsymbol{I} \tag{10.3b}$$

$$q = \sqrt{\frac{3}{2}\mathbf{S} : \mathbf{S}} \tag{10.3c}$$

$$\mathbf{S} = \boldsymbol{\sigma} + p\boldsymbol{I} \tag{10.3d}$$

where $\boldsymbol{\varepsilon}$ is the strain tensor, $\boldsymbol{\sigma}$ is the stress tensor, p is the equivalent hydrostatic pressure, q is the von Mises equivalent stress, $\mathbf{S}$ is the deviatoric stress, and υ is the Poisson's ratio.

Figure 10.1b shows the temperature history at the impact point of the composite. It can be noticed that during the first impact, the temperature rises abruptly, after which the temperature increases gradually in stages. When the stress wave is transmitted to the composite, the temperature increases sharply; when the stress wave is reflected back, the temperature remains essentially constant. This is due to the stress wave propagating to the composite beam during each impact cycle, which converts the plastic deformation energy of the composite into heat energy. As the impact pressure increases, the temperature increases significantly. As the impact energy increases, the energy that can be converted into heat energy also increases. However, the preform structure has a more significant effect on the adiabatic temperature rise. For the 5D30 specimen, the addition of the axial yarn allows more impact energy to be absorbed and increases the damage tolerance, allowing a large amount of energy generated by the impact to be stored as plastic deformation energy, which is further converted into heat [1].

As can be seen in Fig. 10.2a, the ductile damage in the composite is concentrated at the cracking and is greater in the lower half than in the upper half. This is because the strain at the back under tensile stress is greater than the strain at the impact surface under compressive stress, as can be seen in Fig. 10.2b. The plastic strain energy generated on the back of the composite is greater, and the ductility damage is more severe. In addition, the ductile damage of the preform is much greater than that of the resin. It is noteworthy that as the impact pressure increases, the ductile damage of the axial yarn is greater than that of the braided yarn. Considering that the axial yarn has the largest specific energy absorption in Fig. 4.27b, the axial yarn plays an important role in enhancing the damage resistance of the composite. A significant M-shaped boundary in the region of maximum ductile damage is observed in the cross section of the composite beam. This may be due to the fact that the stress wave propagates in the Y-direction in the composite beam under transverse impact loading, while the ductile composites generate the anti-deformation force in the Y-direction to resist deformation. The microscopic forces in the composite interact with each other and partially cancel each other, resulting in the ductile damage distribution.

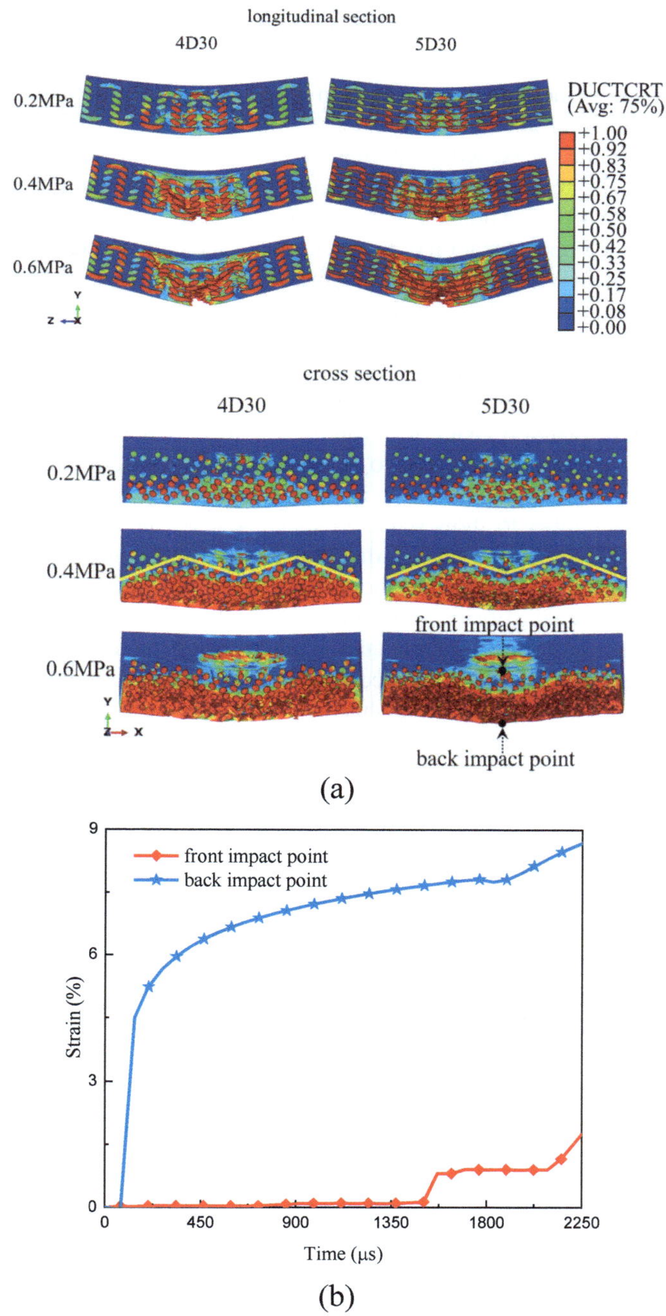

Fig. 10.2 Ductile damage of composites. (**a**) Ductile damage distribution in the sections of composites, (**b**) strain history at the front and back impact points of 5D30 specimen

10.2 Structural Effect at High Temperature

Figure 10.3 shows the damage propagation and temperature evolution of the composite for the first four impact cycles under the impact pressure of 0.6 MPa at 210 °C. It is found that the weakened properties of the epoxy resin at high temperature greatly reduce the resistance to impact loading for the composite beam with the braiding angle of 40°. The braided preform plays a dominant role in the resistance to deformation. The impact location of the preform has the highest stress, and the stress propagates outward with it as the center of the circle, but the stress distribution of the matrix is similar to that of the strip. The adiabatic temperature rise distribution of the matrix is mainly concentrated in the impact zone and the clamping zone. It is known that the adiabatic temperature rise is concentrated on both the front and back sides of the specimen. For the composite beam with the braiding angle of 30°, the deformation resistance is relatively reduced. It can be seen that the maximum stress concentration region changes from circular to cross-shaped. The adiabatic temperature rise distribution of the matrix is similar to its stress distribution, and the heat and stress interact with each other to show scattered spots. And for the composite beam with the braiding angle of 20°, the stress is distributed over almost the entire preform. The stress distribution in the matrix has the same distinct spots. Considering the difference in CTE between the fiber tow and the resin, resulting in the stress concentration, exhibits the phenomenon. Similarly, the adiabatic temperature rise distribution is expanded in an elliptical shape and accompanied by distinct scattered spots [2].

Figure 10.4 shows the temperature history of the resin surface extracted along the X-axis for the first four impact cycles through the impact point under the impact pressure of 0.6 MPa at 210 °C. The extrusion contact between the surface braided yarns and the resin makes the curve fluctuate regularly. Compared to the composite beam with the smaller braiding angle, the preform of the 40° specimen is more tightly interwoven which improves the structural integrity of the composite upon impact. Therefore, from small braiding angle to large braiding angle, this law gradually weakens. The maximum adiabatic temperature rise is concentrated at the impact point, while the temperature on both sides decreases. In addition, we extracted the temperature history of the composite beam at the impact point for three braiding angles. It can be seen from Fig. 10.5 that the temperature increases abruptly at the first impact. For composite beams with large braiding angles, more impact energy is absorbed, increasing the damage tolerance and leading to further conversion of the large amount of energy generated during impact into plastic deformation energy into heat. From the curves, we can see that the adiabatic temperature rise of the resin is largest for the braiding angle of 40°. We also find that the temperature rise at the impact point is not the highest. Figure 10.6 shows the temperature history of the resin surface of the 20° composite beam through the impact point along the Z axis during the third impact cycle under the impact pressure of 0.6 MPa at 210 °C. It can be seen that the maximum temperature rise occurs at the clamping zone. Given that the resin fractures at the clamping zone, the higher the degree of damage, the more heat is generated. Since the resin undergoes physicochemical changes at 210 °C, becoming softer and tougher, tending to rubberize, its

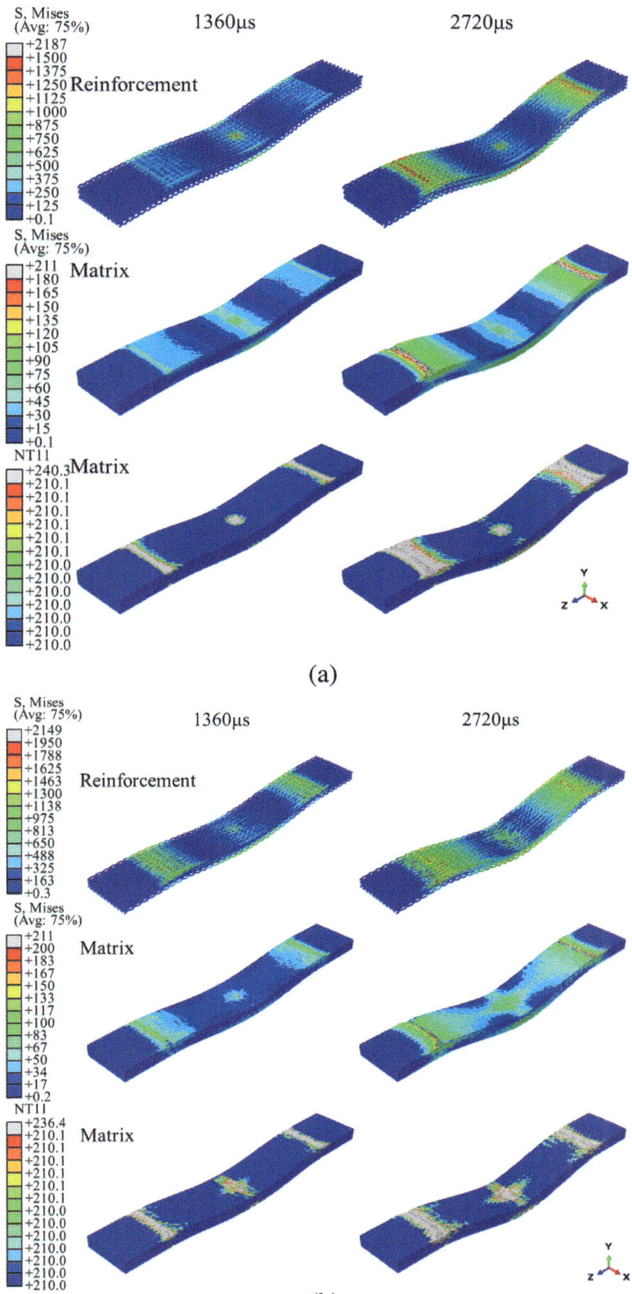

Fig. 10.3 Damage propagation and temperature evolution of composites for the first four impact cycles under the impact pressure of 0.6 MPa at 210 °C. (**a**) 40°, (**b**) 30°, (**c**) 20°

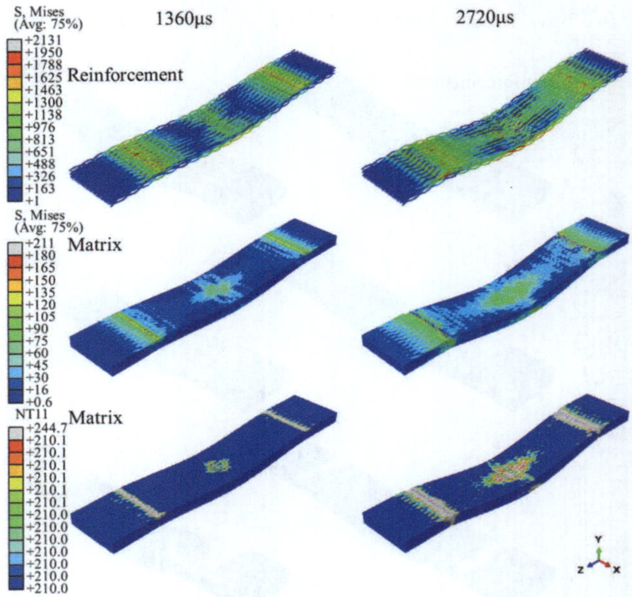

(c)

Fig. 10.3 (continued)

damage state changes from brittle to ductile damage, so the sharp temperature rise occurs in the damage stage, not the elastic-plastic stage [3]. Figure 10.7a shows the stress distribution on the back of the resin during the damage propagation. The stress concentration is particularly pronounced in the direction of braiding angle but tends to be uniform at the braiding angle of 40°. This is because the preform structure of the composite with large braiding angle is denser and the integrity is better during impact. Combined with Fig. 10.7b, it can be found that there is a significant ductile fracture of the resin along the braided yarn, which is consistent with the higher temperature at the extrusion contact between the surface braided yarn and resin. It can be seen that the larger braiding angle not only increases the damage tolerance of the composite but also improves the impact extension characteristics of the cracks.

10.3 Temperature Effect

Figure 10.8 shows the temperature history at the impact point for the composite beam with the braiding angle of 40° at different temperatures. We find that the lower the ambient temperature, the larger the adiabatic temperature rise span. As the stress and modulus of the composite decrease with the increase in ambient temperature, it leads to a decrease in plastic work, which in turn leads to a decrease in temperature rise. All four curves show a sudden increase in temperature at the first impact

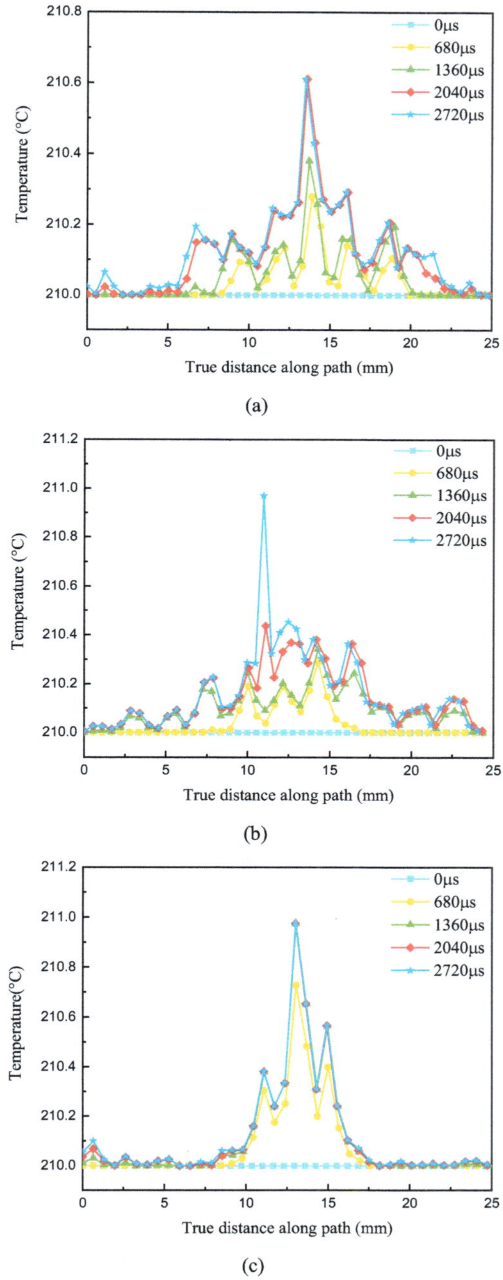

Fig. 10.4 Temperature history of the resin surface extracted along the X-axis through the impact point during the first four impact cycles under the impact pressure of 0.6 MPa at 210 °C. (**a**) 20°, (**b**) 30°, (**c**) 40°

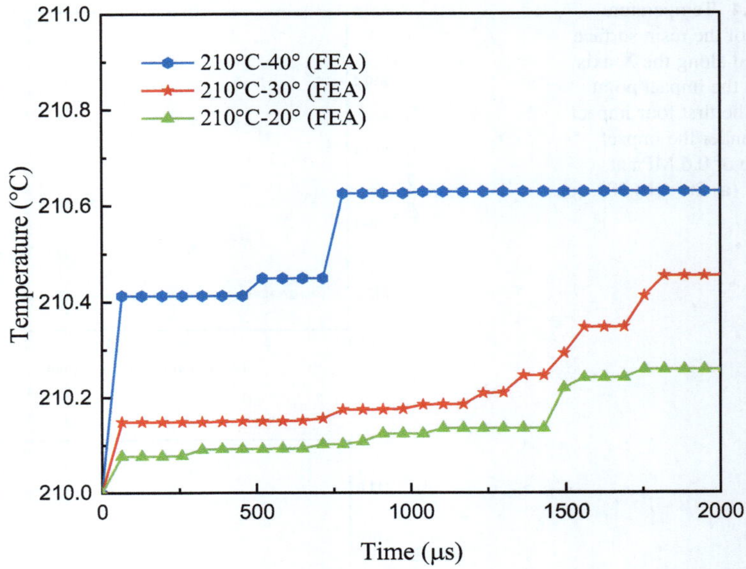

Fig. 10.5 Temperature history at the impact point for 3-D four-directional braided composite beams with different braiding angles

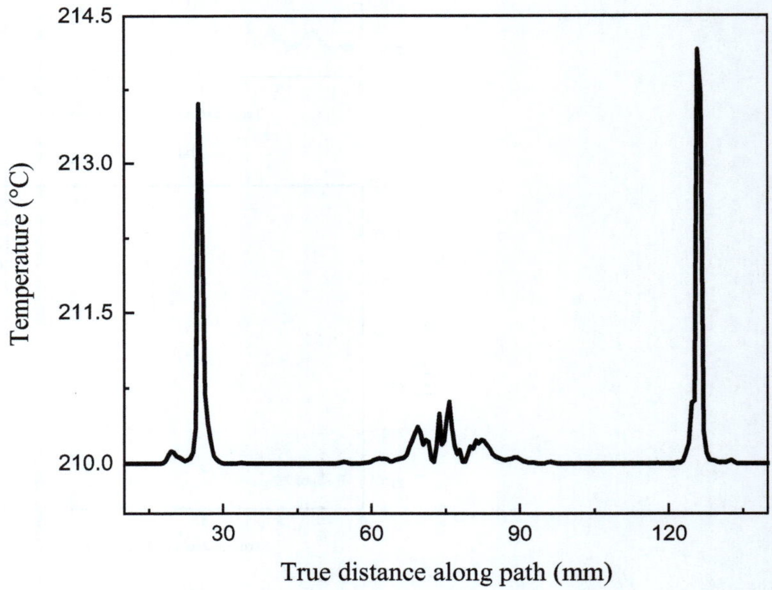

Fig. 10.6 Temperature history of the resin surface of the composite beam with the braiding angle of 20° along the Z-axis through the impact point during the third impact cycle under the impact pressure of 0.6 MPa at 210 °C and 2040 μs

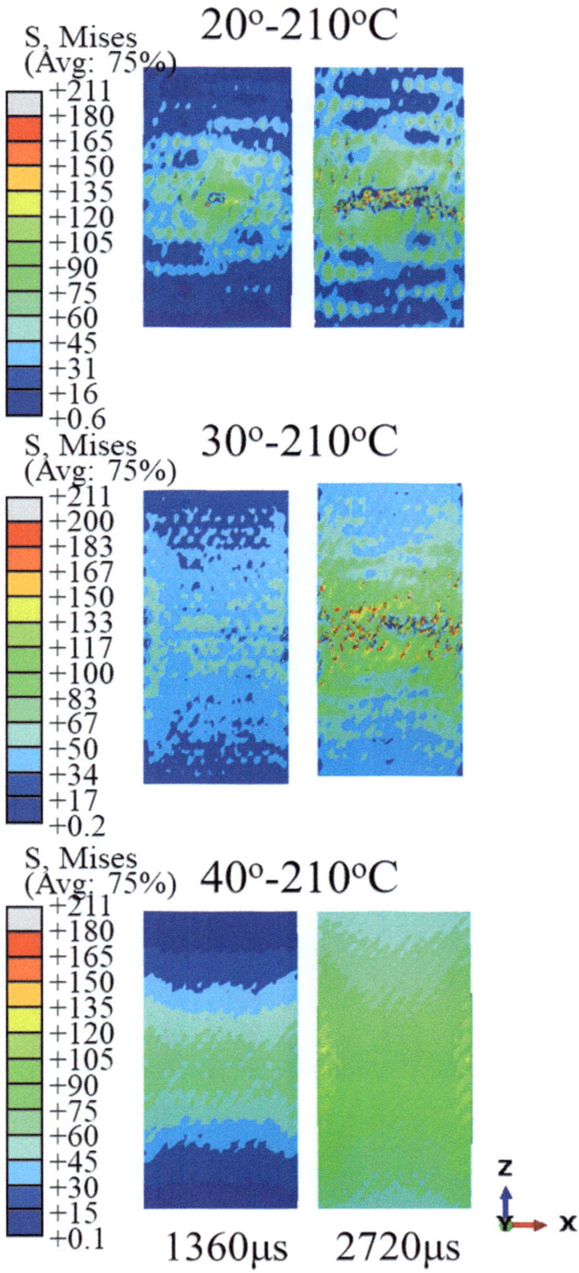

(a)

Fig. 10.7 Damage propagation for the first four impact cycles under the impact pressure of 0.6 MPa at 210 °C. (**a**) Resin back, (**b**) composite back

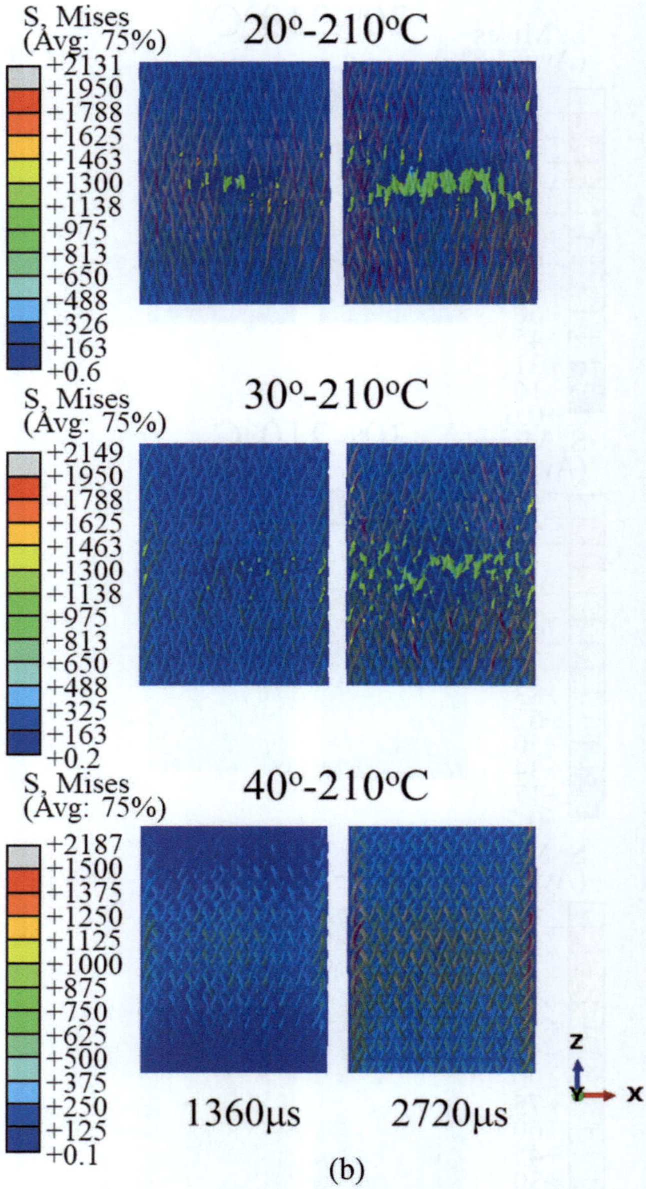

Fig. 10.7 (continued)

contact. When the ambient temperature is 25 °C, the temperature increase in the final stage can be attributed to the brittle fracture of the resin. Due to the stress concentration at the impact zone, the greater plastic work is converted into heat energy to increase the temperature. In addition, Fig. 10.9 shows the damage propagation and

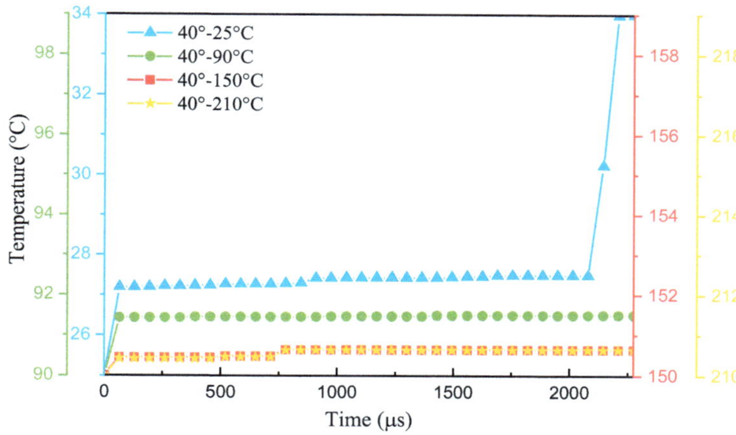

Fig. 10.8 Temperature history at the impact point for composite beams with the braiding angle of 40° at different temperatures

temperature evolution for the cross section of the composite beam with the braiding angle of 40° at 0.6 MPa for the first four impact cycles. The degree of damage decreases with increasing temperature. The stress and temperature are highest at the impact zone before the resin cracks, the maximum value occurs at the fracture surface of the resin after cracking, and the stress decreases after complete failure of the fracture surface. The stress distribution of cross section is X-shaped, which is because the middle part is in equilibrium state under the interaction of compression stress and tensile stress. The higher the stress is, the higher the temperature is, which corresponds to the adiabatic temperature rise distribution of the section. When the temperature (25 and 90 °C) is lower than T_g, the damage stage occurs early, with brittle fracture and higher temperature rise; when the temperature (150 and 210 °C) is higher than T_g, the damage stage occurs later, with ductile fracture and relatively small temperature rise. In summary, the stress distribution in braided composites is influenced by the ambient temperature and braiding angle, while the temperature rise is influenced by the material stress, forming a closed thermo-mechanical coupling cycle [4].

Figure 10.10 shows the damage evolution, adiabatic temperature rise, and stress distribution corresponding to μ-CT images of the sections of 40° specimen at different ambient temperatures. The increase in ambient temperature makes the mechanical properties of the material weaker, and the carbon fiber tows are not significantly affected by the temperature. When the temperature of 150 °C exceeds the glass transition temperature of the resin (T_g is about 110 °C), the mechanical properties of the resin weaken rapidly after the temperature exceeds T_g and the toughness is greatly enhanced. It is observed that the deformation of the specimen along the impact direction increases, the resin on the impact surface peels off and debonds, and the fiber tows on the surface split. There is interfacial cracking on the whole cross section, which is completely different from the cracking of the resin at room temperature. The interfacial damage is found to be distributed over almost the

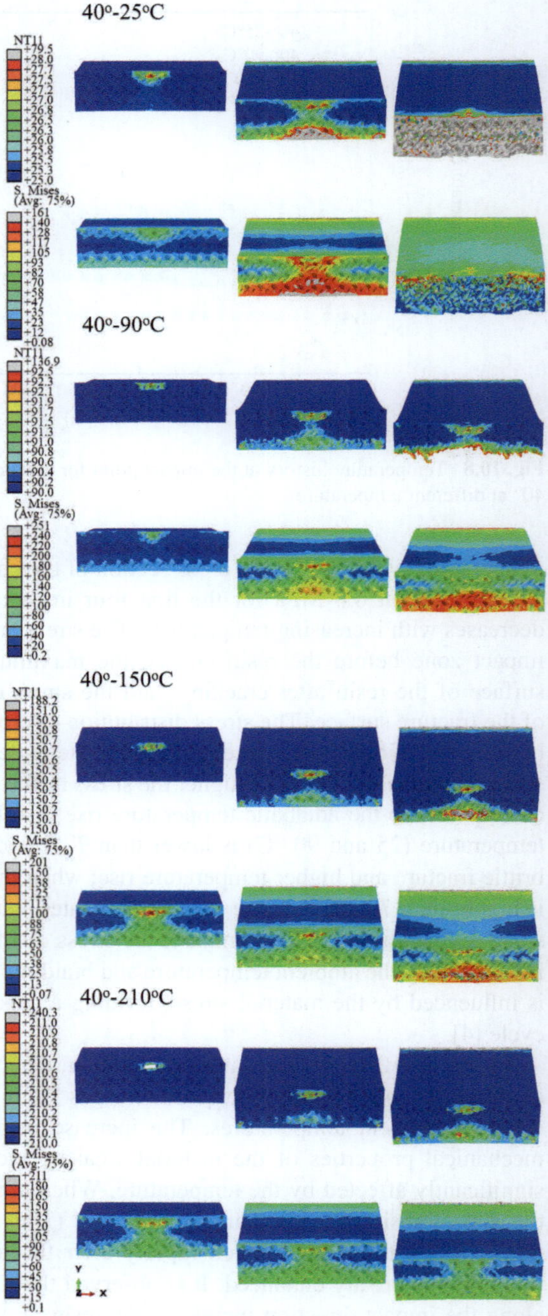

Fig. 10.9 Damage propagation and temperature evolution in the cross section of the composite beam with the braiding angle of 40° at 0.6 MPa for the first four impact cycles

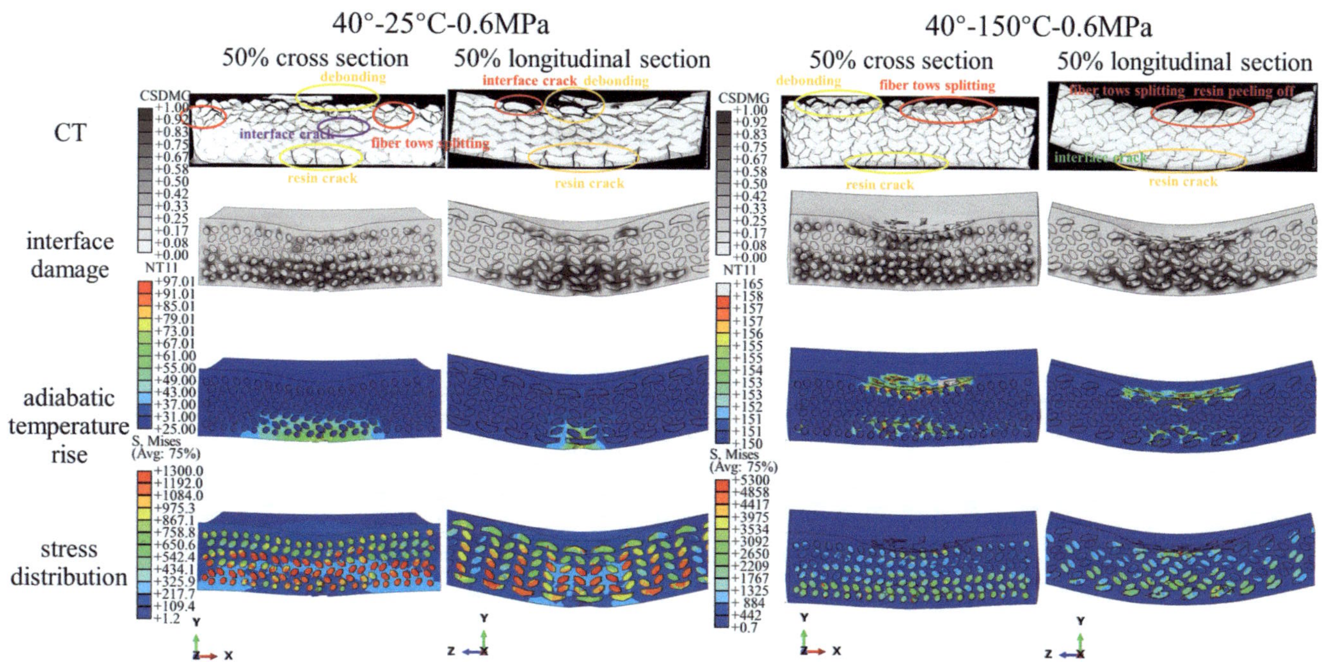

Fig. 10.10 Damage evolution, adiabatic temperature rise, and stress distribution corresponding to μ-CT images of the sections of 40° specimen at different ambient temperatures

entire cross section through the impact direction, and the brittle failure of resin changes to ductile failure with the increase in temperature. The stress carried by the preform is more uniform and significantly larger than that of the resin, and the preform is still the main body that bears the load and resists impact deformation at high temperature. It indicates that the compact meso-structure improves the impact tolerance. The temperature rise of the resin is mainly concentrated near the impact point and the gap between the braided yarns on the back surface and the interior braided yarns. The temperature is as high as 160 °C, which is far higher than the glass transition temperature of the resin. The decrease of the stress and modulus of the resin in the high-temperature region leads to the decrease of the plastic work, which leads to the decrease of the temperature rise and the interfacial bonding strength. Figure 10.11 shows the comprehensive comparison of the interfacial damage, adiabatic temperature rise, and energy absorption in the cross section of 3-D four-directional braided composite under different conditions. For the interfacial damage distribution, the darker the color, the more serious the damage [5].

The failure modes of the specimens are found to change significantly after heating during the impact cycles by finite element calculations. Figure 10.12 shows the damage evolution and stress distribution of the 40° specimens for the first four impact cycles at different temperatures. It is observed from the longitudinal section that the stress concentration distribution of the resin changes significantly with time and the stress concentration region gradually expands from the impact direction (y-direction) to both sides (z-direction) as the impact cycles increase. At room temperature, the stress concentration of the resin leads to the failure and cracking in the fourth impact cycle, and the ductile damage of the resin occurs from the first impact cycle. The impact surface and the back bear large ductile damage, and the ductile damage near the back is serious. However, at 150 °C, the stress concentration of resin is on the upper and lower surfaces, which is significantly different from that at room temperature. The ductile damage of the resin occurs from the second impact cycle, which is concentrated on the impact surface and the back, and does not penetrate the entire section. The increase of temperature leads to the enhancement of material toughness. It is observed from the ductile damage of yarns that the damage degree decreases after temperature rise (Fig. 10.11).

10.4 Low-Temperature Behaviors for the Braided Tubes

10.4.1 Stress-Strain Curves

In order to verify the FEA model for thermo-mechanical coupled meso-structure, the calculated stress-strain curves of the 3-D braided composite tube at −50 and −100 °C with different high strain rates under axial impact compression are extracted and compared with the experimentally obtained stress-strain curves, as shown in Figs. 10.13 and 10.14. The simulated curves agree well with the test curves before reaching the peak stress, and the peak stress and compression modulus of the composite tube predicted by the finite element simulation are in good agreement

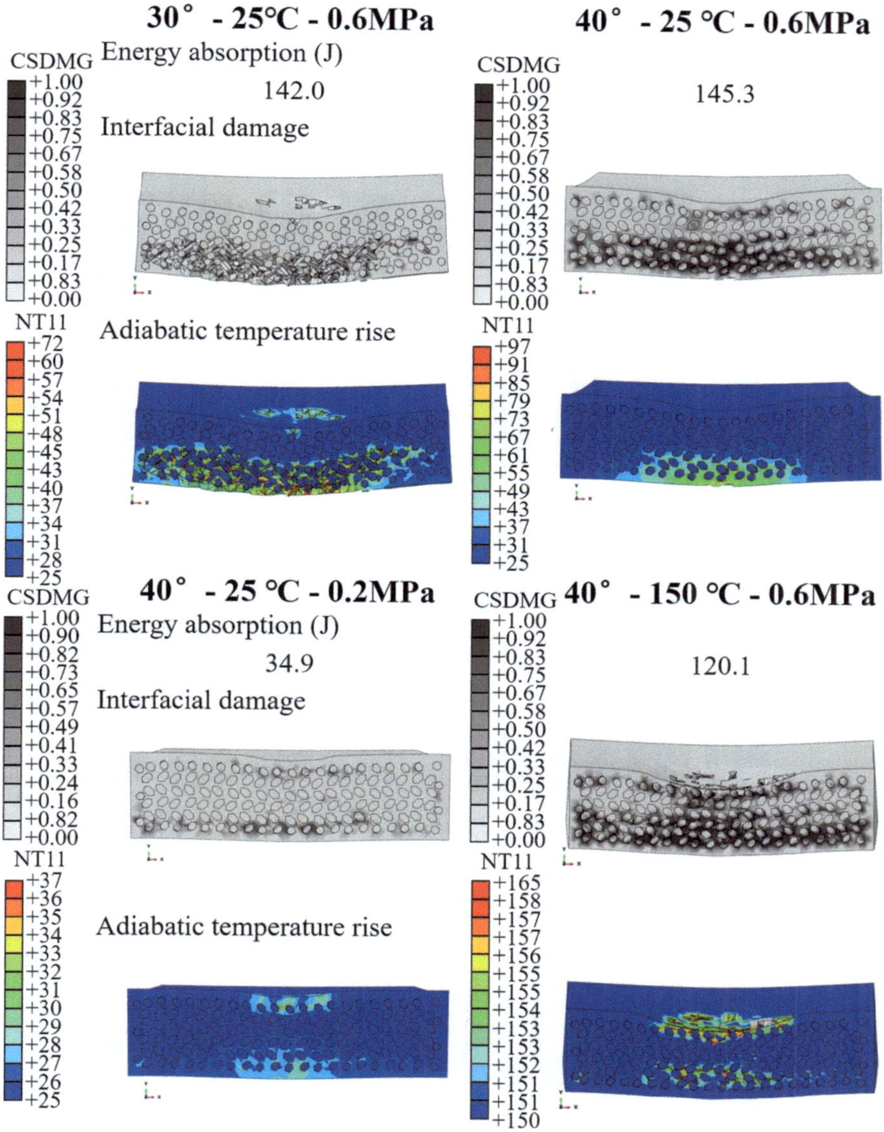

Fig. 10.11 Comparison of interfacial damage, adiabatic temperature rise, and energy absorption in the cross section of 3-D four-directional braided composite beams under different conditions

with the test results, indicating that the finite element model can predict the basic mechanical properties of the composite tube under high strain rate impact compression in low-temperature environment. After the peak stress is reached, there are some differences between the simulated and test curves, especially at 0.7 MPa, where the

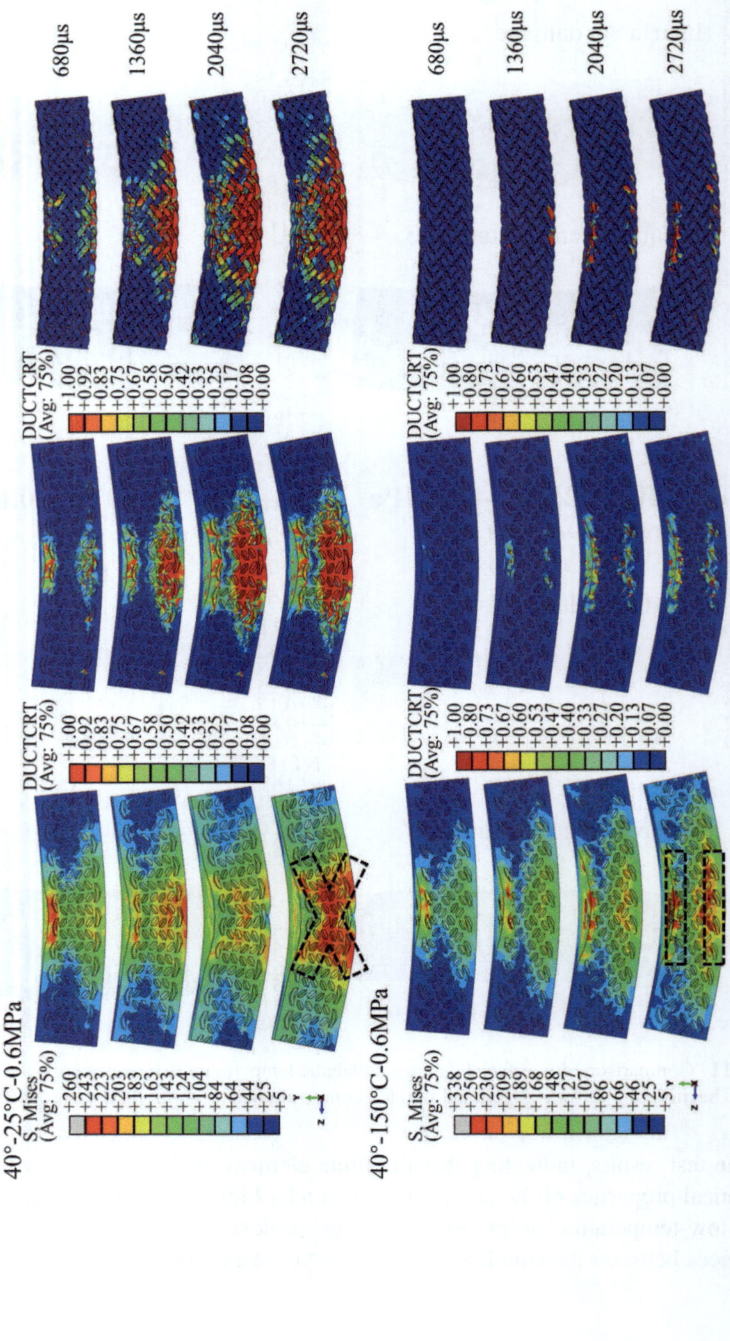

Fig. 10.12 Stress distribution and ductile damage evolution (longitudinal section) of 40° specimen for the first four impact cycles at different temperatures (25 and 150 °C)

Fig. 10.13 Test and FEA stress-strain curves for 3-D braided composite tube at −50 °C

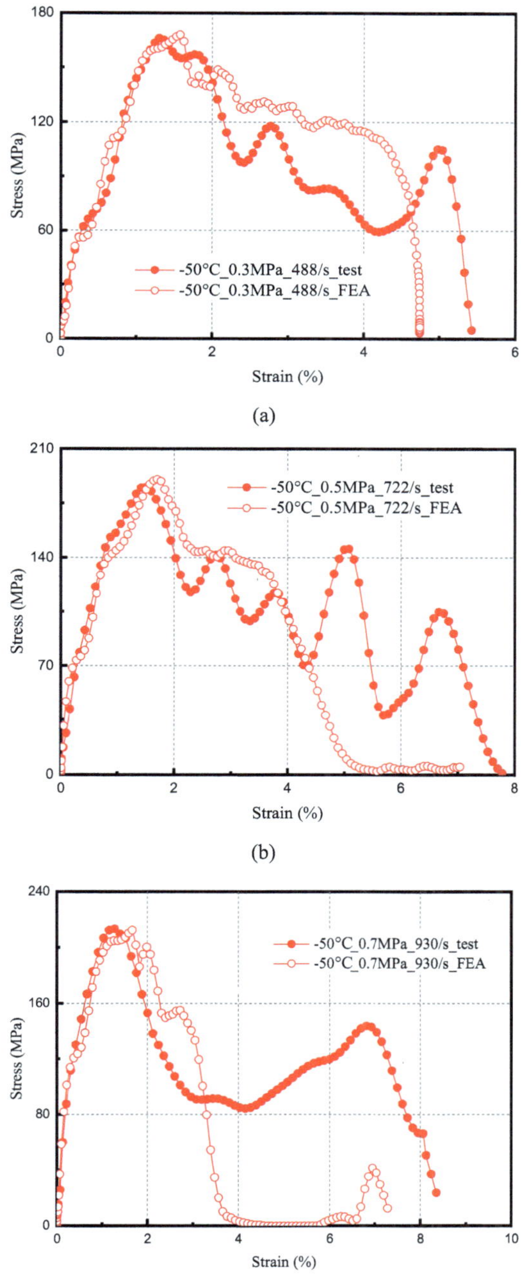

Fig. 10.14 Test and FEA stress-strain curves for 3-D braided composite tube at 100 °C

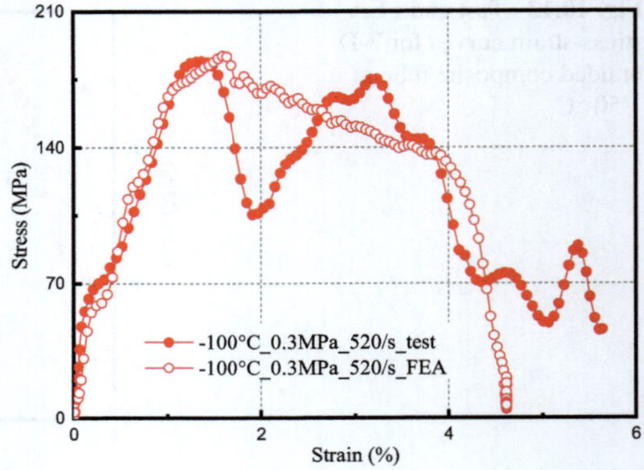

(a)

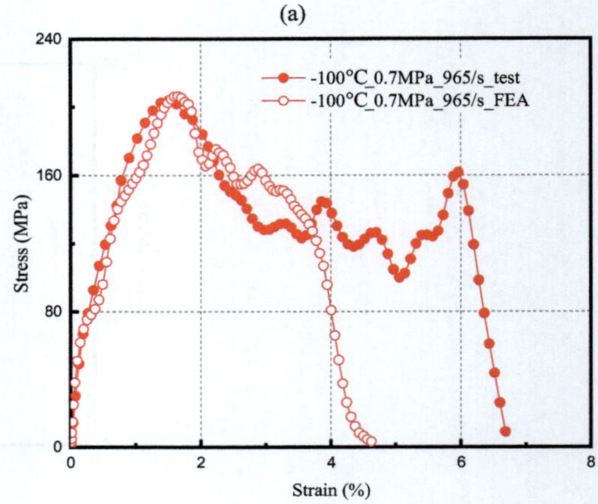

(b)

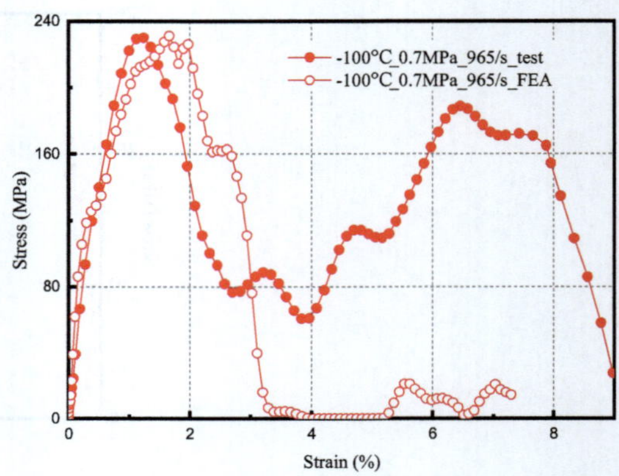

(c)

differences are most obvious. The simulated curve drops directly from the peak stress to zero, without the obvious fluctuation in the latter stage of the test curve. This is due to the fact that in the finite element model, once the material reaches the peak stress, the element will be eliminated, so that the material no longer bears the load. But in the experiment, the material can still bear a certain load after failure, and the material is no longer in force equilibrium after failure.

10.4.2 Thermo-mechanical Coupling Finite Element Analysis

In the process of plastic strain and failure of materials, the failure mechanism could be revealed from the node temperature (NT11) of thermo-mechanical coupled meso-structure finite element model. Taking the impact pressure of 0.3 MPa as an example, in order to analyze the heat generation process and characterize the damage transfer in the resin and preform, five typical points on the stress-strain curves are selected from the low-temperature fields at −50 and −100 °C in Figs. 10.15 and 10.16 to give the temperature rise distributions of the matrix and preform at the corresponding moments (the left half of the model is the matrix and the right half is preform). It can be seen that during the impact compression process, the inelastic thermal temperature rise concentration phenomenon appears first at the flexure of the braided yarn and then is rapidly distributed to the surrounding resin with the structural extrusion deformation of the braided preform. Before the material reaches the maximum stress, the temperature on both the matrix and preform is the same as the ambient temperature. Once the composite tube reaches the maximum stress and fails, it bears an obvious temperature rise. At −50 °C, the temperature rise (9.5 °C) of preform is higher than that (7.0 °C) of matrix; at −100 °C, the temperature rise (8.7 °C) of preform is higher than that (7.3 °C) of matrix. Meanwhile, the distribution and accumulation of inelastic heat lead to the formation of "serrated" shear band in the middle region of the specimen [6].

In order to further investigate the formation of internal inelastic heat under low-temperature field of the braided composite tube, the temperature distribution inside the material and the specific nodal temperature-time curves of the XY section obtained by sequentially intercepting the specimen along the axial direction (Z-direction) of the tube under low-temperature field compression at −50 and −100 °C are presented in Figs. 10.17 and 10.18, respectively. Two nodes (B, A) are selected from the cross section of the yarn and its surrounding resin to characterize the variation of the nodal temperature with the impact compression time. The temperature change of node B is clearly faster than that of node A at both −50 and −100 °C, indicating that the nodal temperature of the braided yarn changes faster than that of the resin. From Figs. 10.17c and 10.18c, it can be seen that the inelastic heat is first generated on the outer surface yarn along the loading direction and, then

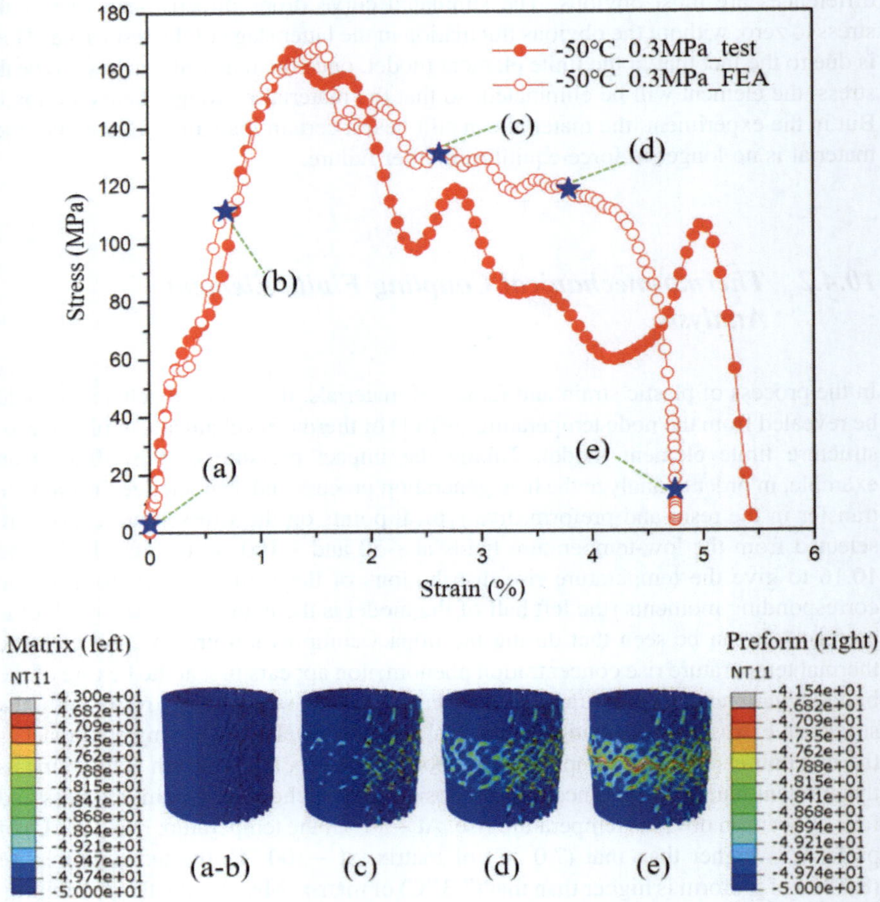

Fig. 10.15 Temperature rise distribution and progressive failure process of matrix and preform at −50 °C

as the stress wave propagates, the inelastic heat is mainly concentrated on the inner and outer surface braided yarns and the resin around them. For the braided preform, the inner and outer surface braided yarns generate more inelastic heat than the internal braided yarns due to more severe damage. For the resin, the closer to the braided yarn, the more serious the damage is, and therefore the more inelastic heat is generated. Also the nonuniform failure of the resin between the braided yarns can be observed from the cross sections inside the specimen.

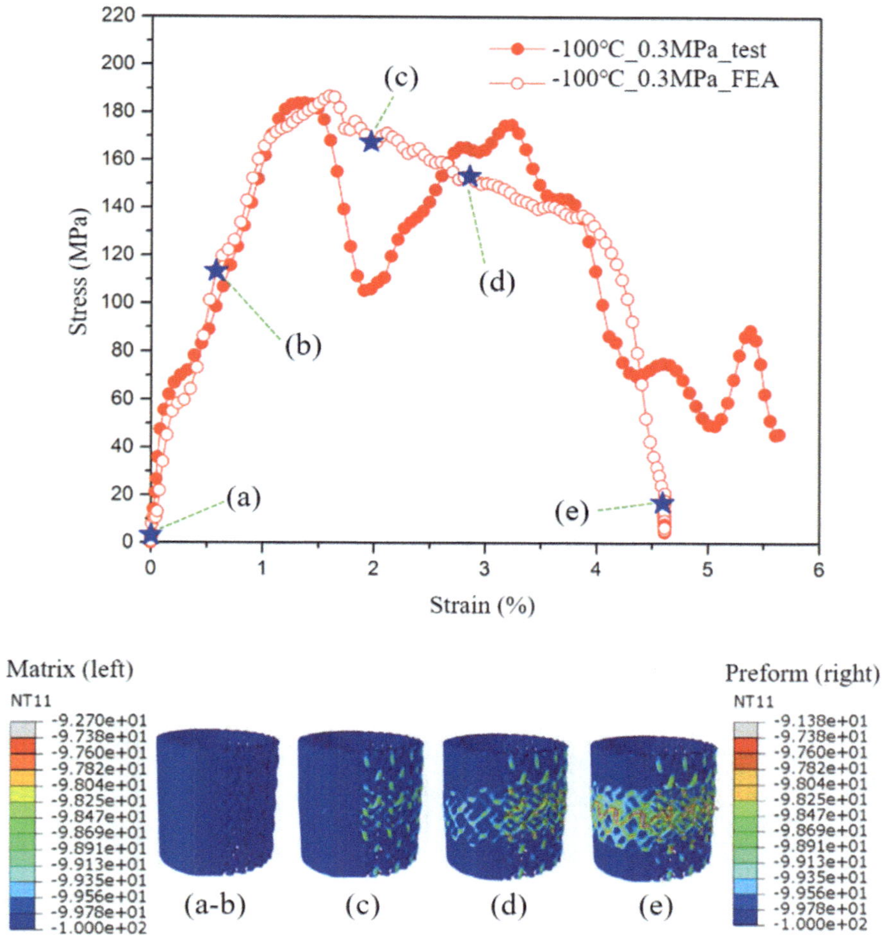

(b)

Fig. 10.16 Temperature rise distribution and progressive failure process of matrix and preform at −100 °C

10.4.3 Thermo-mechanical Coupling Failure

10.4.3.1 Stress Distribution

In order to further investigate the impact compression response and failure mechanism of 3-D braided composite tube at different temperatures and strain rates, the von Mises stress distributions of the final failure morphology of 3-D braided composite tube under axial impact compression at −50 and −100 °C and different strain rates are presented in Fig. 10.19. The stress distributions of the damage

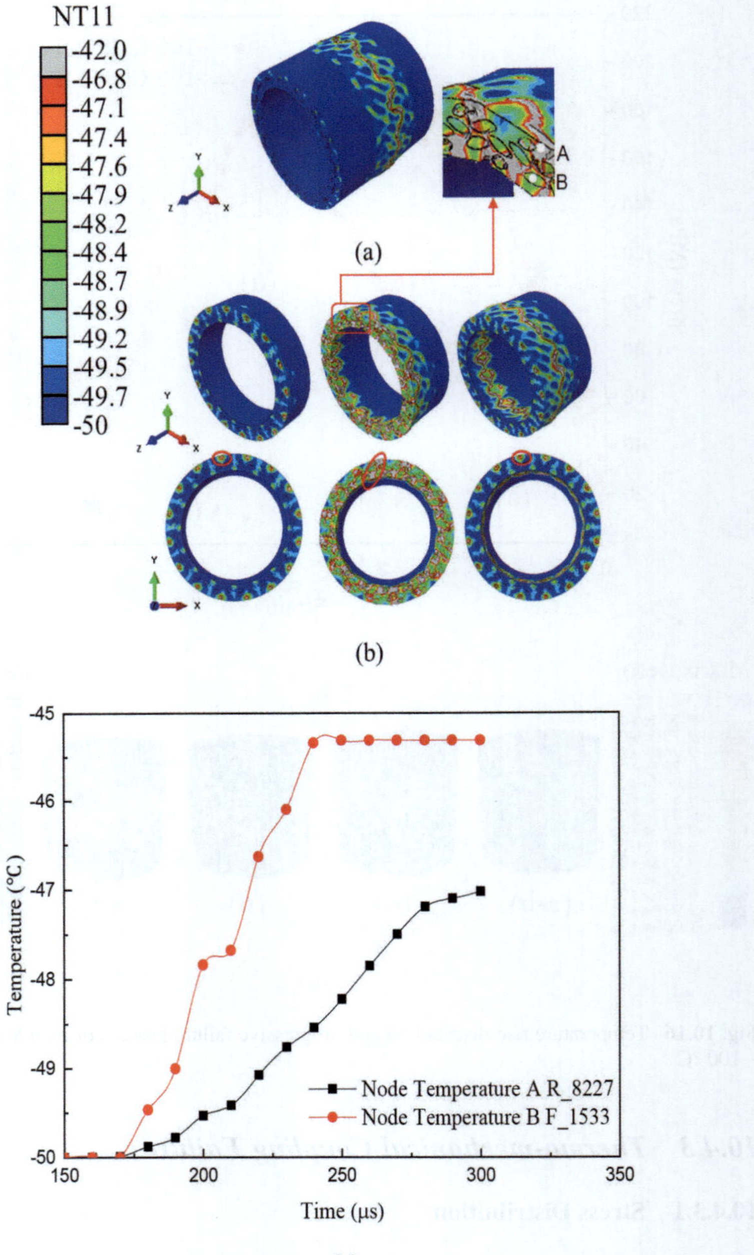

Fig. 10.17 Temperature distribution and nodal temperature-time curves under the impact pressure of 0.3 MPa and the strain rate of 488/s at −50 °C. (**a**) Temperature distribution of the whole specimen, (**b**) temperature distribution of different sections, (**c**) nodal temperature-time curves

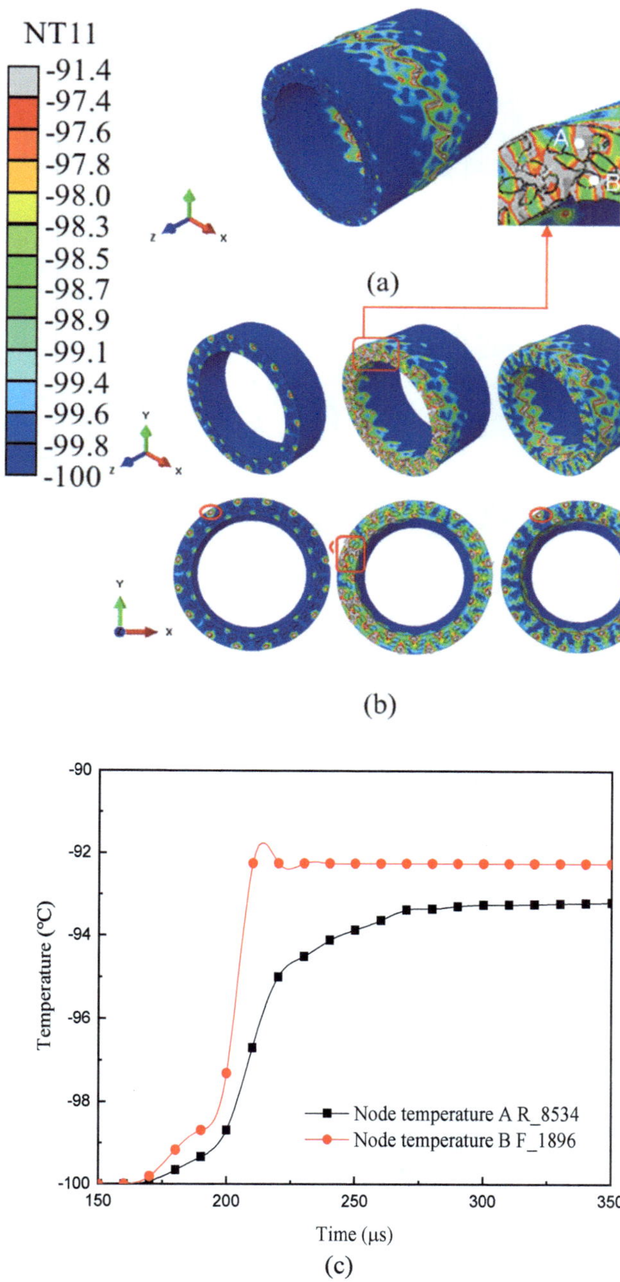

Fig. 10.18 Temperature distribution and nodal temperature-time curves under the impact pressure of 0.3 MPa and the strain rate of 520/s at −100 °C. (**a**) Temperature distribution of the whole specimen, (**b**) temperature distribution of different sections, (**c**) nodal temperature-time curves

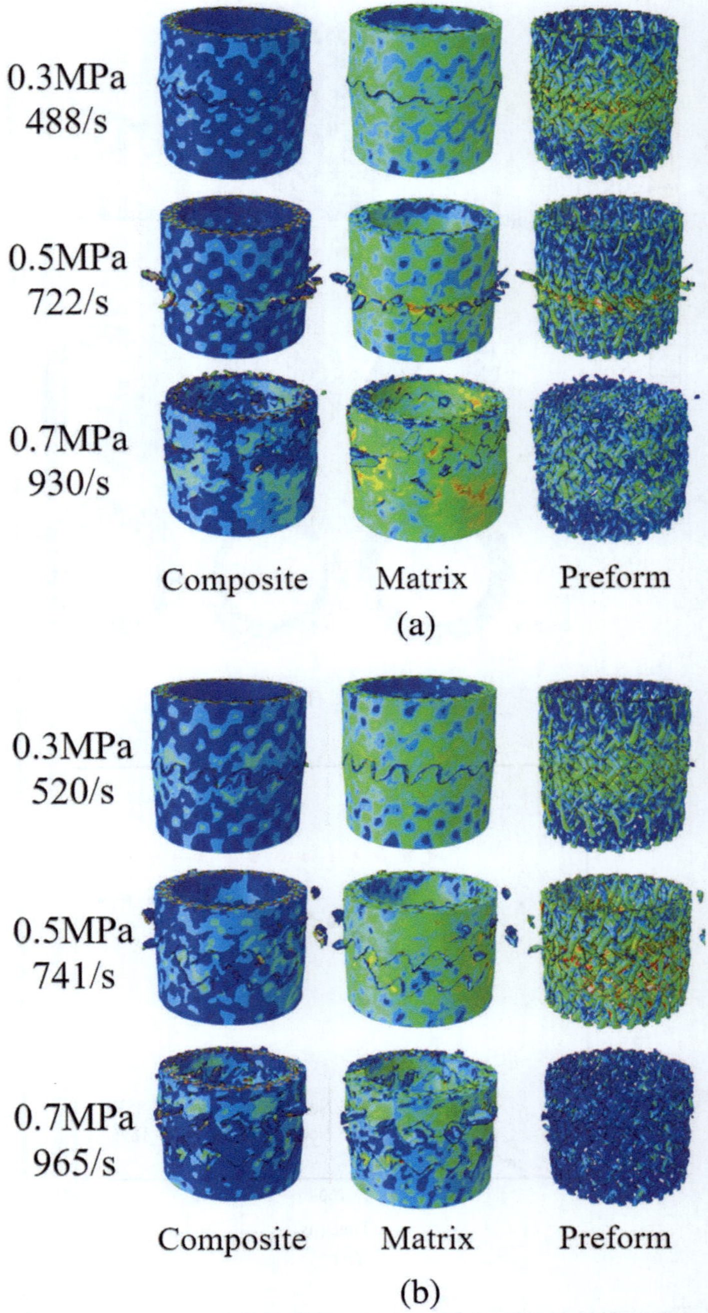

Fig. 10.19 FEA results of thermo-mechanical coupling for different temperature fields and strain rates for 3-D braided composite tube—stress distribution. (**a**) −50 °C, (**b**) −100 °C

morphology of the braided preform and the resin matrix are also presented separately in Fig. 10.19, with higher stress in the 3-D braided preform than in the resin matrix. Although the ideal meso-structure model constructed in this study is not exactly the same as the damage morphology observed in the experiments, the damage morphology of the specimens in the finite element simulations is generally less severe than in the experiments. However, the results of the effects of temperature and strain rate on the finite element model are consistent with the test results given in Chap. 7. The damage morphology of the tube is more severe with increasing strain rate at the same temperature. The resin damage is more serious when the tube is loaded under compression at low-temperature field compared to loading at room temperature (Fig. 9.12a), and more resin fragments can be seen on the surface of the tube. The compression deformation of the specimens is also more pronounced at low-temperature field than at room temperature. Under the strain rate of 488/s at −50 °C, a serrated crack appears in the middle of the composite tube. In addition to the fracture at the serrated crack of the resin, the fiber tows of the preform also fracture at the bending of other fiber tows. As the strain rate increases, cracked resin blocks and fractured fiber tows can be seen around the serrated crack. The serrated cracks in the middle of the composite tube at −100 °C and 520/s are significantly deeper and longer than those at −50 °C and 488/s. Two serrated cracks are found in the specimen at 741/s, and the damage of resin and fiber tows is obvious. The damage occurs at the loaded end of the specimen at 965/s. The resin is crushed and broken, and the braided preform is broken and fractured severely although the structure is intact.

The difference between the damage morphology obtained from finite element simulation and experiment is, on the one hand, due to the fact that the final failure morphology of the material in the experiment is formed under the action of several impact stress waves, while only one impact wave is loaded in the finite element simulation and, on the other hand, due to the fact that perfect bonding of the interface between the resin and the preform is assumed in the finite element model, which to some extent hinders the debonding of the resin during the impact compression.

10.4.3.2 Temperature Rise Distribution

Figure 10.20 shows the temperature rise distributions of the final failure morphology of 3-D braided composite tube under axial impact compression at −50 and −100 °C under different strain rates. The damage evolution of the tube is caused by the generation and accumulation of inelastic heat. Since the damage morphology of 3-D braided composite tube has a strain rate effect, i.e., the damage morphology becomes more severe as the strain rate increases, therefore, the inelastic heat generated by 3-D braided composite tube also increases with increasing strain rate at low-temperature fields. The temperature rise distribution in the middle region of the specimen is serrated, and the temperature rise at the impact loading surface is scattered. Since brittle materials do not show significant thermal phenomena under high strain rate loading, the 3-D braided composite tube is a typical thermoplastic material due to the thermo-mechanical coupled failure mechanism.

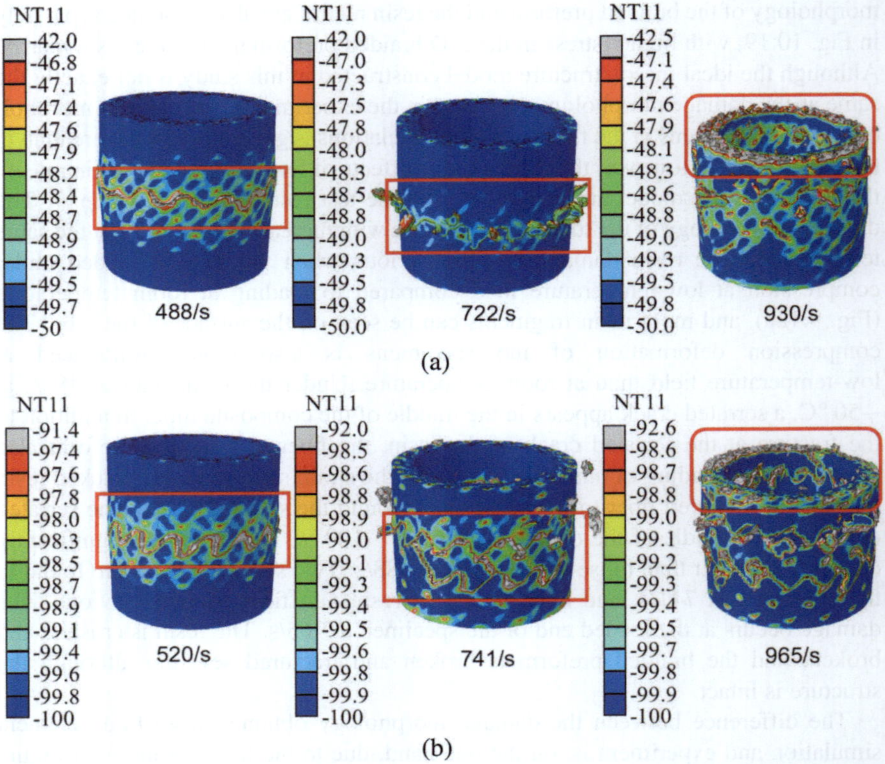

Fig. 10.20 FEA results of thermo-mechanical coupling for different temperature fields and strain rates for 3-D braided composite tubes—temperature distribution. (**a**) −50 °C, (**b**) −100 °C

References

1. Hu Meiqi, Sun Baozhong, Gu Bohong: Microstructure modeling multiple transverse impact damages of 3-D braided composite based on thermo-mechanical coupling approach. Compos. Part B. **214**, 108741 (2021)
2. Hu Meiqi, Liu Shengkai, Zhang Junjie, Wang Lei, Sun Baozhong, Gu Bohong: Multiple transverse impact damage behaviors of 3-D-braided composite beams under room and high temperatures. Int. J. Damage Mech. **29**(5), 715–747 (2020)
3. Pan, Z., Gu, B., Sun, B.: Longitudinal compressive behaviour of 3D braided composite under various temperatures and strain rates. Appl. Phys. A Mater. Sci. Process. **118**, 1315–1337 (2015)
4. Zhang, W., Gu, B., Sun, B.: Thermal-mechanical coupling modeling of 3D braided composite under impact compression loading and high temperature field. Compos. Sci. Technol. **140**, 73–88 (2017)
5. Hu Meiqi, Sun Baozhong, Gu Bohong: Temperature and structure effects on the impact damage distribution of 3D braided composites. Zhongguo Kexue Jishu Kexue/Scientia Sinica Technologica. **51**(1), 108–118 (2021)
6. Xianyan Wu, Qian Zhang, Bohong Gu, Baozhong Sun: Influence of temperature and strain rate on the longitudinal compressive crashworthiness of 3-D braided composite tubes and finite element analysis. Int. J. Damage Mech. **25**(4), 574–607 (2016)

Chapter 11
Summary

11.1 Multiscale Impact Damage Mechanisms of Braided Composite Beams

We have developed a thermo-mechanical coupled elastic-plastic constitutive model based on the microstructure of braided composite. An integrated impact dynamic computational framework has been established to analyze multiple transverse impact damage of 3-D braided composites. The transverse impact deformation and damage behavior of 3-D multidirectional braided composites under four successive stress waves loading at 25, 90, 150, and 210 °C were tested with a modified split Hopkinson pressure bar (SHPB) apparatus attached with a self-designed heating device. The impact processed was recorded with a high-speed camera. The temperature and structural effects on impact damage have been investigated to show the thermo-mechanical coupling behaviors. The microstructure mechanisms of impact damage have been revealed to show the effects of micro-structure and loading rate as well as temperature on the thermo-mechanical coupled damage behavior.

At room temperature, the elastic deformation of the composite beam occurs in the first impact cycle, and the main failure mode in the second impact cycle is plastic deformation accompanied by slight resin cracking. In the third impact cycle, the fibers gradually fracture with the accumulation of plastic deformation. The final failure mode is obtained in the fourth cycle. The braided preform is the main load-bearing component and absorbs most of the impact energy. The peak load and energy absorption gradually decrease with the number of impact cycles, and the displacement and damage area gradually increase.

The meso-structure affects the fracture path, stress distribution, and adiabatic temperature rise distribution and protects the structural integrity of the preform. The preform is the main load-bearing component of the composite and dominates the plastic deformation of the entire composite. As the braiding angle increases, the impact load and energy absorption increase, and the impact displacement and damage area decrease. At room temperature, the main failure modes include resin

M. Hu, B. Gu, *Impact Damages of Braided Composites*, Engineering Materials,
https://doi.org/10.1007/978-981-16-5703-0_11

cracking, fiber breakage under low impact pressure, yarn fracture, and interface debonding at large braiding angles under high impact pressure. As the braiding angle increases, the fiber volume fraction, interlacing points between yarns, and the thickness of the beams all increase. This hinders the propagation of stress waves along the impact direction, and the cracking path is converted from straight line to curve. In addition, the stress propagates along the path of least resistance during extension, leading to inconsistent fracture directions. The maximum stress distribution is mainly X-shaped in the cross section. The results suggest that the high-impact braided composites can be designed with hybrid braided yarns, where the high-strength yarns should be braided in the X-shaped position. Combined with the μ-CT observation results, it is known that for the specimen with the braiding angle of 30°, the adiabatic temperature rise is mainly concentrated at the Y-shaped damage notch formed in the cross section and the difference in stress distribution between the preform and resin decreases after the interface failure; for the specimen with the braiding angle of 40°, the adiabatic temperature rise is mainly concentrated at the local area of tensile failure on the back of the specimen, with more interface cracking area and less failure area. The stress of preform is higher, and the stress difference between the preform and the matrix increases. The resin cracks propagate along the braiding direction. In addition, the toughening mechanism provided by the structure strengthened preform enhances the mechanical properties of the composites and delays the failure time.

In high-temperature environments, lower ambient temperatures and larger braiding angles result in a significant increase in both peak load and energy absorption. The braiding angle has a significant effect on structural integrity and further affects the thermo-mechanical properties. An increase in the braiding angle increases the damage tolerance and crack extension properties of the composite. The composite beams can absorb additional impact energy stored as plastic deformation energy and further converted into heat energy. The difference in CTE between the fiber tows and the resin and the loose structure lead to a scattered distribution of stress. In addition, the elastic-plastic properties of epoxy resins are temperature-dependent. When the ambient temperature is higher than T_g, the epoxy resin tends to the rubber state. As the modulus decreases with increasing ambient temperature, the plastic work and local temperature rise also decrease. The adiabatic temperature rise occurs mainly in the damage stage, not in the elastic-plastic stage. The stress of the material is influenced by the ambient temperature and the braided structure, and the temperature rise is in turn influenced by the stress. As the temperature of the specimen increases, the resin changes from brittle failure to ductile failure. The ductile damage to the resin is gradually transferred from the interior to the upper and lower surfaces of the specimen, and the damage region of preform is also extended from the interior to the upper and lower surfaces. The heat energy transformed by plastic work during impact loading is mainly concentrated near the impact point in the resin and the braided yarn-resin interface region adjacent to the back surface. The local temperature rise leads to a decrease in the interfacial bond strength and severe damage, with the failure mode changing to resin debonding and peeling off and interfacial cracking. The main failure modes below T_g are brittle fracture of the resin and crack extension along the braided yarn direction. The main failure modes above T_g are yarn fracture, ductile fracture of resin, and interfacial debonding. The thermo-mechanical coupling behavior of epoxy resin is the key to optimize the impact resistance at room and high temperatures.

During the impact process, the crack and the impact surface undergo local violent plastic deformation, resulting in the concentration of adiabatic temperature rise and interfacial damage. The resulting heat energy increases the temperature of the local region, leading to thermal expansion of the resin, causing the resin and yarn to extrude each other, and resulting in adiabatic shear damage such as internal stress and local deformation. The high stress concentration and interfacial debonding accelerate the damage development in the local region, converting the generated plastic deformation energy into heat energy to increase the adiabatic temperature rise even more and finally forming a thermo-mechanical coupling closed loop. Using the proposed modeling approach, impact failure and damage tolerance design of various fiber-reinforced composites (e.g., 3-D woven and knitted composites, etc.) can be achieved, and the foundation for designing braided composite engineering structures in aerospace, high-speed vehicles, and impact resistance areas can be laid.

11.2 Transverse Impact Damage Mechanisms of Braided Composite Tubes

The effects of impact pressures (impact velocities) and braiding parameters on impact damage mechanisms were investigated with the SHPB apparatus and FEA. Various braided structural composite specimens were prepared by test methods. Each specimen was tested for transverse impact using SHPB testing technique to obtain load-displacement curves, displacement-time curves, and energy absorption. μ-CT results were performed on the posttest specimens to obtain the internal damage of the material. The full-scale meso-structure finite element model has been established to simulate the transverse impact deformation process of 3-D braided composite tube to further explore the damage failure mechanism of the material meso-structure. The main findings include:

During the transverse impact process, the stress wave is reflected in the incident bar several times, and the specimen is impacted by the stress wave several times. The load-displacement curves and displacement-time curves both show fluctuation. During the impact process, due to the dissipation of the stress wave impact energy and the continuous degradation of the specimen stiffness, the peak load tends to decrease with the increase of the number of impact cycles. With the increase of impact pressure, the peak load, displacement, and energy absorption increase.

The effects of three braiding parameters (braiding angle, number of braided layers, and axial yarn) on the transverse impact mechanical properties of 3-D braided composite tubes include the following: the peak load increases, and the displacement decreases as the braiding angle increases, the number of braided layers increases, or the axial yarns are added. The smaller the braiding angle, the greater the number of braided layers or the addition of axial yarns, and the greater the specific energy absorbed by the specimen.

From the finite element analysis results, it can be seen that the stress is mainly concentrated in the impact zone and the clamping zone, and the 3-D braided preform is the main load-bearing component. The deformation includes the overall bending deformation of the tube, cross-sectional flattening, and local depression in the impact zone, and the damage mainly occurs in the local region subjected to the impact. The damage of the specimen increases with the increase of impact pressure. Different braiding parameters directly affect the internal meso-structure of the composite and thus the stress propagation and failure mode during impact loading. Changes in the braiding angle change the yarn orientation angle on the one hand and the density of interlacing points between the yarns on the other hand. As the braiding angle increases, the fiber orientation gradually deviates from the axis, and the density of interlacing points along the axial direction increases, which is not conducive to stress propagation along the axial direction, while the compact structure makes it more integral and more resistant to impact deformation damage. As the number of braided layers increases, the alternation angle of yarn in the circumferential direction increases, and the number of interlacing points in the thickness direction increases, which also improves the structural integrity of the tube and enhances the impact damage resistance. The impact deformation and damage resistance of 3-D five-directional braided composite tube are much higher than that of 3-D four-directional braided composite tube due to the addition of the axial yarn. At the beginning of the impact, the stress of the axial yarn is rapidly enhanced and propagates along the axial direction much faster than that of the braided yarn, which plays an important role in resisting local damage in the impact zone.

From the μ-CT observations, it can be seen that the main failure modes of 3-D braided composite tube after transverse impact include matrix fragmentation, interface cracking, fiber tow fracture, and internal splitting of fiber tow. Under small impact pressure, matrix fragmentation and interface cracking between fiber tows are dominant, and fiber tows start to fracture as the impact pressure increases. Different braiding parameters directly affect the internal meso-structure of the composite and thus affect the failure mode of the material during impact loading. For the specimens with different braiding angles, the 15° specimen is dominated by transverse fracture of fiber tows in the center of the impact zone, and the failure modes include fiber tow fracture, delamination between fiber tows, and internal splitting of fiber tow. The damage morphologies of the 30° specimen include impact zone depression and formation of an elliptical shear band around the impact zone, and the failure modes mainly include fiber tow fracture, matrix fragmentation, interface cracking, and internal splitting of fiber tow. The 45° specimen is less damaged, and the main damage is the matrix fragmentation between yarns. The main failure modes include interface cracking, matrix cracking, and fiber tow deformation. For specimens with different braided layers, the damage of the two-layer specimen is the most serious. The middle yarns of the impact surface are pulled off, and the whole impact surface is broken inward along the periphery, and the specimen is completely failed.

Compared with the two-layer specimen, both the three-layer and four-layer specimens have smaller depressions in the impact zone, but the damage is more severe inside the entire impact zone. Fragments in the damage zone of the three-layer specimen are loose and dominated by fiber fracture and fiber tow splitting; while the damage zone of the four-layer specimen is compact, dominated by fiber tow extrusion deformation and interface cracking.

11.3 Axial Impact Damages of Braided Composite Tubes

The axial quasi-static and impact compression damage mechanisms of 3-D four-step braided composite tubes with different structural parameters have been investigated at room temperature and low temperature. The structural effects, strain rate effects, and temperature effects have been analyzed.

The compression behavior of 3-D braided composite tubes depends on geometrical structures. The braiding angle, the number of braided layers, and the addition of axial yarn all have significant effects on the mechanical properties of braided composite tubes. When the loading conditions are the same, the compressive strength and compressive stiffness of the composite tube decrease with the increase of the braiding angle. Both increase with the increase of the number of braided layers. The variation of the specific energy absorption of the composite tube is more complicated, which increases with the increase of the braiding angle and the number of braided layers under quasi-static compression. It decreases with the increase of the braiding angle under high strain rate and increases first and then decreases with the increase of the number of braided layers. 3-D five-directional braided composite tubes have higher compressive strength, modulus, and specific energy absorption capacity than 3-D four-directional composite tubes. The damage modes of 3-D four-directional braided composite tubes with different braiding angles and braided layers are in shear damage and compression buckling failure under quasi-static compression. 3-D five-directional braided composite tubes show only slight resin damage at the surface. Under compressive load at high strain rate, the effect of braiding angle on the failure mode of the material is much greater than that of the number of braided layers. When the braiding angle is small, the failure mode of the specimen changes from shear damage to petal crushing failure with the increase of strain rate. When the braiding angle is large, the damage morphology of the specimen maintains the structural integrity, and the failure modes are dominated by resin cracking, fiber pullout, and compression deformation. When the braiding angle is 35°, the two-layer specimens have the largest difference under different strain rates. The failure mode changes from fiber breakage and pullout to resin debonding and almost complete crushing damage with increasing strain rate. With the increase in the number of braided layers, the resistance of the tube to compressive deformation increases, and the failure mode is dominated by resin cracking to maintain the structural integrity. The structural integrity is still maintained at the strain rate of 839/s compared with 3-D four-directional braided composite tube with the same braided layers.

The compressive behavior of 3-D braided composite tube has the strain rate effect. The stress-strain curves of 3-D four- and five-directional braided composite tubes with different braiding angles and different braided layers have strain rate sensitivity. The compressive strength and compressive modulus of the composite tubes increase with increasing strain rate, while the specific energy absorption decreases. For 3-D braided composite tubes with the same braided structural parameters, the higher the strain rate, the more severe the material damage. The specimens with small braiding angles and braided layers (25° and two-layer) both show crushing failure modes at the maximum strain rate.

The compression behavior of 3-D braided composite tube has temperature effect. Carbon fibers have lower thermal expansion coefficient and resistance to ultrahigh temperatures and good resistance to low temperatures and are temperature-insensitive materials, while the epoxy resin hardens during the cooling process from room temperature to −140 °C, and the modulus increases significantly around −100 °C with temperature dependence. At low temperature, 3-D braided composite tube has the same strain rate effect at room temperature and at the same time has sensitive temperature dependence. The compression strength, modulus, and specific energy absorption all increase with decreasing temperature. The strain rate sensitivity of compression strength, compression modulus, and specific energy absorption is greater than that of the temperature. The brittleness of the material increases and the toughness weakens at low temperature. The fracture of the fiber tow is neat, and the resin on the surface of the specimen is peeling off over a large area. The resin damage between the fiber tows is more serious, and the interfacial bond strength between the fiber and resin is enhanced. Both the degree of material damage and deformation become more severe with the increase of strain rate and the decrease of temperature.

Based on the real structure of 3-D braided composite tube, the meso-structure finite element model is established to calculate the mechanical properties, stress propagation process, and damage morphologies. The stress-strain curves and damage morphologies of 3-D braided composite tubes obtained from finite element analyses are in good agreement with the test results, thus verifying the finite element model. The differences in the compressive mechanical response mechanisms of 3-D four- and five-directional braided composite tubes with different braiding angles and number of braided layers are analyzed. The propagation velocity of stress wave on the preform is higher than that on the resin. For the specimens with small braiding angle and more braided layers, the difference of the propagation velocity of the stress wave between the preform and the resin is more obvious. For the 3-D five-directional braided composite tube, the stress wave propagates faster on the axial yarn than the braided yarn, and the axial yarn is the main load-bearing component and reaches the maximum stress first. The initial damage occurs at different times and locations for specimens with different braiding structural parameters under the same loading conditions. The initial damage occurs first on the braided yarn for the three-layer specimens with three different braiding angles, while the damage occurs almost simultaneously on the braided yarn and resin for the two-layer and four-layer specimens.

The thermo-mechanical coupling framework developed in this investigation provides an effective way for showing impact damage mechanisms of the 3-D braided composite beam and tubes. By analyzing the stress-temperature rise-progressive failure process, final failure morphologies, and stress and temperature distribution of the braided preform and resin, it is found that the stress concentration appears first at the flexure of the braided yarn and leads to the inelastic heat generation of the resin due to the compression deformation.

Index

M. Hu, B. Gu, *Impact Damages of Braided Composites*, Engineering Materials,
https://doi.org/10.1007/978-981-16-5703-0

MIX
Papier aus verantwortungsvollen Quellen
Paper from responsible sources
FSC® C105338
FSC
www.fsc.org

If you have any concerns about our products,
you can contact us on
ProductSafety@springernature.com

In case Publisher is established outside the EU,
the EU authorized representative is:
Springer Nature Customer Service Center GmbH
Europaplatz 3, 69115 Heidelberg, Germany

Printed by Libri Plureos GmbH
in Hamburg, Germany